Extending the Linear Model with R:
Generalized Linear, Mixed Effects and Nonparametric Regression Models

---

Julian J. Faraway

Published in 2006 by
Chapman & Hall/CRC
Taylor & Francis Group
6000 Broken Sound Parkway NW, Suite 300
Boca Raton, FL 33487-2742

Printed in the United States of America on acid-free paper
10 9 8 7 6 5 4 3 2 1

International Standard Book Number-10: 1-58488-424-X (Hardcover)
International Standard Book Number-13: 978-1-58488-424-8 (Hardcover)
Library of Congress Card Number 2005054822

Library of Congress Cataloging-in-Publication Data

Faraway, Julian James.
Extending the linear model with R : generalized linear, mixed effects and nonparametric regression models / Julian J. Faraway.
p. cm. -- (Texts in statistical science)
Includes bibliographical references and index.
ISBN 1-58488-424-X
1. Analysis of variance. 2. Regression analysis. 3. R (Computer program language)--Mathematical models. I. Title. II. Series.

QA279.F368 2006
519.5--dc22 2005054822

Visit the Taylor & Francis Web site at
http://www.taylorandfrancis.com

and the CRC Press Web site at
http://www.crcpress.com

Preface

Linear models are central to the practice of statistics. They are part of the core knowledge expected of any applied statistician. Linear models are the foundation of a broad range of statistical methodologies; this book is a survey of techniques that grow from a linear model. Our starting point is the regression model with response y and predictors $x_1, \ldots x_p$. The model takes the form:

$$y = \beta_0 + \beta_1 x_1 + \cdots + \beta_p x_p + \varepsilon$$

where ε is normally distributed. This book presents three extensions to this framework. The first generalizes the y part; the second, the ε part; and the third, the x part of the linear model.

Generalized Linear Models: The standard linear model cannot handle nonnormal responses, y, such as counts or proportions. This motivates the development of generalized linear models that can represent categorical, binary and other response types.

Mixed Effect Models: Some data has a grouped, nested or hierarchical structure. Repeated measures, longitudinal and multilevel data consist of several observations taken on the same individual or group. This induces a correlation structure in the error, ε. Mixed effect models allow the modeling of such data.

Nonparametric Regression Models: In the linear model, the predictors, x, are combined in a linear way to model the effect on the response. Sometimes this linearity is insufficient to capture the structure of the data and more flexibility is required. Methods such as additive models, trees and neural networks allow a more flexible regression modeling of the response that combine the predictors in a nonparametric manner.

This book aims to provide the reader with a well-stocked toolbox of statistical methodologies. A practicing statistician needs to be aware of and familiar with the basic use of a broad range of ideas and techniques. This book will be a success if the reader is able to recognize and get started on a wide range of problems. However, the breadth comes at the expense of some depth. Fortunately, there are book-length treatments of topics discussed in every chapter of this book, so the reader will know where to go next if needed.

R is a free software environment for statistical computing and graphics. It runs on a wide variety of platforms including the Windows, Linux and Macintosh operating systems. Although there are several excellent statistical packages, only R is both free and possesses the power to perform the analyses demonstrated in this book. While it is possible in principle to learn statistical methods from purely theoretical expositions, I believe most readers learn best from the demonstrated interplay of theory and practice. The data analysis of real examples is woven into this book and all the R commands necessary to reproduce the analyses are provided.

Prerequisites: Readers should possess some knowledge of linear models. The first chapter provides a review of these models. This book can be viewed as a sequel to *Linear Models with R*, Faraway (2004). Even so there are plenty of other good books on linear models such as Draper and Smith (1998) or Weisberg (2005), that would provide ample grounding. Some knowledge of likelihood theory is also very useful. An outline is provided in Appendix A, but this may be insufficient for those who have never seen it before. A general knowledge of statistical theory is also expected concerning such topics as hypothesis tests or confidence intervals. Even so, the emphasis in this text is on application, so readers without much statistical theory can still learn something here.

This is not a book about learning R, but the reader will inevitably pick up the language by reading through the example data analyses. Readers completely new to R will benefit from studying an introductory book such as Dalgaard (2002) or one of the many tutorials available for free at the R website. Even so, the book should be intelligible to a reader without prior knowledge of R just by reading the text and output. R skills can be further developed by modifying the examples in this book, trying the exercises and studying the help pages for each command as needed. There is a large amount of detailed help on the commands available within the software and there is no point in duplicating that here. Please refer to Appendix B for details on obtaining and installing R along with the necessary add-on packages and data necessary for running the examples in this text. S-plus derives from the same `S` language as R, so many of the commands in this book will work. However, there are some differences in the syntax and the availability of add-on packages, so not everything here will work in S-plus.

The website for this book is at `people.bath.ac.uk/jjf23/ELM` where data described in this book appears. Updates and errata will also appear there.

Thanks to the builders of R without whom this book would not have been possible.

Contents

CHAPTER 1

Introduction

This book is about extending the linear model methodology using R statistical software. Before setting off on this journey, it is worth reviewing both linear models and R. We shall not attempt a detailed description of linear models; the reader is advised to consult texts such as Faraway (2004), Draper and Smith (1998) or Weisberg (2005). However, we will review the main points. Also, we do not intend this as a self-contained introduction to R as this may be found in books such as Dalgaard (2002) or Maindonald and Braun (2003) or from guides obtainable from the R website. Even so, a reader unfamiliar with R should be able to follow the analysis to follow and learn a little R in the process without further preparation.

Let's consider an example. The 2000 United States Presidential election generated much controversy, particularly in the state of Florida where there were some difficulties with the voting machinery. In Meyer (2002), data on voting in the state of Georgia is presented and analyzed. Let's take a look at this data using R. Please refer to Appendix B for details on obtaining and installing R along with the necessary add-on packages and data for running the examples in this text. R commands are typed at the command prompt: >. We start by loading the package of datasets that are used in this book:

```
> library(faraway)
```

Please remember that every time you want to access a dataset specific to this book, you will need to type the `library(faraway)` command. Since you might start a new session at any point in this book, in future we will simply assume that you type this first. If you forget, you will receive an error message notifying you that the data could not be found.

Next we load the dataset with the Georgia voting information:

```
> data(gavote)
```

The `data` command loads the particular dataset into R. In R, the object containing the data is called a *dataframe*. In most installations of R, this `data` step will be unnecessary as the datasets will be silently accessed using a process called *lazy loading*. However, we will retain this command throughout this book as a marker to indicate that we intend to use a particular dataset in R. Rather than typing the command, you might regard it as a reminder to consult the help page for the dataset. We can obtain definitions of the variables and more information about the dataset using the `help` command:

```
> help(gavote)
```

You can use the `help` command to learn more about any of the commands we use. For example, to learn about the `quantile` command:

```
> help(quantile)
```

If you do not already know or guess the name of the command you need, use:

```
> help.search("quantiles")
```

to learn about all commands that refer to quantiles. More generally, use:

```
> help.start()
```

to browse through the documentation.

We can examine the contents of the dataframe simply by typing its name:

```
> gavote
              equip   econ perAA rural    atlanta   gore   bush
APPLING       LEVER   poor 0.182 rural notAtlanta   2093   3940
ATKINSON      LEVER   poor 0.230 rural notAtlanta    821   1228
....
```

We have deleted most of the output although this dataset is small enough to be comfortably examined in its entirety. Sometimes, we simply want to look at the first few cases. The head command is useful for this:

```
> head(gavote)
         equip   econ perAA rural    atlanta gore bush other
APPLING  LEVER   poor 0.182 rural notAtlanta 2093 3940    66
ATKINSON LEVER   poor 0.230 rural notAtlanta  821 1228    22
BACON    LEVER   poor 0.131 rural notAtlanta  956 2010    29
BAKER    OS-CC   poor 0.476 rural notAtlanta  893  615    11
BALDWIN  LEVER middle 0.359 rural notAtlanta 5893 6041   192
BANKS    LEVER middle 0.024 rural notAtlanta 1220 3202   111
         votes ballots
APPLING   6099    6617
ATKINSON  2071    2149
BACON     2995    3347
BAKER     1519    1607
BALDWIN  12126   12785
BANKS     4533    4773
```

The cases in this dataset are the counties of Georgia and the variables are (in order) the type of voting equipment used, the economic level of the county, the percentage of African Americans, whether the county is rural or urban, whether the county is part of the Atlanta metropolitan area, the number of voters for Al Gore, the number of voters for George Bush, the number of voters for other candidates, the number of votes cast, and ballots issued.

A potential voter goes to the polling station where it is determined whether he or she is registered to vote. If so, a ballot is issued. However, a vote is not recorded if the person fails to vote for President, votes for more than one candidate or the equipment fails to record the vote. For example, we can see that in Appling county, $6617 - 6099 = 518$ ballots did not result in votes for President. This is called the *undercount*. The purpose of our analysis will be to determine what factors affect the undercount. We will not attempt a full and conclusive analysis here because our main purpose is to illustrate the use of linear models and R. The reader is invited to fill in some of the gaps in the analysis.

Initial data analysis: The first stage in any data analysis should be an initial graphical and numerical look at the data. A compact numerical overview is:

```
> summary(gavote)
   equip         econ          perAA            rural
 LEVER:74   middle:69   Min.   :0.000   rural:117
 OS-CC:44   poor  :72   1st Qu.:0.112   urban: 42
 OS-PC:22   rich  :18   Median :0.233
 PAPER: 2               Mean   :0.243
 PUNCH:17               3rd Qu.:0.348
                        Max.   :0.765
       atlanta          gore              bush
 Atlanta   : 15   Min.   :   249   Min.   :   271
 notAtlanta:144   1st Qu.:  1386   1st Qu.:  1804
                  Median :  2326   Median :  3597
                  Mean   :  7020   Mean   :  8929
                  3rd Qu.:  4430   3rd Qu.:  7468
                  Max.   :154509   Max.   :140494
     other          votes           ballots
 Min.   :   5   Min.   :   832   Min.   :   881
 1st Qu.:  30   1st Qu.:  3506   1st Qu.:  3694
 Median :  86   Median :  6299   Median :  6712
 Mean   : 382   Mean   : 16331   Mean   : 16927
 3rd Qu.: 210   3rd Qu.: 11846   3rd Qu.: 12251
 Max.   :7920   Max.   :263211   Max.   :280975
```

For the categorical variables, we get a count of the number of each type that occur. We notice, for example, that only two counties used a paper ballot. This will make it difficult to estimate the effect of this particular voting method on the undercount. For the numerical variables, we have six statistics that are sufficient to get a rough idea of the distributions. In particular, we notice that the number of ballots cast ranges over orders of magnitudes. This suggests that we should consider the relative, rather than the absolute, undercount. We create this new relative undercount variable, where we specify the variables using the `dataframe$variable` syntax:

```
> gavote$undercount <- (gavote$ballots-gavote$votes)/gavote$ballots
> summary(gavote$undercount)
   Min. 1st Qu.  Median    Mean 3rd Qu.    Max.
 0.0000  0.0278  0.0398  0.0438  0.0565  0.1880
```

We see that the undercount ranges from zero up to as much as 19%. The mean across counties is 4.38%. Note that this is not the same thing as the overall relative undercount which is:

```
> sum(gavote$ballots-gavote$votes)/sum(gavote$ballots)
[1] 0.03518
```

Graphical summaries are also valuable in gaining an understanding of the data. Considering just one variable at a time, histograms are a well-known way of examining the distribution of a variable:

```
> hist(gavote$undercount,main="Undercount",xlab="Percent Undercount")
```

The plot is shown in the left panel of Figure 1.1. A histogram is a fairly crude estimate

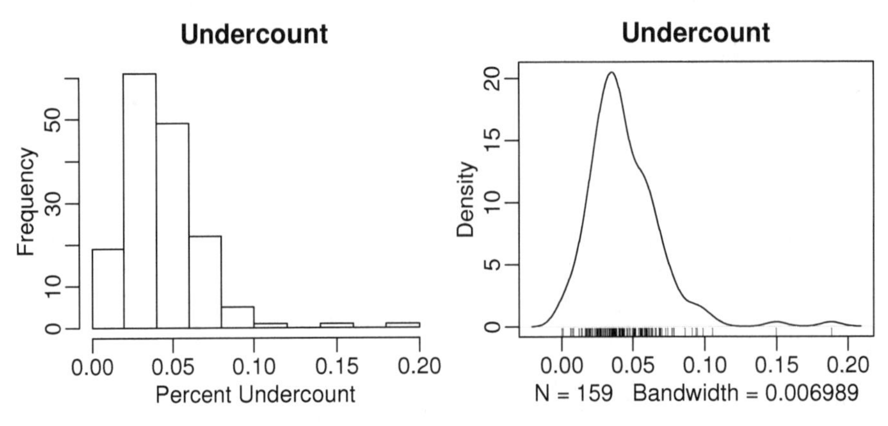

Figure 1.1 *Histogram of the undercount is shown on the left and a density estimate with a data rug is shown on the right.*

of the density of the variable that is sensitive to the choice of bins. A kernel density estimate can be viewed as a smoother version of a histogram that is also a superior estimate of the density. We have added a "rug" to our display that makes it possible to discern the individual data points:

```
> plot(density(gavote$undercount),main="Undercount")
> rug(gavote$undercount)
```

We can see that the distribution is slightly skewed and that there are two outliers in the right tail of the distribution. Such plots are invaluable in detecting mistakes or unusual points in the data. Categorical variables can also be graphically displayed. The pie chart is a popular method. We demonstrate this on the types of voting equipment:

```
> pie(table(gavote$equip),col=gray(0:4/4))
```

The plot is shown in the first panel of Figure 1.2. We have used shades of gray for the slices of the pie because this is a monochrome book. If you omit the `col` argument, you will see a color plot by default. Of course, a color plot is usually preferable, but bear in mind that most photocopying machines and many laser printers are black and white only, so a good greyscale plot is still needed. An alternative plot is a Pareto chart which is simply a bar plot with categories in descending order of frequency:

```
> barplot(sort(table(gavote$equip),decreasing=TRUE),las=2)
```

The plot is shown in the second panel of Figure 1.2. The `las=2` argument means that the bar labels are printed vertically as opposed to horizontally, ensuring that there is enough room for them to be seen. The Pareto chart (or just a bar plot) is superior to the pie chart because the size of the categories are easier to comprehend in the former plot.

Two dimensional plots are also very helpful. A scatterplot is the obvious way to depict two quantitative variables. Let's see how the proportion voting for Gore relates to the proportion of African Americans:

```
> gavote$pergore <- gavote$gore/gavote$votes
> plot(pergore ~ perAA, gavote, xlab="Proportion African American",
```

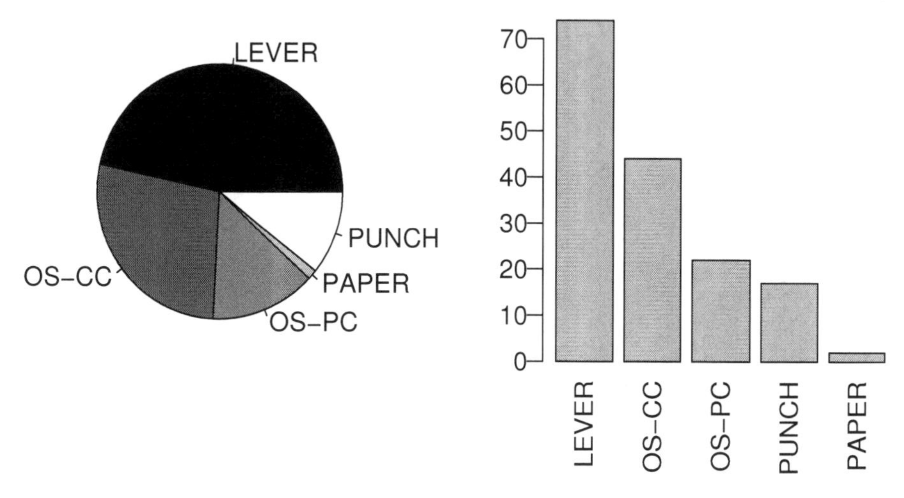

Figure 1.2 *Pie chart of the voting equipment frequencies is shown on the left and a Pareto chart on the right.*

```
  ylab="Proportion for Gore")
```

The plot, seen in the first panel of Figure 1.3, shows a strong correlation between these variables. This is an *ecological* correlation because the data points are aggregated across counties. The plot, in and of itself, does not prove that individual African Americans were more likely to vote for Gore, although we know this to be true from other sources. We could also compute the proportion of voters for Bush, but this is, not surprisingly, strongly negatively correlated with the proportion of voters for Gore. We do not need both variables as the one explains the other. We will use the proportion for Gore in the analysis to follow although one could just as well replace this with the proportion for Bush. We will not consider the proportion for other voters as this has little effect on our conclusions. The reader may wish to verify this.

Side-by-side boxplots are a good way of displaying the relationship between qualitative and quantitative variables:

```
> plot(undercount ~ equip, gavote, xlab="", las=3)
```

The plot, shown in the second panel of Figure 1.3, shows no major differences in undercount for the different types of equipment. Two outliers are visible for the optical scan-precinct count (OS-PC) method. Plots of two qualitative variables are generally not worthwhile unless both variables have more than three or four levels. The `xtabs()` function is useful for compiling cross-tabulations:

```
> xtabs(~ atlanta + rural, gavote)
            rural
atlanta       rural urban
  Atlanta         1    14
  notAtlanta    116    28
```

We see that just one county in the Atlanta area is classified as rural.

Correlations are the standard way of numerically summarizing the relationship

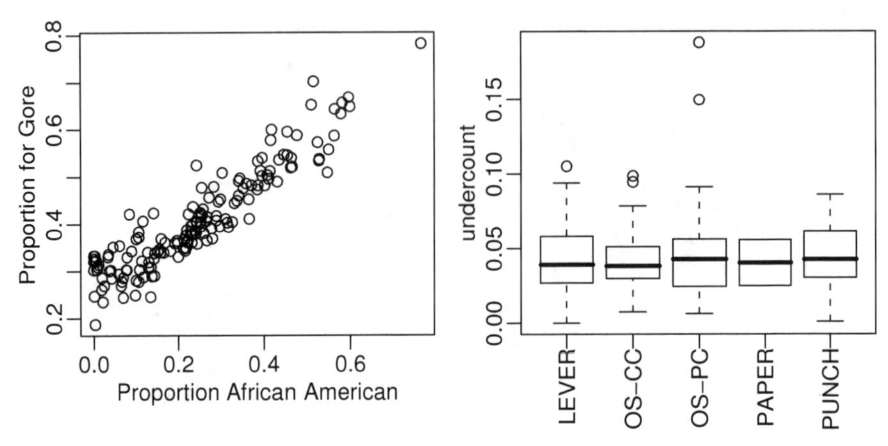

Figure 1.3 *A scatterplot plot of proportions of Gore voters and African Americans by county is shown on the left. Boxplots showing the distribution of the undercount by voting equipment are shown on the right.*

between quantitative variables. However, not all the variables in our dataframe are quantitative or immediately of interest. First we construct a vector using `c()` of length three which contains the indices of the variables of interest. We select these columns from the dataframe and compute the correlation. The syntax for selecting rows and/or columns is `dataframe[rows,columns]` where rows and/or columns are vectors of indices. In this case, we want all the rows, so we omit that part of the construction:

```
> nix <- c(3,10,11,12)
> cor(gavote[,nix])
               perAA   ballots undercount  pergore
perAA       1.000000  0.027732    0.22969 0.921652
ballots     0.027732  1.000000   -0.15517 0.095617
undercount  0.229687 -0.155172    1.00000 0.218765
pergore     0.921652  0.095617    0.21877 1.000000
```

We see some mild correlation between some of the variables except for the Gore — African Americans correlation which we know is large from the previous plot.

Defining a linear model: We shall try to describe this data with a linear model which takes the form:

$$Y = \beta_0 + \beta_1 X_1 + \beta_2 X_2 + \cdots + \beta_{p-1} X_{p-1} + \varepsilon$$

where β_i, $i = 0, 1, 2, \ldots, p-1$ are unknown *parameters*. β_0 is called the *intercept* term. The *response* is Y and the *predictors* are $X_1, \ldots, X_{p-1}$. The predictors may be the original variables in the dataset or transformations or combinations of them. The error ε represents the difference between what is explained by the systematic part of the model and what is observed. ε may include measurement error although it is often due to the effect of unincluded or unmeasured variables.

The regression equation is more conveniently written as:

$$y = X\beta + \varepsilon$$

where, in terms of the n data points, $y=(y_1,\ldots,y_n)^T$, $\varepsilon=(\varepsilon_1,\ldots,\varepsilon_n)^T$, $\beta=(\beta_0,\ldots,\beta_{p-1})^T$ and:

$$X=\begin{pmatrix} 1 & x_{11} & x_{12} & \ldots & x_{1,p-1} \\ 1 & x_{21} & x_{22} & \ldots & x_{2,p-1} \\ \vdots & & & & \vdots \\ 1 & x_{n1} & x_{n2} & \ldots & x_{n,p-1} \end{pmatrix}$$

The column of ones incorporates the intercept term. The *least squares* estimate of β, called $\hat{\beta}$ minimizes:

$$\sum \varepsilon_i^2 = \varepsilon^T\varepsilon = (y-X\beta)^T(y-X\beta)$$

Differentiating with respect to β and setting to zero, we find that $\hat{\beta}$ satisfies:

$$X^TX\hat{\beta}=X^Ty$$

These are called the *normal equations*.

Fitting a linear model: Linear models in R are fit using the `lm` command. For example, suppose we model the undercount as the response and the proportions of Gore voters and African Americans as predictors:

```
> lmod <- lm(undercount ~ pergore + perAA, gavote)
```

This corresponds to the linear model formula:

$$\texttt{undercount} = \beta_0 + \beta_1\texttt{pergore} + \beta_2\texttt{perAA} + \varepsilon$$

R uses the *Wilkinson–Rogers* notation of Wilkinson and Rogers (1973). For a straightforward linear model, such as here, we see that it corresponds to just dropping the parameters from the mathematical form. The intercept is included by default.

We can obtain the least squares estimates of β, called the regression coefficients, $\hat{\beta}$ by:

```
> coef(lmod)
(Intercept)      pergore       perAA
   0.032376     0.010979    0.028533
```

The construction of the least squares estimates do not require any assumptions about ε. If we are prepared to assume that the errors are at least independent and have equal variance, then the *Gauss–Markov* theorem tells us that the least squares estimates are the best linear unbiased estimates. Although it is not necessary, we might further assume that the errors are normally distributed, we might compute the maximum likelihood estimate (MLE) of β (see Appendix A for more MLEs). For the linear models, these MLEs are identical with the least squares estimates. However, we shall find that, in some of the extension of linear models considered later in this book, an equivalent notion to least squares is not suitable and that likelihood methods must be used. This issue does not arise with the standard linear model.

The predicted or fitted values are $\hat{y}=X\hat{\beta}$ while the residuals are $\hat{\varepsilon}=y-X\hat{\beta}=y-\hat{y}$. We can compute these as:

```
> predict(lmod)
 APPLING ATKINSON    BACON    BAKER  BALDWIN    BANKS
0.041337 0.043291 0.039618 0.052412 0.047955 0.036016
...
```

```
> residuals(lmod)
   APPLING   ATKINSON       BACON       BAKER     BALDWIN
 0.0369466 -0.0069949  0.0655506  0.0023484   0.0035899
...
```

where the ellipsis indicates that (much of) the output has been omitted.

It is useful to have some notion of how well the model fits the data. The residual sum of squares (RSS) is $\hat{\varepsilon}^T\hat{\varepsilon}$. This can be computed as:

```
> deviance(lmod)
[1] 0.09325
```

The term *deviance* is a more general measure of fit than RSS, which we will meet again in chapters to follow. For linear models, the deviance is the RSS.

The *degrees of freedom* for a linear model is the number of cases minus the number of coefficients or:

```
> df.residual(lmod)
[1] 156
> nrow(gavote)-length(coef(lmod))
[1] 156
```

Let the variance of the error be σ^2, then σ is estimated by the residual standard error computed from $\sqrt{(\mathrm{RSS}/\mathrm{df})}$. For our example, this works out to be:

```
> sqrt(deviance(lmod)/df.residual(lmod))
[1] 0.024449
```

Although several useful regression quantities are stored in the `lm` model object (which we called `lmod` in this instance), we can compute several more using the `summary` command on the model object. For example:

```
> lmodsum <- summary(lmod)
> lmodsum$sigma
[1] 0.024449
```

R is an object-oriented language. One important feature of such a language is that *generic* functions, such as `summary`, recognize the type of object being passed to it and behave appropriately. We used `summary` for dataframes above and now for linear models. `residuals` is another generic function and we shall see how it can be applied to many model types and return appropriately defined residuals.

The deviance measures how well the model fits in an absolute sense, but it does not tell us how well the model fits in a relative sense. The popular choice is R^2, called the *coefficient of determination* or *percentage of variance explained*:

$$R^2 = 1 - \frac{\sum(\hat{y}_i - y_i)^2}{\sum(y_i - \bar{y})^2} = 1 - \frac{\mathrm{RSS}}{\mathrm{TSS}}$$

where $\mathrm{TSS} = \sum(y_i - \bar{y})^2$ and stands for total sum of squares. This can be most conveniently extracted as:

```
> lmodsum$r.squared
[1] 0.053089
```

We see that R^2 is only about 5% which indicates that this particular model does not fit so well. An appreciation of what constitutes a good value of R^2 varies according to the application. Another way to think of R^2 is the (squared) correlation between the predicted values and the response:

```
> cor(predict(lmod),gavote$undercount)^2
[1] 0.053089
```

R^2 suffers as a criterion for choosing models among those available because it can never decrease when you add a new predictor to the model. This means that it will favor the largest models. The adjusted R^2 makes allowance for the fact a larger model also uses more parameters. It is defined as:

$$R_a^2 = 1 - \frac{RSS/(n-p)}{TSS/(n-1)}$$

Adding a predictor will only increase R_a^2 if it has some predictive value. Furthermore, minimizing $\hat{\sigma}^2$ means maximizing R_a^2 over a set of possible linear models. The value can be extracted as:

```
> lmodsum$adj.r.squared
[1] 0.040949
```

One advantage of R over many statistical packages is that we can extract all these quantities individually for subsequent calculations in a convenient way. However, if we simply want to see the regression output printed in a readable way, we use the `summary`:

```
> summary(lmod)
Residuals:
     Min       1Q   Median       3Q      Max
-0.04601 -0.01500 -0.00354  0.01178  0.14244

Coefficients:
            Estimate Std. Error t value Pr(>|t|)
(Intercept)   0.0324     0.0128    2.54    0.012
pergore       0.0110     0.0469    0.23    0.815
perAA         0.0285     0.0307    0.93    0.355

Residual standard error: 0.0244 on 156 degrees of freedom
Multiple R-Squared: 0.0531,Adjusted R-squared: 0.0409
F-statistic: 4.37 on 2 and 156 DF,  p-value: 0.0142
```

We have already separately computed many of the quantities given above.

Qualitative predictors: The addition of qualitative variables requires the introduction of dummy variables. Two-level variables are easy to code; consider the rural/urban indicator variable. We can code this using a dummy variable d:

$$d = \begin{cases} 0 & \text{rural} \\ 1 & \text{urban} \end{cases}$$

This is the default coding used in R. Zero is assigned to the level which is first alphabetically, unless something is done to change this (perhaps using the `relevel` command). If we add this variable to our model, it would now be:

$$\texttt{undercount} = \beta_0 + \beta_1 \texttt{pergore} + \beta_2 \texttt{perAA} + \beta_3 \texttt{d} + \varepsilon$$

So β_3 would now represent the difference between the undercount in an urban county and a rural county. Codings other than 0-1 could be used although the interpretation of the associated parameter would not be quite as straightforward.

A more extensive use of dummy variables is needed for factors with $k > 2$ levels. Let B be an $n \times k$ dummy variable matrix where $B_{ij} = 1$ if case i falls in class j and is zero otherwise. The coding is determined by a *contrast matrix* C which has dimension $k \times (k-1)$. The contribution of the factor to the model matrix X is then given by BC.

Consider the voting equipment, which is a five-level factor. The default choice for R is *treatment coding*. The contrast matrix, C that describes this coding, where the columns represent the dummy variables and the rows represent the levels, is:

```
> contr.treatment(5)
  2 3 4 5
1 0 0 0 0
2 1 0 0 0
3 0 1 0 0
4 0 0 1 0
5 0 0 0 1
```

This treats level one (LEVER in this example) as the standard level to which all other levels are compared. Each parameter for the dummy variable then represents the difference between the given level and the first level. Other codings, which define different C matrices, are used, such as Helmert or sum contrasts, but the treatment coding is generally the easiest to interpret.

Interactions between variables can be added to the model by taking the columns of the model matrix X that correspond to the two variables. Call these submatrices A and B having r and s columns, respectively. We then form a new matrix by multiplying every column of A elementwise with every column of B. This new matrix will have rs columns with typical element $a_{ij}b_{ik}$ where $i = 1, \ldots, n$, $j = 1, \ldots, r$ and $k = 1, \ldots, s$.

Interpretation: Let's add some qualitative variables to the model to see how the terms can be interpreted. We have centered the `pergore` and `perAA` terms by their means for reasons that will become clear:

```
> gavote$cpergore <- gavote$pergore - mean(gavote$pergore)
> gavote$cperAA <- gavote$perAA - mean(gavote$perAA)
> lmodi <- lm(undercount ~ cperAA+cpergore*rural+equip, gavote)
> summary(lmodi)
Coefficients:
                     Estimate Std. Error t value Pr(>|t|)
(Intercept)           0.04330    0.00284   15.25  < 2e-16
cperAA                0.02826    0.03109    0.91   0.3648
cpergore              0.00824    0.05116    0.16   0.8723
ruralurban           -0.01864    0.00465   -4.01 0.000096
equipOS-CC            0.00648    0.00468    1.39   0.1681
equipOS-PC            0.01564    0.00583    2.68   0.0081
equipPAPER           -0.00909    0.01693   -0.54   0.5920
equipPUNCH            0.01415    0.00678    2.09   0.0387
cpergore:ruralurban -0.00880     0.03872   -0.23   0.8205
```

```
Residual standard error: 0.0233 on 150 degrees of freedom
Multiple R-Squared: 0.17,Adjusted R-squared: 0.125
F-statistic: 3.83 on 8 and 150 DF,  p-value: 0.0004
```

Consider a rural county which has an average proportion of Gore voters and an average proportion of African Americans where lever machines are used for voting. Because rural and lever are the reference levels for the two qualitative variables, there is no contribution to the predicted undercount from these terms. Furthermore, because we have centered the two quantitative variables at their mean values, these terms also do not enter into the prediction. Notice the worth of the centering because otherwise we would need to set these variables to zero to get them to drop out of the prediction equation; zero is not a typical value for these predictors. Given that all the other terms are dropped, the predicted undercount is just given by the intercept, which is 4.33%.

The interpretation of the coefficients can now be made relative to this baseline. We see that, with all other predictors unchanged, except using optical scan with precinct count (OS-PC), the predicted undercount increases by 1.56%. The other equipment methods can be similarly interpreted. Notice that we need to be cautious about the interpretation. Given two counties with the same values of the predictors, except having different voting equipment, we would *predict* the undercount to be 1.56% higher for the OS-PC county compared to the lever county. However, we would not go so far as to say that if we went to a county with lever equipment and changed it to OS-PC that this would *cause* the undercount to increase by 1.56%.

With all other predictors held constant, we would predict the undercount to increase by 2.83% going from a county with no African Americans to all African American. Sometimes a one-unit change in a predictor is too large or too small, prompting a rescaling of the interpretation. For example, we might predict a 0.283% increase in the undercount for a 10% increase in the proportion of African Americans. Of course, this interpretation should not be taken too literally. We already know that the proportion of African Americans and Gore voters is strongly correlated so that an increase in the proportion of one would lead to an increase in the proportion of the other. This is the problem of *collinearity* which makes interpretation of regression coefficients more difficult. Furthermore, the proportion of African Americans is likely to be associated with other socioeconomic variables which might also be related to the undercount. This further hinders the possibility of a causal conclusion.

The interpretation of the `rural` and `pergore` cannot be done separately as there is an interaction term between these two variables. We see that for an average number of Gore voters, we would predict a 1.86%-lower undercount in an urban county compared to a rural county. In a rural county, we predict a 0.08% increase in the undercount as the proportion of Gore voters increases by 10%. In an urban county, we predict a $(0.00824 - 0.00880) * 10 = -0.0056\%$ increase in the undercount as the proportion of Gore voters increases by 10%. Since the increase is by a negative amount, this is actually a decrease. This illustrates the potential pitfalls in interpreting the effect of a predictor in the presence of an interaction. We cannot give a simple stand-alone interpretation of the effect of the proportion of Gore voters. The effect is to increase the undercount in rural counties and to decrease it, if only very slightly, in urban counties.

Hypothesis testing: We often wish to determine the significance of one, some or all of the predictors in a model. If we assume that the errors are independent and identically normally distributed, there is a very general testing procedure that may be used. Suppose we compare two models, a larger model Ω and a smaller model ω contained within that can be represented as a linear restriction on the parameters of the larger model. Most often, the predictors in ω are just a subset of the predictors in Ω.

Now suppose that the dimension (or number of parameters) of Ω is p and the dimension of ω is q, then, assuming that the smaller model ω is correct, the F-statistic is:

$$F = \frac{(\text{RSS}_\omega - \text{RSS}_\Omega)/(p-q)}{\text{RSS}_\Omega/(n-p)} \sim F_{p-q,n-p}$$

Thus we would reject the null hypothesis that the smaller model is correct if $F > F^{(\alpha)}_{p-q,n-p}$.

For example, we might compare the two linear models we have fit so far. The smaller model has just `pergore` and `perAA` while the larger model adds `rural` and `equip` along with an interaction. We may compute the F-test as:

```
> anova(lmod,lmodi)
Analysis of Variance Table

Model 1: undercount ~ pergore + perAA
Model 2: undercount ~ cperAA + cpergore * rural + equip
  Res.Df    RSS  Df Sum of Sq    F Pr(>F)
1    156 0.0932
2    150 0.0818   6    0.0115 3.51 0.0028
```

It does not matter that the variables have been centered in the larger model but not in the smaller model, because the centering makes no difference to the RSS. The p-value here is small indicating the null hypothesis of preferring the smaller model should be rejected.

One common F-test is the comparison of the current model to the null model, which is the model with no predictors and just an intercept term. This corresponds to the question of whether any of the variables have predictive value. For the larger model above, we can see that this F-statistic is 3.83 on 8 and 150 degrees of freedom with a p-value of 0.0004. We can see clearly that at least some of the predictors have some significance.

Another common need is to test specific predictors in the model. It is possible to use the general F-testing method: fit a model with the predictor and without the predictor and compute the F-statistic. It is important to know what other predictors are also included in the models and the results may differ if these are also changed. An alternative approach is to use a t-statistic for testing the hypothesis:

$$t_i = \hat{\beta}_i / se(\hat{\beta}_i)$$

and check for significance using a t-distribution with $n-p$ degrees of freedom. This approach will produce exactly the same p-value as the F-testing method. For example, in the larger model above, the test for the significance of the proportion of African Americans gives a p-value of 0.3648. This indicates that this predictor is not statistically significant after adjusting for the effect of the other predictors on the response.

We would usually avoid using the t-tests for the levels of qualitative predictors with more than two levels. For example, if we were interested in testing the effects of the various voting equipment, we would need to fit a model without this predictor and compute the corresponding F-test. A comparison of all models with one predictor less than the larger model may be obtained conveniently as:

```
> drop1(lmodi,test="F")
Single term deletions

Model:
undercount ~ cperAA + cpergore * rural + equip
               Df Sum of Sq      RSS   AIC F value Pr(F)
<none>                      0.081775 -1186
cperAA          1  0.000451 0.082226 -1187    0.83 0.365
equip           4  0.005444 0.087219 -1184    2.50 0.045
cpergore:rural  1  0.000028 0.081804 -1188    0.05 0.821
```

We see that the equipment is barely statistically significant in that the p-value is just less than the traditional 5% significance level. You will also notice that only the interaction term `cpergore:rural` is considered and not the corresponding main effects terms, `cpergore` and `rural`. This demonstrates respect for the *hierarchy principle* which demands that all lower-order terms corresponding to an interaction be retained in the model. In this case, we see that the interaction is not significant, but a further step would now be necessary to test the main effects.

There are numerous difficulties with interpreting the results of hypothesis tests and the reader is advised to avoid taking the results too literally before understanding these problems.

Confidence intervals for β may be constructed using:

$$\hat{\beta}_i \pm t_{n-p}^{(\alpha/2)} se(\hat{\beta}_i)$$

where $t_{n-p}^{(\alpha/2)}$ is the upper $\alpha/2^{th}$ quantile of a t distribution with $n-p$ degrees of freedom. A convenient way of computing the 95% confidence intervals in R is:

```
> confint(lmodi)
                          2.5 %     97.5 %
(Intercept)          0.03768844  0.0489062
cperAA              -0.03317106  0.0896992
cpergore            -0.09284293  0.1093166
ruralurban          -0.02782090 -0.0094523
equipOS-CC          -0.00276464  0.0157296
equipOS-PC           0.00412523  0.0271540
equipPAPER          -0.04253684  0.0243528
equipPUNCH           0.00074772  0.0275515
cpergore:ruralurban -0.08529909  0.0677002
```

Confidence intervals have a duality with the corresponding t-tests in that if the p-value is greater than 5%, zero will fall in the interval and vice versa. Confidence intervals give a range of plausible values for the parameter and are more useful for judging the size of the effect of the predictor than a p-value which merely indicates statistical significance, not necessarily practical significance. These intervals are individually correct, but there is not a 95% chance that the true parameter values fall in all the intervals. This problem of *multiple comparisons* is particularly acute for the voting equipment, where five levels leads to 10 possible pairwise comparisons, more than just the four shown here.

Diagnostics: The validity of the inference depends on the assumptions concerning the linear model. One part of these assumptions is that the systematic form of the model $EY = X\beta$ is correct; we have included all the right variables and transformed and combined them correctly. Another set of assumptions concerns the random part of the model: ε. We require that the errors have equal variance, be uncorrelated and have a normal distribution. We are also interested in detecting points, called *outliers*, that are unusual in that they do not fit the model that seems adequate for the rest of the data. Ideally, we would like each case to have an equal contribution to the fitted model; yet sometimes a few points have a much larger effect than others. Such points are called *influential*.

Diagnostic methods can be graphical or numerical. We generally prefer graphical methods because they tend to be more versatile and informative. It is virtually impossible to verify that a given model is exactly correct. The purpose of the diagnostics is more to check whether the model is not grossly wrong. Indeed, a successful data analyst should pay more attention to avoiding big mistakes than optimizing the fit.

A collection of four useful diagnostics can be simply obtained with:

```
> plot(lmodi)
```

as can be seen in Figure 1.4. The plot in the upper-left panel shows the residuals plotted against the fitted values. The plot can be used to detect lack of fit. If the residuals show some curvilinear trend, this is a sign that some change to the model is required, often a transformation of one of the variables. In this instance, there is no sign of such a problem. The plot is also used to check the constant variance assumption on the errors. In this case, it seems the variance is roughly constant as the fitted values vary. Assuming symmetry of the errors, we can effectively double the resolution by plotting the absolute value of the residuals against the fitted values. As it happens $|\hat{\varepsilon}|$ tends to be rather skewed and is better to use $\sqrt{\hat{\varepsilon}}$. Such a plot is shown in the lower-left panel, confirming what we have already observed about the constancy of the variance. Notice that a few larger residuals have been labeled.

The residuals can be assessed for normality using a *QQ plot*. This compares the residuals to "ideal" normal observations. We plot the sorted residuals against $\Phi^{-1}(\frac{i}{n+1})$ for $i = 1, \ldots, n$. This can be seen in the upper-right panel of Figure 1.4. In this plot, the points follow a linear trend (except for one or two cases), indicating that normality is a reasonable assumption. If we observe a curve, this indicates skewness, suggesting a possible transformation of the response, while two tails of points diverging from linearity would indicate a long-tailed error, suggesting that we should consider robust fitting methods. Particularly for larger datasets, the normality assumption is not

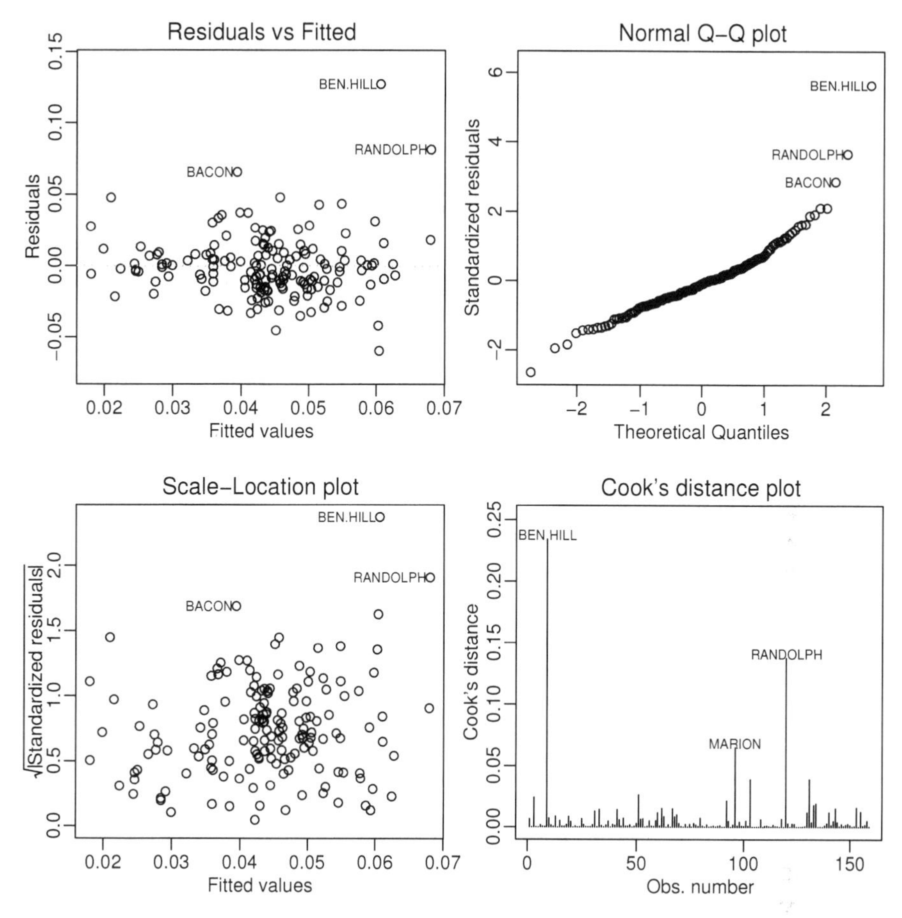

Figure 1.4 *Diagnostics obtained from plotting the model object.*

crucial, as the inference will be approximately correct in spite of the nonnormality. Only a clear deviation from normality should necessarily spur some action to change the model.

The Cook statistics are a popular influence diagnostic because they reduce the information to a single value for each case. They are defined as:

$$D_i = \frac{(\hat{y} - \hat{y}_{(i)})^T (\hat{y} - \hat{y}_{(i)})}{p\hat{\sigma}^2}$$

They represent a scaled measure of the change in the fit if the single case is dropped from the dataset. An index plot of the Cook statistics for the current model is given in the lower-right panel of Figure 1.4. We can see that there are a couple of cases that stick out and we should investigate more closely the influence of these points. We can pick out the top two influential cases with:

```
> gavote[cooks.distance(lmodi) > 0.1,]
```

```
         equip econ perAA rural    atlanta gore bush other votes
BEN.HILL OS-PC poor 0.282 rural notAtlanta 2234 2381    46  4661
RANDOLPH OS-PC poor 0.527 rural notAtlanta 1381 1174    14  2569
         ballots undercount pergore cpergore   cperAA
BEN.HILL    5741    0.18812 0.47930 0.070975 0.039019
RANDOLPH    3021    0.14962 0.53756 0.129241 0.284019
```

Notice how we can select a subset of a dataframe using a logical expression. Here we ask for all rows in the dataframe that have Cook statistics larger than 0.1. We see that these are the same two counties that stuck out in the boxplots seen in Figure 1.3.

The fitted values can be written as $X\hat{\beta} = X(X^TX)^{-1}X^Ty = Hy$ where the *hat-matrix* $H = X(X^TX)^{-1}X^T$. $h_i = H_{ii}$ are called *leverages* and are useful diagnostics. For example, since var $\hat{\varepsilon}_i = \sigma^2(1-h_i)$, a large leverage, h_i, will tend to make var $\hat{\varepsilon}_i$ small. The fit will be "forced" close to y_i. It is useful to examine the leverages to determine which cases have the power to be influential. Points on the boundary of the predictor space will have the most leverage.

A useful technique for judging whether some cases in a set of positive observations are unusually extreme is the half-normal plot. Here we plot the sorted values against $\Phi^{-1}(\frac{n+i}{2n+1})$ which represent the quantiles of the upper half of a standard normal distribution. We are usually not looking for a straight line relationship since we do not necessarily expect a positive normal distribution for quantities like the leverages. We are looking for outliers, which will be apparent as points that diverge substantially from the rest of the data. Here is the half-normal plot of the leverages:

```
> halfnorm(influence(lmodi)$hat)
```

The plot, seen in the left panel of Figure 1.5, shows two points with much higher leverage than the rest. These points are:

```
> gavote[influence(lmodi)$hat>0.3,]
           equip econ perAA rural    atlanta gore bush other
MONTGOMERY PAPER poor 0.243 rural notAtlanta 1013 1465    31
TALIAFERRO PAPER poor 0.596 rural notAtlanta  556  271     5
           votes ballots undercount pergore   cpergore
MONTGOMERY  2509    2573   0.024874 0.40375 -0.0045753
TALIAFERRO   832     881   0.055619 0.66827  0.2599475
```

These are the only two counties that use a paper ballot, so they will be the only cases that determine the coefficient for paper. This is sufficient to give them high leverage as the remaining predictor values are all unremarkable. Note that these counties were not identified as influential — having high leverage alone is not necessarily enough to be influential.

Partial residual plots display $\hat{\varepsilon} + \hat{\beta}_i x_i$ against x_i. To see the motivation, look at the response with the predicted effect of the other X removed:

$$y - \sum_{j \neq i} x_j\hat{\beta}_j = \hat{y} + \hat{\varepsilon} - \sum_{j \neq i} x_j\hat{\beta}_j = x_i\hat{\beta}_i + \hat{\varepsilon}$$

The partial residual plot for `cperAA` is shown in the right panel of Figure 1.5:

```
> termplot(lmodi,partial=TRUE,terms=1)
```

The line is the least squares fit to the data on this plot as well as having the same

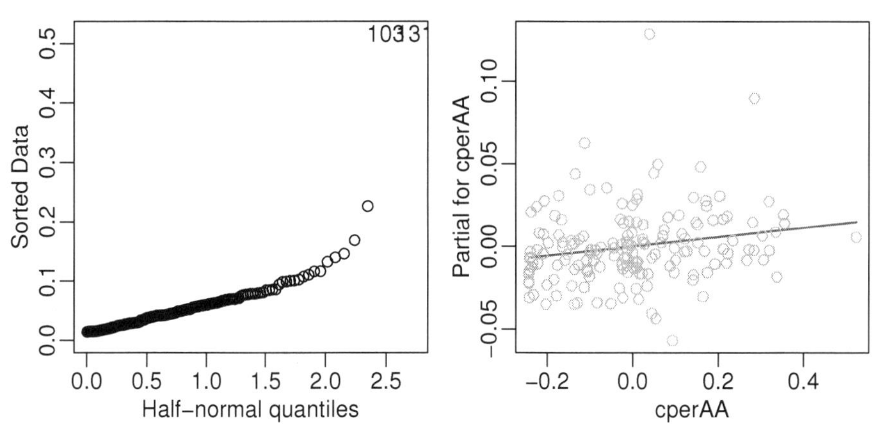

Figure 1.5 *Half-normal plot of the leverages is shown on the left and a partial residual plot for the proportion of African Americans is shown on the right.*

slope as the `cperAA` term in the current model. This plot gives us a snapshot of the marginal relationship between this predictor and the response. In this case, we see a linear relationship indicating that it is not worthwhile seeking transformations. Furthermore, there is no sign that a few points are having undue influence on the relationship.

Robust regression: Least squares works well when there are normal errors, but performs poorly for long-tailed errors. We have identified a few potential outliers in the current model. One approach is to simply eliminate the outliers from the dataset and then proceed with least squares. This approach is satisfactory when we are convinced that the outliers represent truly incorrect observations, but even then, detecting such cases is not always easy as multiple outliers can mask each other. However, in other cases, outliers are real observations. Sometimes, removing these cases simply creates other outliers. A generally better approach is to use a robust alternative to least squares that downweights the effect of larger errors. The Huber method is the default choice of the `rlm` function, which is part of the MASS package of Venables and Ripley (2002):

```
> library(MASS)
> rlmodi <- rlm(undercount ~ cperAA+cpergore*rural+equip, gavote)
> summary(rlmodi)
Coefficients:
                    Value  Std. Error t value
(Intercept)          0.041  0.002      17.866
cperAA               0.033  0.025       1.290
cpergore            -0.008  0.042      -0.197
ruralurban          -0.017  0.004      -4.406
equipOS-CC           0.007  0.004       1.802
equipOS-PC           0.008  0.005       1.695
equipPAPER          -0.006  0.014      -0.427
equipPUNCH           0.017  0.006       3.072
```

```
cpergore:ruralurban  0.007  0.032      0.230

Residual standard error: 0.0172 on 150 degrees of freedom
```

Inferential methods are more difficult to apply when robust estimation methods are used, hence there is less in this output than for the corresponding `lm` output above. The most interesting change is that the coefficient for OS-PC is now about half the size. Recall that, using the treatment coding, this represents the difference between OS-PC and the reference lever method. There is some fluctuation in the other coefficients, but not enough to change our impression of the important effects. The robust fit here has reduced the effect of the two outlying counties.

Weighted least squares: The sizes of the counties in this dataset vary greatly with the number of ballots cast in each county ranging from 881 to 280,975. We might expect the proportion of undercounted votes to be more variable in smaller counties than larger ones. Since the responses from the larger counties might be more precise, one might think they should count for more in the fitting of the model. This effect can be achieved by the use of weighted least squares where we attempt to minimize $\sum w_i \varepsilon_i^2$. The appropriate choice for the weights w_i is to set them to be inversely proportional to var y_i.

Now var y for a binomial proportion is inversely proportional to the group size, in this case, the number of ballots. This suggests setting the weights proportional to the number of ballots:

```
> wlmodi <- lm(undercount ~ cperAA+cpergore*rural+equip,
  gavote, weights=ballots)
```

This results in a fit that is substantially different from the unweighted fit. It is dominated by the data from a few large counties.

However, the variation in the response is likely to be caused by more than just binomial variation due to the number of ballots. There are likely to be other variables that affect the response in a way that is not proportional to ballot size. Consider the relative size of these effects. Even for the smallest county, assuming an average undercount rate, the standard deviation using the binomial is:

```
> sqrt(0.035*(1-0.035)/881)
[1] 0.0061917
```

which is much smaller than the residual standard error of 0.0233. The effects will be substantially smaller for other counties. So since the other sources of variation dominate, we would recommend leaving this particular model unweighted.

Transformation: Models can sometimes be improved by transforming the variables. Ideas for transformations can come from several sources. One method is to search through a family of possible transformations looking for the best fit. An example of this approach is the Box–Cox method of selecting a transformation on the response variable. Alternatively, the diagnostic plots for the current model can suggest transformations that might improve the fit or ameliorate an apparent violations of the assumptions. In other situations, transformations may be motivated by theories concerning the relationship between the variables or to aid the interpretation of the model.

For this dataset, transformation of the response is problematic for both technical and interpretational reasons. The minimum undercount is exactly zero which precludes directly applying some popular transformations such as the log or inverse. An arbitrary fix for this problem is to add a small amount (say 0.005 here) to the response which would enable the use of all power transformations. The application of the Box–Cox method, using the `boxcox` function from the `MASS` package, suggests a square root transformation of the response. However, it is difficult to give an interpretation to the regression coefficients with this transformation on the response. Other than no transformation at all, a logged response does allow a simple interpretation. For an untransformed response, the coefficients represent addition to the undercount whereas for a logged response, the coefficients can be interpreted as multiplying the response. So we see that, although transformations of the response might sometimes improve the fit, they can lead to difficulties with interpretation and so should be applied with care. Another point to consider is that if the untransformed response was normally distributed, it will not be so after transformation. This suggests considering nonnormal, continuous responses as seen in Section 7.1, for example.

Transformations of the predictors are less problematic. Let's first consider the proportion of African Americans predictor in the current model. Polynomials provide a commonly used family of transformations. The use of orthogonal polynomials is recommended as these a more numerically stable and make it easier to select the correct degree:

```
> plmodi <- lm(undercount ~ poly(cperAA,4)+cpergore*rural+equip, gavote)
> summary(plmodi)
Coefficients:
                      Estimate Std. Error t value Pr(>|t|)
(Intercept)            0.04346    0.00288   15.12  < 2e-16
poly(cperAA, 4)1       0.05226    0.06939    0.75   0.4526
poly(cperAA, 4)2      -0.00299    0.02613   -0.11   0.9091
poly(cperAA, 4)3      -0.00536    0.02427   -0.22   0.8254
poly(cperAA, 4)4      -0.01651    0.02420   -0.68   0.4961
cpergore               0.01315    0.05693    0.23   0.8176
ruralurban            -0.01913    0.00474   -4.03 0.000088
equipOS-CC             0.00644    0.00472    1.36   0.1746
equipOS-PC             0.01559    0.00588    2.65   0.0089
equipPAPER            -0.01027    0.01720   -0.60   0.5514
equipPUNCH             0.01405    0.00687    2.05   0.0425
cpergore:ruralurban   -0.01054    0.04136   -0.25   0.7993

Residual standard error: 0.0235 on 147 degrees of freedom
Multiple R-Squared: 0.173,Adjusted R-squared: 0.111
F-statistic: 2.79 on 11 and 147 DF,  p-value: 0.00254
```

The hierarchy principle requires that we avoid eliminating lower-order terms of a variable when high-order terms are still in the model. From the output, we see that the fourth-order term is not significant and can be eliminated. With standard polynomials, the elimination of one term would cause a change in the values of the remaining coefficients. The advantage of the orthogonal polynomials is that the coefficients for the lower-order terms do not change as we change the maximum degree of the model. Here we see that all the terms of cperAA are not significant and all can be removed. Some insight into the relationship may be gained by plotting the fit on top of the partial residuals:

```
> termplot(plmodi,partial=TRUE,terms=1)
```

The plot, seen in the first panel of Figure 1.6, shows that the quartic polynomial is not so different from a constant fit, explaining the lack of significance.

Polynomial fits become less attractive with higher-order terms. The fit is not local in the sense that a point in one part of the range of the variable affects the fit across the whole range. Furthermore, polynomials tend to have rather oscillatory fits and extrapolate poorly. A more stable fit can be had using splines, which are piecewise polynomials. Various types of splines are available and they typically have the local fit and stable extrapolation properties. We demonstrate the use of cubic B-splines here:

```
> library(splines)
> blmodi <- lm(undercount ~ cperAA+bs(cpergore,4)+rural+equip, gavote)
```

Because the spline fit for cperAA was very similar to orthogonal polynomials, we consider cpergore here for some variety. Notice that we have eliminated the interaction with rural for simplicity. The complexity of the B-spline fit may be controlled by specifying the degrees of freedom. We have used four here. The nature of the fit can be seen in the second panel of Figure 1.6:

```
> termplot(blmodi,partial=TRUE,terms=2)
```

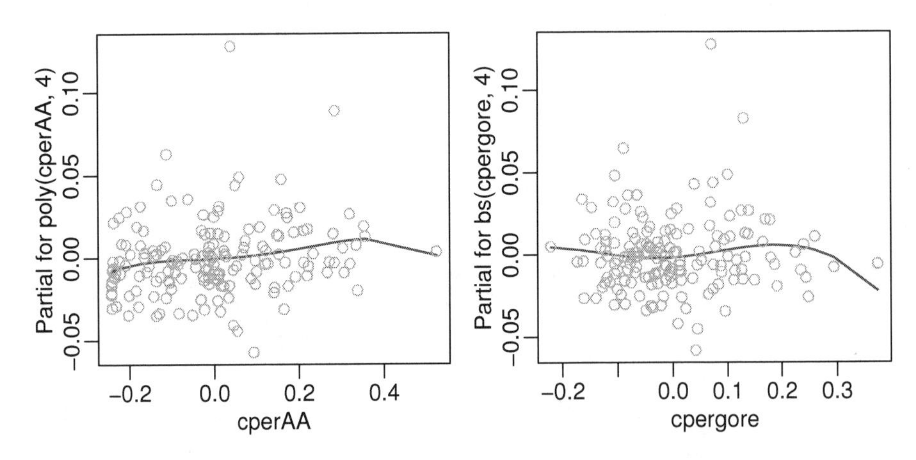

Figure 1.6 *Partial fits using orthogonal polynomials for* cperAA *(shown on the left) and cubic B-splines for* cpergore *(shown on the right).*

We see that the curved fit is not much different from a constant.

Variable selection: One theoretical view of the problem of variable selection is that one subset of the available variables represents the correct model for the data and that any method should be judged by its success in identifying this correct model. While this may be a tempting world in which to test competing variable selection methods, it seems unlikely to match with reality. Even if we believe that a correct model even exists, it is more than likely that we will not have recorded all the relevant variables or not have chosen the correct transformations or functional form for the model amongst the set we choose to consider. We might then retreat from the initial goal and hope to identify the best model from the available set. Even then, we would need to define what is meant by best.

Linear modeling serves two broad goals. Some build linear models for the purposes of prediction — they expect to observe new X and wish to predict y, along with measures of uncertainty in the prediction. Prediction performance is improved by removing variables that contribute little or nothing to the model. We can define a criterion for prediction performance and search for the model that optimizes that criterion. One such criterion is the adjusted R^2 previously mentioned. The `regsubsets` function in the `leaps` package implements this search. For problems involving a moderate number of variables, it is possible to exhaustively search all possible models for the best. As the number of variables increases, exhaustive search become prohibitive and various stepwise methods must be used to search the model space. The implementation also has the disadvantage that it can only be applied to quantitative predictors.

Another popular criterion is the Akaike Information Criterion or AIC defined as:

$$\text{AIC} = -2 \text{ maximum log likelihood} + 2p$$

where p is the number of parameters. This criterion has the advantage of generality and can be applied far beyond normal linear models. The `step` command implements a stepwise search strategy through the space of possible models. It does allow qualitative variables and respects the hierarchy principle. We start by defining a rather large model:

```
> biglm <- lm(undercount ~ (equip+econ+rural+atlanta)^2+
   (equip+econ+rural+atlanta)*(perAA+pergore), gavote)
```

This model includes up to all two-way interactions between the qualitative variables along with all two-way interaction between a qualitative and a quantitative variable. All main effects are included. The `step` command sequentially eliminates terms to minimize the AIC:

```
> smallm <- step(biglm,trace=F)
```

The resulting model includes interactions between `equip` and `econ`, `econ` and `perAA`, and `rural` and `perAA`, together with the associated main effects. The `trace=F` arguments blocks the large amount of intermediate model information that we would otherwise see.

Linear modeling is also used to try to understand the relationship between the variables — we want to develop an explanation for the data. For this dataset, we are much more interested in explanation than prediction. However, the two goals are not mutually exclusive and often the same methods are used for variable selection in both cases. Even so, when explanation is the goal, it may be unwise to rely on completely automated variable selection methods. For example, the proportion of voters for Gore was eliminated from the model by the AIC-based `step` method and yet we know this variable to be strongly correlated with the proportion of African Americans which is in the model. It would be rash to conclude that the latter variable is important and the former is not — the two are intertwined. Researchers interested in explaining the relationship may prefer a more manual variable selection approach that takes into account background information and is geared toward the substantive questions of interest.

The other major class of variable selection methods is based on testing. We can use F-tests to compare larger models with smaller nested models. A stepwise testing approach can then be applied to select a model. The consensus view among statisticians is that this is an inferior method to variable selection compared to the criterion-based methods. Nevertheless, testing-based methods are still useful, particularly when under manual control. They have the advantage of applicability across a wide class of models where tests have been developed. They allow the user to respect restrictions of hierarchy and situations where certain variables must be included for explanatory purposes. Let's compare the AIC-selected models above to models with one fewer term:

```
> drop1(smallm,test="F")
Single term deletions

Model:
undercount ~ equip + econ + rural + perAA + equip:econ + equip:perAA +
    rural:perAA
            Df Sum of Sq    RSS   AIC F value  Pr(F)
<none>                   0.0536 -1231
equip:econ   6    0.0075 0.0612 -1222    3.25 0.0051
equip:perAA  4    0.0068 0.0605 -1220    4.43 0.0021
rural:perAA  1    0.0010 0.0546 -1230    2.65 0.1060
```

We see that the `rural:perAA` can be dropped. A subsequent test reveals that `rural` can also be removed. This gives us a final model of:

```
> finalm <- lm(undercount~equip + econ  + perAA + equip:econ
  + equip:perAA, gavote)
> summary(finalm)
Coefficients: (2 not defined because of singularities)
                    Estimate Std. Error t value Pr(>|t|)
(Intercept)          0.04187    0.00503    8.33  6.5e-14
equipOS-CC          -0.01133    0.00737   -1.54  0.12670
equipOS-PC           0.00858    0.01118    0.77  0.44429
equipPAPER          -0.05843    0.03701   -1.58  0.11669
equipPUNCH          -0.01575    0.01875   -0.84  0.40218
econpoor             0.02027    0.00553    3.67  0.00035
econrich            -0.01697    0.01239   -1.37  0.17313
perAA               -0.04204    0.01659   -2.53  0.01239
equipOS-CC:econpoor -0.01096    0.00988   -1.11  0.26922
equipOS-PC:econpoor  0.04838    0.01380    3.51  0.00061
equipPAPER:econpoor       NA         NA      NA       NA
equipPUNCH:econpoor -0.00356    0.01243   -0.29  0.77492
equipOS-CC:econrich  0.00228    0.01538    0.15  0.88246
equipOS-PC:econrich -0.01332    0.01705   -0.78  0.43615
equipPAPER:econrich       NA         NA      NA       NA
```

```
equipPUNCH:econrich  0.02003    0.02200     0.91  0.36405
equipOS-CC:perAA     0.10725    0.03286     3.26  0.00138
equipOS-PC:perAA    -0.00591    0.04341    -0.14  0.89198
equipPAPER:perAA     0.12914    0.08181     1.58  0.11668
equipPUNCH:perAA     0.08685    0.04650     1.87  0.06388

Residual standard error: 0.02 on 141 degrees of freedom
Multiple R-Squared: 0.428,Adjusted R-squared: 0.359
F-statistic:  6.2 on 17 and 141 DF,  p-value: 1.32e-10
```

Because there are only two paper-using counties, there is insufficient data to estimate the interaction terms involving paper. This model output is difficult to interpret because of the interaction terms.

Conclusion: Let's attempt an interpretation of this final model. Certainly we should explore more models and check more diagnostics, so our conclusions can only be tentative. The reader is invited to investigate other possibilities.

To interpret interactions, it is often helpful to construct predictions for all the levels of the variables involved. Here we generate all combinations of `equip` and `econ` for a median proportion of `perAA`:

```
> pdf <- data.frame(econ=rep(levels(gavote$econ),5),
  equip=rep(levels(gavote$equip),rep(3,5)),perAA=0.233)
```

We now compute the predicted undercount for all 15 combinations and display the result in a table:

```
> pp <- predict(finalm,new=pdf)
> xtabs(round(pp,3) ~ econ + equip, pdf)

         equip
econ      LEVER  OS-CC  OS-PC  PAPER  PUNCH
  middle  0.032  0.046  0.039  0.004  0.037
  poor    0.052  0.055  0.108  0.024  0.053
  rich    0.015  0.031  0.009 -0.013  0.040
```

We can see that the undercount is lower in richer counties and higher in poorer counties. The amount of difference depends on the voting system. Of the three most commonly used voting methods, the LEVER method seems best. It is hard to separate the two optical scan methods, but there is clearly a problem with the precinct count in poorer counties, which is partly due to the two outliers we observed earlier. We notice one impossible prediction — a negative undercount in rich paper-using counties, but given the absence of such data (there were no such counties), we are not too disturbed.

We use the same approach to investigate the relationship between the proportion of African Americans and the voting equipment. We set the proportion of African Americans at three levels — the first quartile, the median and the third quartile and then compute the predicted undercount for all types of voting equipment. We set the `econ` variable to middle:

```
> pdf <- data.frame(econ=rep("middle",15),equip=rep(levels(gavote$equip),
  rep(3,5)),perAA=rep(c(.11,0.23,0.35),5))
> pp <- predict(finalm,new=pdf)
```

We create a three-level factor for the three levels of `perAA` to aid the construction of the table:

```
> propAA <- gl(3,1,15,labels=c("low","medium","high"))
> xtabs(round(pp,3) ~ propAA + equip,pdf)
        equip
propAA    LEVER  OS-CC  OS-PC  PAPER  PUNCH
  low     0.037  0.038  0.045 -0.007  0.031
  medium  0.032  0.046  0.039  0.003  0.036
  high    0.027  0.053  0.034  0.014  0.042
```

We see that the effect of the proportion of African Americans on the undercount is mixed. High proportions are associated with higher undercounts for OS-CC and PUNCH and associated with lower undercounts for LEVER and OS-PC.

In summary, we have found that the economic status of a county is the clearest factor determining the proportion of undercounted votes, with richer counties having lower undercounts. The type of voting equipment and the proportion of African Americans do have some impact on the response, but the direction of the effects are not simply stated. We would like to emphasize again that this dataset deserves further analysis before any definitive conclusions are drawn.

Exercises

Since this is a review chapter, it is best to consult the recommended background texts for specific questions on linear models. However, it is worthwhile gaining some practice using R on some real data. Your data analysis should consist of:

- An initial data analysis that explores the numerical and graphical characteristics of the data.
- Variable selection to choose the best model.
- An exploration of transformations to improve the fit of the model.
- Diagnostics to check the assumptions of your model.
- Some predictions of future observations for interesting values of the predictors.
- An interpretation of the meaning of the model with respect to the particular area of application.

There is always some freedom in deciding which methods to use, in what order to apply them, and how to interpret the results. So there may not be one clear right answer and good analysts may come up with different models.

Here are some datasets which should provide some good practice at building linear models:

1. The `swiss` data — use `Fertility` as the response.
2. The `rock` data — use `perm` as the response.
3. The `mtcars` data — use `mpg` as the response.
4. The `attitude` data — use `rating` as the response.
5. The `prostate` data — use `lpsa` as the response.
6. The `teengamb` data — use `gamble` as the response.

CHAPTER 2

Binomial Data

2.1 Challenger Disaster Example

In January 1986, the space shuttle Challenger exploded shortly after launch. An investigation was launched into the cause of the crash and attention focused on the rubber O-ring seals in the rocket boosters. At lower temperatures, rubber becomes more brittle and is a less effective sealant. At the time of the launch, the temperature was 31°F. Could the failure of the O-rings have been predicted? In the 23 previous shuttle missions for which data exists, some evidence of damage due to blow by and erosion was recorded on some O-rings. Each shuttle had two boosters, each with three O-rings. For each mission, we know the number of O-rings out of six showing some damage and the launch temperature. This is a simplification of the problem — see Dalal, Fowlkes, and Hoadley (1989) for more details.

Let's start our analysis with R. For help in obtaining R and installing the necessary add-on packages and datasets, please see Appendix B. First we load the data. To do this, you will first need to load the `faraway` package using the `library` command as seen in here. You will need to do this in every session that you run examples from this book. If you forget, you will receive a warning message about the data not being found. We then plot the proportion of damaged O-rings against temperature in Figure 2.1:

```
> library(faraway)
> data(orings)
> plot(damage/6 ~ temp, orings, xlim=c(25,85), ylim = c(0,1),
  xlab="Temperature", ylab="Prob of damage")
```

We are interested in how the probability of failure in a given O-ring is related to the launch temperature and predicting that probability when the temperature is 31°F. A naive approach, based on linear models, simply fits a line to this data:

```
> lmod <- lm(damage/6 ~ temp, orings)
> abline(lmod)
```

The fit is shown in Figure 2.1. There are several problems with this approach. Most obviously from the plot, it can predict probabilities greater than one or less than zero. One might suggest truncating predictions outside the range to zero or one as appropriate, but it does not seem credible that these probabilities would be exactly zero or one, in this particular example or many others.

We might consider the number of damage incidents to be binomially distributed. For a linear model, we require the errors to be approximately normally distributed for accurate inference. However, for a binomial with only six trials, the normal approx-

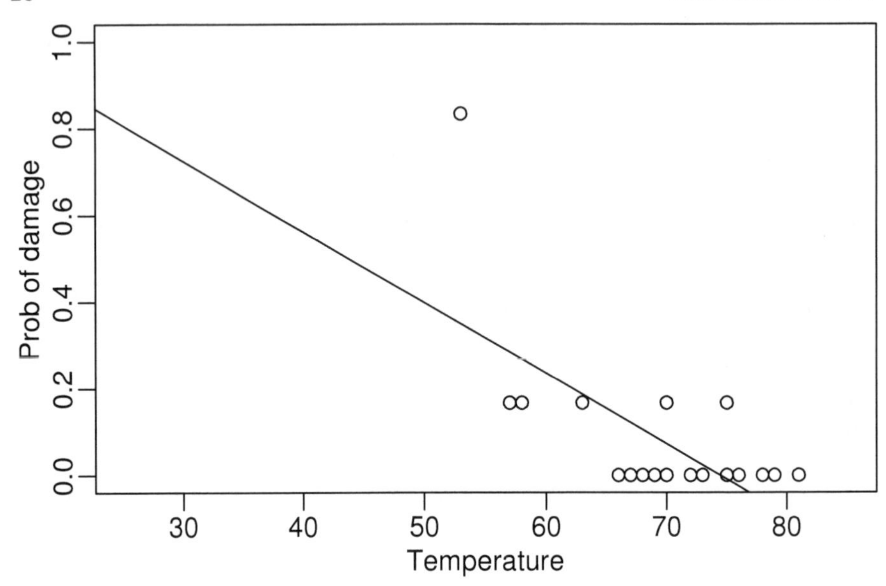

Figure 2.1 *Damage to O-rings in 23 space shuttle missions as a function of launch temperature. Least squares fit line is shown.*

imation is too much of a stretch. Furthermore, the variance of a binomial variable is not constant which violates another crucial assumption of the linear model.

The standard linear model is clearly not directly suitable here. Although, we could use transformation and weighting to correct some of these problems, it is better to develop a model that is directly suited for binomial data.

2.2 Binomial Regression Model

Suppose the response variable Y_i for $i = 1, \ldots, n_i$ is binomially distributed $B(n_i, p_i)$ so that:

$$P(Y_i = y_i) = \binom{n_i}{y_i} p_i^{y_i} (1 - p_i)^{n_i - y_i}$$

We further assume that the Y_i are independent. The individual trials that compose the response Y_i are all subject to the same q predictors $(x_{i1}, \ldots, x_{iq})$. The group of trials is known as a *covariate class*. We need a model that describes the relationship of $x_1, \ldots, x_q$ to p. Following the linear model approach, we construct a *linear predictor*:

$$\eta_i = \beta_0 + \beta_1 x_{i1} + \cdots + \beta_q x_{iq}$$

Since the linear predictor can accommodate quantitative and qualitative predictors with the use of dummy variables and also allows for transformations and combinations of the original predictors, it is very flexible and yet retains interpretability. This notion that we can express the effect of the predictors on the response solely through the linear predictor is important. The idea can be extended to models for other types of response and is one of the defining features of the wider class of generalized linear models (GLMs) discussed in Chapter 6.

We have already seen above that setting $\eta_i = p_i$ is not appropriate because we require $0 \leq p_i \leq 1$. Instead we shall use a *link function* g such that $\eta_i = g(p_i)$. For this application, we shall need g to be monotone and be such that $0 \leq g^{-1}(\eta) \leq 1$ for any η. There are three common choices:

1. Logit: $\eta = \log(p/(1-p))$.
2. Probit: $\eta = \Phi^{-1}(p)$ where Φ is the normal cumulative distribution function.
3. Complementary log-log: $\eta = \log(-\log(1-p))$.

The idea of the link function is also one of the central ideas of generalized linear models. It is used to link the linear predictor to the mean of the response in the wider class of models.

We will compare these three choices of link function later, but first we estimate the parameters of the model. We shall use the method of maximum likelihood; see Appendix A for a brief introduction to this method. The log-likelihood using the logit link is given by:

$$l(\beta) = \sum_{i=1}^{n} \left[y_i\eta_i - n_i \log(1 + e_i^{\eta}) + \log \binom{n_i}{y_i} \right]$$

We can maximize this to obtain the maximum likelihood estimates $\hat{\beta}$ and use the standard theory to obtain approximate standard errors. An algorithm to perform the maximization will be discussed in Chapter 6.

We use R to estimate the regression parameters for the Challenger data:

```
> logitmod <- glm(cbind(damage,6-damage) ~ temp, family=binomial, orings)
> summary(logitmod)
Deviance Residuals:
   Min       1Q  Median       3Q      Max
-0.953  -0.735  -0.439  -0.208   1.957

Coefficients:
            Estimate Std. Error z value Pr(>|z|)
(Intercept)  11.6630     3.2963    3.54    4e-04
temp         -0.2162     0.0532   -4.07  4.8e-05

(Dispersion parameter for binomial family taken to be 1)

    Null deviance: 38.898  on 22  degrees of freedom
Residual deviance: 16.912  on 21  degrees of freedom
AIC: 33.67

Number of Fisher Scoring iterations: 6
```

For binomial response data, we need two pieces of information about the response values — y and n. In R, one way of doing this is to form a two-column matrix with the first column representing the number of "successes" y and the second column the number of "failures" $n-y$. We have specified that the response is binomially distributed. The default choice of link is the logit — other choices need to be specifically stated as we shall see shortly. This default choice is sometimes called *logistic regression*. The regression coefficients are given in the output — $\hat{\beta}_0 = 11.6$ and $\hat{\beta}_1 = -0.216$ along with their respective standard errors. The rest of the output will be explained shortly.

We show the logit fit to the data as seen in Figure 2.2:

```
> plot(damage/6 ~ temp, orings, xlim=c(25,85), ylim = c(0,1),
  xlab="Temperature", ylab="Prob of damage")
> x <- seq(25,85,1)
> lines(x,ilogit(11.6630-0.2162*x))
```

Notice how the logit fit tends asymptotically toward zero and one at high and low temperatures, respectively. The fitted values never actually reach zero or one, so the model never predicts anything to completely certain or completely impossible. Now

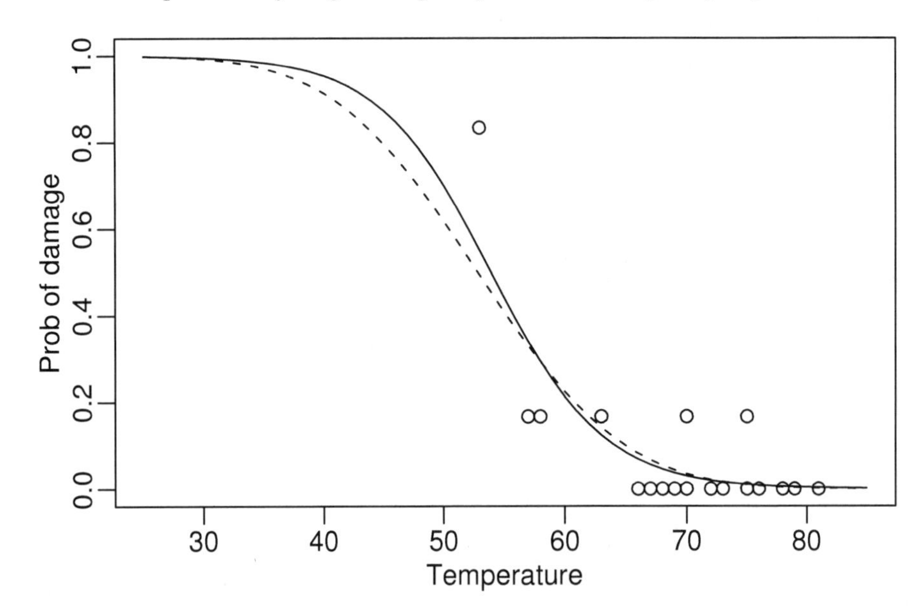

Figure 2.2 *Logit (solid line) and probit (dashed line) fits to the Challenger data*

compare this to the probit fit:

```
> probitmod <- glm(cbind(damage,6-damage) ~ temp,
  family=binomial(link=probit), orings)
> summary(probitmod)
Coefficients:
            Estimate Std. Error z value Pr(>|z|)
(Intercept)   5.5915     1.7105    3.27   0.0011
temp         -0.1058     0.0266   -3.98  6.8e-05

(Dispersion parameter for binomial family taken to be 1)
```

```
    Null deviance: 38.898  on 22  degrees of freedom
Residual deviance: 18.131  on 21  degrees of freedom
AIC: 34.89
```

Although the coefficients seem quite different, the fits are similar, particularly in the range of the data, as seen in Figure 2.2:

```
> lines(x,pnorm(5.5915-0.1058*x),lty=2)
```

We can predict the response at 31°F for both models:

```
> ilogit(11.6630-0.2162*31)
[1] 0.99304
> pnorm(5.5915-0.1058*31)
[1] 0.9896
```

We see a very high probability of damage with either model although we still need to develop some inferential techniques before we leap to conclusions.

2.3 Inference

Consider two models, a larger model with l parameters and likelihood L_L and a smaller model with s parameters and likelihood L_S where the smaller model represents a linear subspace (a linear restriction on the parameters) of the larger model. Likelihood methods suggest the likelihood ratio statistic:

$$2\log\frac{L_L}{L_S} \tag{2.1}$$

as an appropriate test statistic for comparing the two models. Now suppose we choose a saturated larger model — such a model typically has as many parameters as cases and has fitted values $\hat{p}_i = y_i/n_i$. In such a case, the test statistic becomes:

$$D = 2\sum_{i=1}^{n}\{y_i \log y_i/\hat{y}_i + (n_i - y_i)\log(n_i - y_i)/(n_i - \hat{y}_i)\}$$

where $\hat{y}_i$ are the fitted values from the smaller model. Now since the saturated model fits as well as any model can fit, the *deviance* D measures how close the (smaller) model comes to perfection. Thus deviance is a measure of goodness of fit. In the output for the models above, the `Residual deviance` is the deviance for the current model while the `Null deviance` is the deviance for a model with no predictors and just an intercept term.

Provided that Y is truly binomial and that the n_i are relatively large, the deviance is approximately χ^2 distributed with $n - s$ degrees of freedom if the model is correct. Thus we can use the deviance to test whether the model is an adequate fit. For the logit model of the Challenger data, we may compute:

```
> pchisq(deviance(logitmod),df.residual(logitmod),lower=FALSE)
[1] 0.71641
```

Since this p-value is well in excess of 0.05, we may conclude that this model fits sufficiently well. Of course, this does not mean that this model is correct or that a simpler model might not also fit adequately. Even so, for the null model:

```
> pchisq(38.9,22,lower=FALSE)
[1] 0.014489
```

we see that the fit is inadequate, so we cannot ascribe the response to simple variation not dependent on any predictor. Note that a χ^2_d variable has mean d and standard deviation $\sqrt{2d}$ so that it is often possible to quickly judge whether a deviance is large or small without explicitly computing the p-value. If the deviance is far in excess of the degrees of freedom, the null hypothesis can be rejected.

The χ^2 distribution is only an approximation that becomes more accurate as the n_i increase. For the case, $n_i = 1$, when $y_i = 0$ or 1, in other words, a binary response, the deviance reduces to:

$$-2\sum_{i=1}^{n}\{\hat{p}_i \text{logit}(\hat{p}_i) + \log(1-\hat{p}_i)\}$$

For a deviance to measure fit, it has to compare the fitted values $\hat{p}_i$ to the data y_i, but here we have only a function of $\hat{p}_i$. Thus this deviance does not assess goodness of fit and furthermore, it is not even approximately χ^2 distributed. Other methods must be used to judge goodness of fit for binary data — for example, the Hosmer-Lemeshow test described in Hosmer and Lemeshow (2000).

The approximation is very poor for small n_i. Although it is not possible to say exactly how large n_i should be for an adequate approximation, $n_i \geq 5$ has often been suggested. Permutation or bootstrap methods might be considered as an alternative.

We can also use the deviance to compare two nested models. The test statistic in (2.1) becomes $D_S - D_L$. This test statistic is asymptotically distributed χ^2_{l-s}, assuming that the smaller model is correct and the distributional assumptions hold. We can use this to test the significance of temperature by computing the difference in the deviances between the model with and without temperature. The model without temperature is just the null model and the difference in degrees of freedom or parameters is one:

```
> pchisq(38.9-16.9,1,lower=FALSE)
[1] 2.7265e-06
```

Since the p-value is so small, we conclude that the effect of launch temperature is statistically significant. An alternative to this test is the z-value, which is $\hat{\beta}/se(\hat{\beta})$, here -4.07 with a p-value of 4.8e-05. In contrast to the normal (Gaussian) linear model, these two statistics are not identical. In this particular example, there is no practical difference, but in some cases, especially with sparse data, the standard errors can be overestimated and so the z-value is too small and the significance of an effect could be missed. This is known as the Hauck–Donner effect — see Hauck and Donner (1977). So the deviance-based test is preferred.

Again, there are concerns with the accuracy of the approximation, but the test involving differences of deviances is generally more accurate than the goodness of fit test involving a single deviance.

Confidence intervals for the regression parameters may be constructed using normal approximations for the parameter estimates. A $100(1-\alpha)\%$ confidence interval for β_i would be:

$$\hat{\beta}_i \pm z^{\alpha/2} se(\hat{\beta}_i)$$

where $z^{\alpha/2}$ is a quantile from the normal distribution. Thus a 95% confidence interval for β_1 in our model would be:

```
> c(-0.2162-1.96*0.0532,-0.2162+1.96*0.0532)
[1] -0.32047 -0.11193
```

It is also possible to construct a profile likelihood-based confidence interval:

```
> library(MASS)
> confint(logitmod)
Waiting for profiling to be done...
               2.5 %   97.5 %
(Intercept)  5.57543 18.73812
temp        -0.33267 -0.12018
```

It is important to load the MASS package or the default `confint` method for ordinary linear models will be used (which will not be quite right). The profile likelihood method is generally preferable for the same Hauck–Donner reasons discussed above although it is more work to compute.

Although we have only computed results for the logit link, the same methods would apply for the probit or any other link.

2.4 Tolerance Distribution

Suppose that students answers questions on a test and that a specific student has an aptitude T. A particular question might have difficulty d_i and the student will get the answer correct only if $T > d_i$. Now if we consider d_i fixed and $T \sim N(\mu, \sigma^2)$, then the probability that a randomly selected student will get the answer wrong is:

$$p_i = P(T \leq d_i) = \Phi((d_i - \mu)/\sigma)$$

So

$$\Phi^{-1}(p_i) = -\mu/\sigma + d_i/\sigma$$

If we set $\beta_0 = -\mu/\sigma$ and $\beta_1 = 1/\sigma$, we now have a probit regression model. So we see that the probit link can be naturally motivated by the existence of a normally distributed *tolerance distribution* T. The term arose from toxicity studies where the aptitude of the subject would be replaced with the tolerance of the insect.

The logit model arises from a logistically distributed tolerance distribution. The logistic and normal density are very similar in the mid-range, but differ more in a relative sense in the tails. The complementary log-log is similarly associated with an extreme value distribution.

2.5 Interpreting Odds

Odds are sometimes a better scale than probability to represent chance. They arose as a way to express the payoffs for bets. An *evens* bet means that the winner gets paid an equal amount to that staked. A 3–1 *against* bet would pay \$3 for every \$1 bet while a 3–1 *on* bet would pay only \$1 for every \$3 bet. If these bets are *fair* in the sense that a bettor would break even in the long-run average, then we can make a correspondence to probability. Let p be the probability and o be the odds, where we represent 3–1 against as 1/3 and 3–1 on as 3, then the following relationships hold:

$$\frac{p}{1-p} = o \qquad\qquad p = \frac{o}{1+o}$$

One mathematical advantage of odds is that they are unbounded above which makes them more convenient for some modeling purposes.

Odds also form the basis of a subjective assessment of probability. Some probabilities are determined from considerations of symmetry or long-term frequencies, but such information is often unavailable. Individuals may determine their subjective probability for events by considering what odds they would be prepared to offer on the outcome. Under this theory, other potential persons would be allowed to place bets for or against the event occurring. Thus the individual would be forced to make an honest assessment of probability to avoid financial loss.

If we have two covariates x_1 and x_2, then the logistic regression model is:

$$\log(\text{odds}) = \log\left(\frac{p}{1-p}\right) = \beta_0 + \beta_1 x_1 + \beta_2 x_2$$

Now β_1 can be interpreted as follows: a unit increase in x_1 with x_2 held fixed increases the log-odds of success by β_1 or increases the odds of success by a factor of $\exp\beta_1$. Of course, the usual interpretational difficulties regarding causation apply as in standard regression. No such simple interpretation exists for other links such as the probit.

An alternative notion to odds-ratio is relative risk. Suppose the probability of "success" in the presence of some condition is p_1 and p_2 in its absence. The relative risk is p_1/p_2. For rare outcomes, the relative risk and the odds ratio will be very similar, but for larger probabilities, there may be substantial differences. There is some debate over which is the more intuitive way of expressing the effect of some condition.

Consider the data shown in Table 2.1 from a study on infant respiratory disease, namely the proportions of children developing bronchitis or pneumonia in their first year of life by type of feeding and sex, which may be found in Payne (1987):

	Bottle Only	Some Breast with Supplement	Breast Only
Boys	77/458	19/147	47/494
Girls	48/384	16/127	31/464

Table 2.1 *Incidence of respiratory disease in infants to the age of 1 year.*

We can recover the layout above with the proportions as follows:

```
> data(babyfood)
> xtabs(disease/(disease+nondisease)~sex+food,babyfood)
      food
sex      Bottle   Breast    Suppl
  Boy  0.16812 0.095142 0.12925
  Girl 0.12500 0.066810 0.12598
```

Fit and examine the model:

```
> mdl <- glm(cbind(disease,nondisease) ~ sex + food, family=binomial,
  babyfood)
> summary(mdl)
Coefficients:
            Estimate Std. Error z value Pr(>|z|)
(Intercept)   -1.613      0.112  -14.35  < 2e-16
sexGirl       -0.313      0.141   -2.22    0.027
foodBreast    -0.669      0.153   -4.37  1.2e-05
foodSuppl     -0.173      0.206   -0.84    0.401

(Dispersion parameter for binomial family taken to be 1)

    Null deviance: 26.37529  on 5  degrees of freedom
Residual deviance:  0.72192  on 2  degrees of freedom
AIC: 40.24
```

The χ^2 approximation can be expected to be accurate here due to the large covariate class sizes. Is there a `sex`-by-`food` interaction? Notice that a model with the interaction effect would be saturated with deviance and degrees of freedom zero, so we can look at the residual deviance of this model to test for an interaction effect. A deviance of 0.72 is not at all large for two degrees of freedom, so we may conclude that there is no evidence of an interaction effect. This means that we may interpret the main effects separately.

We can test for the significance of the main effects:

```
> drop1(mdl,test="Chi")
Single term deletions

Model:
cbind(disease, nondisease) ~ sex + food
       Df Deviance  AIC  LRT Pr(Chi)
<none>          0.7 40.2
sex     1       5.7 43.2  5.0   0.026
food    2      20.9 56.4 20.2 4.2e-05
```

The `drop1` function tests each predictor relative to the full. We see that both predictors are significant in this sense. Now consider the interpretation of the coefficients, starting with the effect of breast feeding:

```
> exp(-0.669)
[1] 0.51222
```

We see that breast feeding reduces the odds of respiratory disease to 51% of that for bottle feeding. We could compute a confidence interval by figuring the standard error on the odds scale; however, we get better coverage properties by computing the interval on the log-odds scale and then transforming the endpoints as follows:

```
> exp(c(-0.669-1.96*0.153,-0.669+1.96*0.153))
[1] 0.37951 0.69134
```

Notice that the interval is asymmetric about the estimated effect of 0.512. Confidence intervals can also be computed using profile likelihood methods:

```
> library(MASS)
```

```
> exp(confint(mdl))
Waiting for profiling to be done...
              2.5 %  97.5 %
(Intercept) 0.15920 0.24743
sexGirl     0.55362 0.96292
foodBreast  0.37819 0.68952
foodSuppl   0.55552 1.24643
```

which gives a slightly wider interval. This latter result is usually more reliable although it makes little difference for this data.

As an aside, note that for small values of ε, we have:

$$\log(x(1+\varepsilon)) = \log x + \log(1+\varepsilon) \approx \log x + \varepsilon$$

This approximation is reasonable for values $-0.25 < \varepsilon < 0.25$. So, for example, given the observed supplement coefficient of -0.173, we can approximate the reduction in odds as about 17% relative to bottle feeding. The exact figure is:

```
> 1-exp(-0.173)
[1] 0.15886
```

that is about 16%. So the approximation is only good for a quick sense of the effect, but an exact calculation is necessary for results that will be presented to others.

Here we see that breast-fed and to a lesser extent supplement-fed babies are less vulnerable to respiratory disease. We also see that boys are more vulnerable than girls. We should be careful about making any general conclusions from this data without knowing how it was collected. In particular, the decision to breast feed is almost certainly related to other socioeconomic factors and we would need to investigate whether it is these rather than the breast feeding that is responsible for the reduction in the incidence of respiratory disease.

2.6 Prospective and Retrospective Sampling

In *prospective* sampling, the predictors are fixed and then the outcome is observed. In other words, in the infant respiratory disease example shown in Table 2.1, we would select a sample of newborn girls and boys whose parents had chosen a particular method of feeding and then monitor them for their first year. This is also called a *cohort study*.

In *retrospective* sampling, the outcome is fixed and then the predictors are observed. Typically, we would find infants coming to a doctor with a respiratory disease in the first year and then record their sex and method of feeding. We would also obtain a sample of respiratory disease-free infants and record their information. How these samples are obtained is important — we require that the probability of inclusion in the study is independent of the predictor values. This is also called a *case-control study*.

Since the question of interest is how the predictors affect the response, prospective sampling seems to be required. Let's focus on just boys who are breast or bottle fed. The data we need is:

```
> babyfood[c(1,3),]
```

```
  disease nondisease sex   food
1      77        381 Boy Bottle
3      47        447 Boy Breast
```

- Given the infant is *breast* fed, the log-odds of having a respiratory disease are $\log 47/447 = -2.25$
- Given the infant is *bottle* fed, the log-odds of having a respiratory disease are $\log 77/381 = -1.60$

The difference between these two log-odds, $\Delta = -1.60 - -2.25 = 0.65$, represents the increased risk of respiratory disease incurred by bottle feeding relative to breast feeding. This is the log-odds ratio.

Now suppose that this had been a retrospective study — we could compute the log-odds of feeding type given respiratory disease status and then find the difference. Notice that this would give the same result because:

$$\Delta = \log 77/47 - \log 381/447 = \log 77/381 - \log 47/447 = 0.65$$

This shows that a retrospective design is as effective as a prospective design for estimating Δ.

Retrospective designs are cheaper, faster and more efficient, so it is convenient that the same result may be obtained from the prospective study. This manipulation is not possible for other links. The downside to retrospective studies is that they are typically less reliable than prospective studies. Retrospective studies rely on historical records which may be of unknown accuracy and completeness. They may also rely on the memory of the subject which may be unreliable.

In most practical situations, we will also need to account for the effects of covariates X. Let π_0 be the probability that an individual is included in the study if they do *not* have the disease, while let π_1 be the probability of inclusion if they do have the disease. For a prospective study, $\pi_0 = \pi_1$ because we have no knowledge of the outcome, while for a retrospective study typically π_1 is much greater than π_0. Suppose that for given x, $p^*(x)$ is the conditional probability that an individual has the disease given that he or she was included in the study, while $p(x)$ is the unconditional probability that he or she has the disease as we would obtain from a prospective study. Now by Bayes theorem:

$$p^*(x) = \frac{\pi_1 p(x)}{\pi_1 p(x) + \pi_0(1 - p(x))}$$

which can be rearranged to show that:

$$\text{logit}(p^*(x)) = \log \frac{\pi_1}{\pi_0} + \text{logit}(p(x))$$

So the only difference between the retrospective and the prospective study would be the difference in the intercept: $\log(\pi_1/\pi_0)$. Generally π_1/π_0 would not be known, so we would not be able to estimate β_0, but knowledge of the other β would be most important since this can be used to assess the *relative* effect of the covariates. We could not, however, estimate the absolute effect. This does not work for other links such as the probit.

2.7 Choice of Link Function

We must choose a link function to specify a binomial regression model. It is usually not possible to make this choice based on the data alone. For regions of moderate p, that is not close to zero or one, the link functions we have proposed are quite similar and so a very large amount of data would be necessary to distinguish between them. Larger differences are apparent in the tails, but for very small p, one needs a very large amount of data to obtain just a few successes, making it expensive to distinguish between link functions in this region. So usually, the choice of link function is made based on assumptions derived from physical knowledge or simple convenience. We now look at some of the advantages and disadvantages of the three proposed link functions and what motivates the choice.

Bliss (1935) analyzed some data on the numbers of insects dying at different levels of insecticide concentration. We fit all three link functions:

```
> data(bliss)
> bliss
  dead alive conc
1    2    28    0
2    8    22    1
3   15    15    2
4   23     7    3
5   27     3    4
> modl <- glm(cbind(dead,alive) ~ conc, family=binomial, data=bliss)
> modp <- glm(cbind(dead,alive) ~ conc, family=binomial(link=probit),
  data=bliss)
> modc <- glm(cbind(dead,alive) ~ conc, family=binomial(link=cloglog),
  data=bliss)
```

We start by considering the fitted values:

```
> fitted(modl)
        1        2        3        4        5
0.089172 0.238323 0.500000 0.761677 0.910828
```

or from `predict(modl,type="response")`. These are constructed using linear predictor, η:

```
> coef(modl)[1]+coef(modl)[2]*bliss$conc
[1] -2.3238 -1.1619  0.0000  1.1619  2.3238
```

Alternatively, these values may be obtained from `modl$linear.predictors` or `predict(modl)`. The fitted values are then:

```
> ilogit(modl$lin)
        1        2        3        4        5
0.089172 0.238323 0.500000 0.761677 0.910828
```

Notice the need to distinguish between predictions in the scale of the response and the link. Now compare the logit, probit and complementary log-log fits:

```
> cbind(fitted(modl),fitted(modp),fitted(modc))
      [,1]     [,2]    [,3]
1 0.089172 0.084242 0.12727
2 0.238323 0.244873 0.24969
```

```
3 0.500000 0.498272 0.45459
4 0.761677 0.752396 0.72177
5 0.910828 0.914411 0.93277
```

These are not very different, but now look at a wider range:

```
> x <- seq(-2,8,0.2)
> pl <- ilogit(modl$coef[1]+modl$coef[2]*x)
> pp <- pnorm(modp$coef[1]+modp$coef[2]*x)
> pc <- 1-exp(-exp((modc$coef[1]+modc$coef[2]*x)))
> plot(x,pl,type="l",ylab="Probability",xlab="Dose")
> lines(x,pp,lty=2)
> lines(x,pc,lty=5)
```

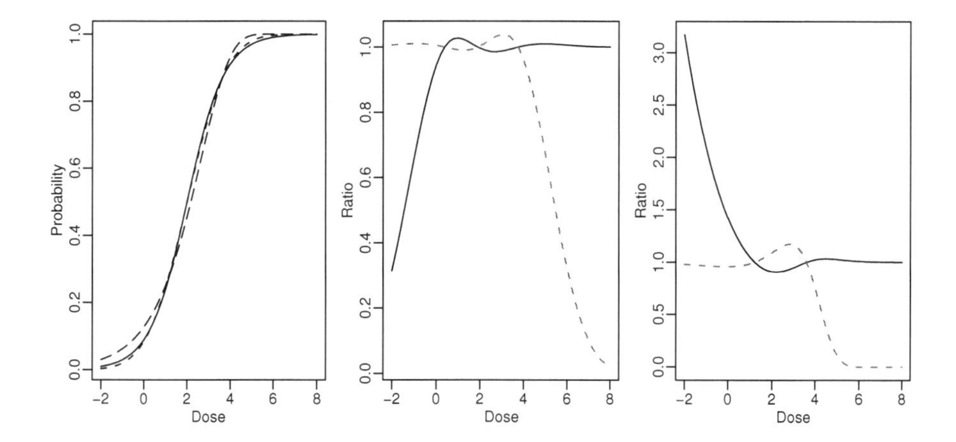

Figure 2.3 *Probit, logit and complementary log-log compared. The fitted probabilities are shown on the left. The logit fit is shown with a solid line, the probit is shown by a dotted line and the complementary log-log by a dashed line. In the central plot, the ratio of probit to logit probabilities in both tails is shown. The lower tail ratio is given by the solid line while the upper tail ratio is given by the dotted line. In the plot on the right the same information is shown for the ratio of the complementary log-log to the logit. The data range from 0 to 4. We see that the links are similar in this range and only begin to diverge as we extrapolate.*

The lines in the left panel of Figure 2.3 do not seem very different, but look at the relative differences:

```
> matplot(x,cbind(pp/pl,(1-pp)/(1-pl)),type="l",xlab="Dose",ylab="Ratio")
> matplot(x,cbind(pc/pl,(1-pc)/(1-pl)),type="l",xlab="Dose",ylab="Ratio")
```

as they appear in the second and third panels of Figure 2.3. We see that the probit and logit differ substantially in the tails. The same phenomenon is observed for the complementary log-log. This is problematic since the former plot indicates it would be difficult to distinguish between the two using the data we have. This is an issue in trials of potential carcinogens and other substances that must be tested for possible harmful effects on humans. Some substances are highly poisonous in that their effects become immediately obvious at doses that might normally be experienced in the environment. It is not difficult to detect such substances. However, there are other substances whose harmful effects only become apparent at large dosages where the observed probabilities are sufficiently larger than zero to become estimable without immense sample sizes. In order to estimate the probability of a harmful effect at a low dose, it would be necessary to select an appropriate link function and yet the data for high dosages will be of little help in doing this. As Paracelsus (1493–1541) said, "All substances are poisons; there is none which is not a poison. The right dose differentiates a poison."

A good example of this problem is asbestos. Information regarding the harmful effects of asbestos derives from historical studies of workers in industries exposed to very high levels of asbestos dust. However, we would like to know the risk to individuals exposed to low levels of asbestos dust such as those found in old buildings. It is virtually impossible to accurately determine this risk. We cannot accurately measure exposure or outcome. This is not to argue that nothing should be done, but that decisions should be made in recognition of the uncertainties.

In summary, the default choice is the logit link. There are three advantages: it leads to simpler mathematics due the intractability of Φ; it is easier to interpret using odds and it allows easier analysis of retrospectively sampled data.

2.8 Estimation Problems

Estimation using the Fisher scoring algorithm, described in Section 6.2, is usually fast. However, difficulties can sometimes arise. When convergence fails, it is sometimes due to a problem exhibited by the following dataset. Urinary androsterone (androgen) and etiocholanolone (estrogen) values were recorded from 26 healthy males by Margolese (1970). The data were also analyzed by Hand (1981). We start by plotting the data as shown in Figure 2.4:

```
> data(hormone)
> plot(estrogen ~ androgen,data=hormone,pch=as.character(orientation))
```

We now fit a binomial model to see if the orientation can be predicted from the two hormone values. Notice that when the response is binary, we can use it directly as the response variable in the `glm` function:

```
> modl <- glm(orientation ~ estrogen + androgen, hormone, family=binomial)
Warning messages:
1: Algorithm did not converge in: glm.fit(x = X, y = Y,
   weights = weights, start = start, etastart = etastart,
2: fitted probabilities numerically 0 or 1 occurred in:
   glm.fit(x = X, y = Y, weights = weights, start = start,
   etastart = etastart,
```

We see that there were problems with the convergence. A look at the summary reveals further evidence:

```
> summary(modl)
Coefficients:
            Estimate Std. Error  z value Pr(>|z|)
(Intercept)    -84.5   136095.1 -0.00062        1
estrogen       -90.2    75911.0 -0.00119        1
```

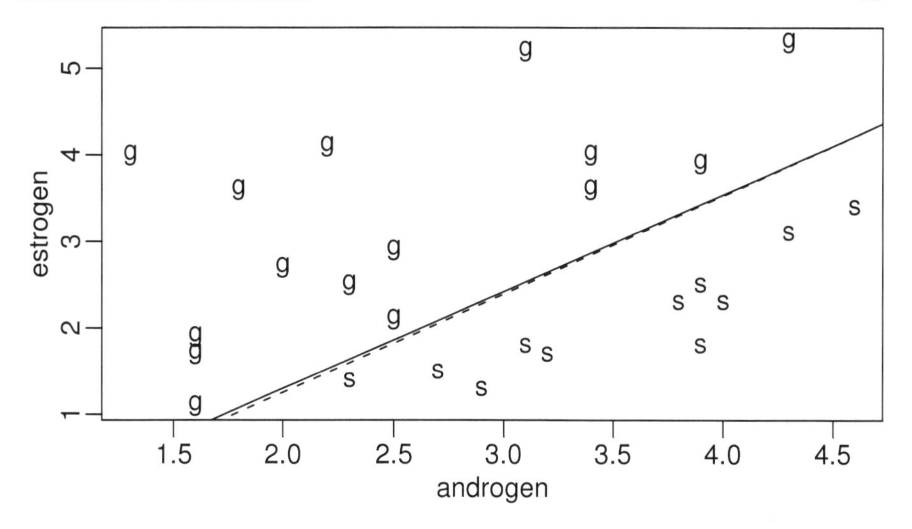

Figure 2.4 *Levels of androgen and estrogen for 15 homosexual (g) and 11 heterosexual (s) males. Solid line shows predictions from* `glm` *fit that correspond to $p = 1/2$. The dotted line is equivalent from* `brglm`.

```
androgen        100.9    92755.6  0.00109        1

(Dispersion parameter for binomial family taken to be 1)

    Null deviance: 3.5426e+01  on 25  degrees of freedom
Residual deviance: 2.3229e-09  on 23  degrees of freedom
AIC: 6

Number of Fisher Scoring iterations: 25
```

Notice that the residual deviance is extremely small indicating a very good fit and yet none of the predictors are significant due to the high standard errors. We see that the maximum default number of iterations (25) has been reached. A look at the data reveals the reason for this. We see that the two groups are *linearly separable* so that a perfect fit is possible. We can compute the line separating the groups by finding the line that corresponds to $p = 1/2$ which is when the logit is zero:

```
> abline(-84.5/90.2,100.9/90.2)
```

We suffer from an embarrassment of riches in this example — we can fit the data perfectly. Unfortunately, this results in unstable estimates of the parameters and their standard errors and would (probably falsely) suggest that perfect predictions can be made. An alternative fitting approach might be considered in such cases called *exact logistic regression*. See Cox (1970) and the work of Cyrus Mehta, for example: Mehta and Patel (1995). Currently, there are no comprehensive packages for such exact methods in R, although it is available in products such as LogExact©.

An alternative to exact methods is the bias reduction method of Firth (1993). For the MLE, $E\hat{\beta} \neq \beta$ and indeed a sensible unbiased estimator would be difficult to obtain. Firth's method removes the $O(1/n)$ term from the asymptotic bias of estimated coefficients. These estimates have the advantage of always being finite:

```
> library(brglm)
> modb <- brglm(orientation ~ estrogen + androgen, hormone,
  family=binomial)
> summary(modb)
Coefficients:
            Estimate Std. Error z value Pr(>|z|)
(Intercept)    -3.65       2.91   -1.25    0.210
estrogen       -3.59       1.50   -2.39    0.017
androgen        4.07       1.62    2.51    0.012

   Null deviance: 28.9231  on 25  degrees of freedom
Residual deviance:  3.7008  on 23  degrees of freedom
Penalized deviance: 4.1837
```

We can see that this results in significant predictors which we expect given Figure 2.4. Although the fit appears, judging from the coefficients, to be different from the `glm` result, it is effectively very close as we can see by plotting the line corresponding to $p = 1/2$:

```
> abline(-3.65/3.59,4.07/3.59,lty=2)
```

Instability in parameter estimation will also occur in datasets that approach linear separability. Care will be needed in such cases.

2.9 Goodness of Fit

The deviance is one measure of how well the model fits the data, but there are alternatives. The Pearson's X^2 statistic takes the general form:

$$X^2 = \sum_{i=1}^{n} \frac{(O_i - E_i)^2}{E_i}$$

where O_i is the observed counts and E_i are the expected counts for case i. For a binomial response, we count the number of successes for which $O_i = y_i$ while $E_i = n_i\hat{p}_i$ and failures for which $O_i = n_i - y_i$ and $E_i = n_i(1-\hat{p}_i)$ which results in:

$$X^2 = \sum_{i=1}^{n} \frac{(y_i - n_i\hat{p}_i)^2}{n_i\hat{p}_i(1-\hat{p}_i)}$$

If we define *Pearson residuals* as:

$$r_i^P = (y_i - n_i\hat{p}_i)/\sqrt{\operatorname{var}\hat{y}_i}$$

which can be viewed as a type of standardized residual, then $X^2 = \sum_{i=1}^{n}(r_i^P)^2$. So the Pearson's X^2 is analogous to the residual sum of squares used in normal linear models.

The Pearson X^2 will typically be close in size to the deviance and can be used in the same manner. Alternative versions of the hypothesis tests described above might use the X^2 in place of the deviance with the same approximate null distributions.

However, some care is necessary because the model is fit to minimize the deviance and not the Pearson's X^2. This means that it is possible, although unlikely, that the X^2 could increase as a predictor is added to the model. X^2 can be computed like this:

```
> modl <- glm(cbind(dead,alive) ~ conc, family=binomial, data=bliss)
> sum(residuals(modl,type="pearson")^2)
[1] 0.36727
> deviance(modl)
[1] 0.37875
```

As can be seen, there is little difference here between X^2 and the deviance.

The proportion of variance explained or R^2 is a popular measure of fit for normal linear models. We might consider applying the same concept to binomial regression models by using the proportion of deviance explained. However, a better statistic is due to Nagelkerke (1991):

$$R^2 = \frac{1-(\hat{L}_0/\hat{L})^{2/n}}{1-\hat{L}_o^{2/n}} = \frac{1-\exp((D-D_{null})/n)}{1-\exp(-D_{null}/n)}$$

where n is the number of binary observations and $\hat{L}_0$ is the maximized likelihood under the null. The numerator can be seen as a ratio of the relative likelihood with the $1/n$ power having the effect of a geometric mean on the observations. The denominator simply normalizes so that $0 \leq R^2 \leq 1$. For example, for the Bliss insect data, the R^2 is:

```
> (1-exp((modl$dev-modl$null)/150))/(1-exp(-modl$null/150))
[1] 0.99532
```

Notice that we have used $n = 150$ as there are 5 covariate class with 30 observations each. We can see that this is a very good fit.

2.10 Prediction and Effective Doses

Sometimes we wish to predict the outcome for given values of the covariates. For binary data this will mean estimating the probability of success. For given covariates x_0, $\hat{\eta} = x_0\hat{\beta}$ with variance given by $x_0^T(X^TWX)^{-1}x_0$. Approximate confidence intervals may be obtained using a normal approximation. To get an answer in the probability scale, it will be necessary to transform back using the inverse of the link function. We predict the response for the insect data:

```
> data(bliss)
> modl <- glm(cbind(dead,alive) ~ conc, family=binomial,data=bliss)
> lmodsum <- summary(modl)
```

We show how to predict the response at dose of 2.5:

```
> x0 <- c(1,2.5)
> eta0 <- sum(x0*coef(modl))
> ilogit(eta0)
[1] 0.64129
```

A 64% predicted chance of death at this dose — now compute a 95% confidence interval (CI) for this probability. First, extract the variance matrix of the coefficients:

```
> (cm <- lmodsum$cov.unscaled)
            (Intercept)      conc
(Intercept)    0.174630 -0.065823
conc          -0.065823  0.032912
```

The standard error on the logit scale is then:

```
> se <- sqrt( t(x0) %*% cm %*% x0)
```

so the CI on the probability scale is:

```
> ilogit(c(eta0-1.96*se,eta0+1.96*se))
[1] 0.53430 0.73585
```

A more direct way of obtaining the same result is:

```
> predict(modl,newdata=data.frame(conc=2.5),se=T)
$fit
[1] 0.58095

$se.fit
[1] 0.2263
> ilogit(c(0.58095-1.96*0.2263,0.58095+1.96*0.2263))
[1] 0.53430 0.73585
```

Note that in contrast to the linear regression situation, there is no distinction possible between confidence intervals for a future observation and those for the mean response. Now we try predicting the response probability at the low dose of -5:

```
> x0 <- c(1,-5)
> se <- sqrt( t(x0) %*% cm %*% x0)
> eta0 <- sum(x0*lmod$coef)
> ilogit(c(eta0-1.96*se,eta0+1.96*se))
[1] 2.3577e-05 3.6429e-03
```

This is not a wide interval in absolute terms, but in relative terms, it certainly is. The upper limit is about 100 times larger than the lower limit.

Logistic regression models have been widely used for classification purposes. Depending on whether $\hat{p}$ is greater or less than 0.5, the case may be classified as a success or failure. In cases where the losses due to misclassification are not symmetrical, such as in disease diagnosis, critical values other than 0.5 should be used. Another example is in credit scoring. When financial institutions decide whether to make a loan, it is helpful to estimate the probability that a given borrower will default. A logistic regression model is one way in which this probability can be estimated using past financial data.

When there is a single (continuous) covariate or when other covariates are held fixed, we sometimes wish to estimate the value of x corresponding to a chosen p. For example we may wish to determine which dose, x, will lead to a probability of success p. ED50 stands for the *effective dose* for which there will be a 50% chance of success. When the objective is to kill the subjects or determine toxicity, as when using insecticides, the term LD50 would be used. LD stands for *lethal dose*. Other percentiles are also of interest. For a logit link, we can set $p = 1/2$ and then solve for x to find:

$$\widehat{\mathrm{ED50}} = -\hat{\beta}_0/\hat{\beta}_1$$

Using the Bliss data, the LD50 is:

```
> (ld50 <- -lmod$coef[1]/lmod$coef[2])
 (Intercept)
           2
```

To determine the standard error, we can use the delta method. The general expression for the variance of $g(\hat{\theta})$ for multivariate θ is given by

$$\text{var } g(\hat{\theta}) \approx g'(\hat{\theta})^T \text{var } \hat{\theta} g'(\hat{\theta})$$

which, in this example, works out as:

```
> dr <- c(-1/lmod$coef[2],lmod$coef[1]/lmod$coef[2]^2)
> sqrt(dr %*% lmodsum$cov.un %*% dr)[,]
[1] 0.17844
```

So the 95% CI is given by:

```
> c(2-1.96*0.178,2+1.96*0.178)
[1] 1.6511 2.3489
```

Other levels may be considered — the effective dose x_p for probability of success p is:

$$x_p = \frac{\text{logit}(p) - \beta_0}{\beta_1}$$

So, for example:

```
> ed90 <- (logit(0.9)-lmod$coef[1])/lmod$coef[2]
> ed90
 (Intercept)
     3.8911
```

More conveniently, we may use the `dose.p` function in the `MASS` package:

```
> library(MASS)
> dose.p(lmod,p=c(0.5,0.9))
             Dose       SE
p = 0.5: 2.0000 0.17844
p = 0.9: 3.8911 0.34499
```

2.11 Overdispersion

If the binomial GLM model specification is correct, we expect that the residual deviance will be approximately distributed χ^2 with the appropriate degrees of freedom. Sometimes, we observe a deviance that is much larger than would be expected if the model were correct. We must then determine which aspect of the model specification is incorrect.

The most common explanation is that we have the wrong structural form for the model. We have not included the right predictors or we have not transformed or combined them in the correct way. We have a number of ways of determining the importance of potential additional predictors and diagnostics for determining better transformations — see Section 6.4. Suppose, however, that we are able to exclude this explanation. This is difficult to achieve, but when we have only one or two predictors, it is feasible to explore the model space quite thoroughly and be sure that there is not a plausible superior model formula.

Another common explanation for a large deviance is the presence of a small number of outliers. Fortunately, these are easily checked using diagnostic methods explained more fully in Section 6.4. When larger numbers of points are identified as outliers, they become unexceptional, and we might more reasonably conclude that there is something amiss with the error distribution.

Sparse data can also lead to large deviances. In the extreme case of a binary response, the deviance is not even approximately χ^2. In situations where the group sizes are simply small, the approximation is poor. Because we cannot judge the fit using the deviance, we shall exclude this case from further consideration in this section.

Having excluded these other possibilities, we might explain a large deviance by deficiencies in the random part of the model. A binomial distribution for Y arises when the probability of success p is independent and identical for each trial within the group. If the group size is m, then $\text{var } Y = mp(1-p)$ if the binomial assumptions are correct. However, the assumptions are broken, the variance may be greater. This is *overdispersion*. In rarer cases, the variance is less and *underdispersion* results.

There are two main ways that overdispersion can arise — the independent or identical assumptions can be violated. We look at the constant p assumption first. It is easy to see how there may be some unexplained heterogeneity within a group that might lead to some variation in p. For example, in the shuttle disaster case study of Section 2.1, the position of the O-ring on the booster rocket may have some effect on the failure probability. Yet this variable was not recorded and so we cannot include it as a predictor. Heterogeneity can also result from clustering. Suppose a population is divided into clusters, so that when you take a sample, you actually get a sample of clusters. This would be common in epidemiological applications.

Let the sample size be m, the cluster size be k and the number of clusters be $l = m/k$. Let the number of successes in cluster i be $Z_i \sim B(k, p_i)$. Now suppose that p_i is a random variable such that $Ep_i = p$ and $\text{var } p_i = \tau^2 p(1-p)$. Let the total number of successes be $Y = Z_1 + \cdots + Z_l$. Then:

$$EY = \sum EZ_i = \sum_{i=1}^{l} kp = mp$$

as in the standard case, but:

$$\text{var } Y = \sum \text{var } Z_i = \sum \{E(\text{var } (Z_i|p_i)) + \text{var } (E(Z_i|p_i))\} = (1+(k-1)\tau^2)mp(1-p)$$

So Y is overdispersed since $1+(k-1)\tau^2 \geq 1$. Notice that in the sparse case, $m = 1$, and this problem cannot arise.

Overdispersion can also result from dependence between trials. If the response has a common cause, say a disease is influenced by genes, the responses will tend to be positively correlated. For example, subjects in human or animal trials may be influenced in their response by other subjects. If the food supply is limited, the probability of survival of an animal may be increased by the death of others. This circumstance would result in underdispersion.

The simplest approach for modeling overdispersion is to introduce an additional dispersion parameter, σ^2. In the standard binomial case $\sigma^2 = \phi = 1$. We now let σ^2 vary and estimate using the data. Notice the similarity to linear regression. The dispersion parameter may be estimated using:

$$\hat{\sigma}^2 = \frac{X^2}{n-p}$$

Using the deviance in place of the Pearson's X^2 is not recommended as it may not be consistent. The estimation of β is unaffected since σ^2 does not change the mean response but:

$$\hat{\text{var}}\hat{\beta} = \hat{\sigma}^2 (X^T \hat{W} X)^{-1}$$

So we need to scale up the standard errors by a factor of $\hat{\sigma}$.

We cannot use the difference in deviances when comparing models, because the test statistic will be distributed $\sigma^2\chi^2$. Since σ^2 is not known and must be estimated in the overdispersion situation, an F-statistic must be used:

$$F = \frac{(D_{small} - D_{large})/(df_{small} - df_{large})}{\hat{\sigma}^2}$$

This statistic is only an approximately F distributed, in contrast to the Gaussian case.

This dispersion parameter method is only appropriate when the covariate classes are roughly equal in size. If not, more sophisticated methods should be used. One such approach uses the beta-binomial distribution where we assume that p follows a beta distribution. This approach is discussed in Williams (1982) and Crowder (1978) and can be implemented using the `aod` package in R.

In Manly (1978), an experiment is reported where boxes of trout eggs were buried at five different stream locations and retrieved at four different times, specified by the number of weeks after the original placement. The number of surviving eggs was recorded. The box was not returned to the stream. The data is also analyzed by Hinde and Demetrio (1988). We can construct a tabulation of the data by:

```
> data(troutegg)
> ftable(xtabs(cbind(survive,total) ~ location+period, troutegg))
                 survive total
location period
1        4            89    94
         7            94    98
         8            77    86
         11          141   155
2        4           106   108
         7            91   106
         8            87    96
         11          104   122
3        4           119   123
         7           100   130
         8            88   119
         11           91   125
4        4           104   104
         7            80    97
         8            67    99
```

```
         11          111  132
5        4            49   93
         7            11  113
         8            18   88
         11            0  138
```

Notice that in one case, all the eggs survive, while in another, none of the eggs survive. We now fit a binomial GLM for the two main effects:

```
> bmod <- glm(cbind(survive,total-survive) ~ location+period,
  family=binomial,troutegg)
> bmod
Coefficients:
(Intercept)    location2    location3    location4    location5
      4.636       -0.417       -1.242       -0.951       -4.614
    period7      period8     period11
     -2.170       -2.326       -2.450

Degrees of Freedom: 19 Total (i.e. Null);  12 Residual
Null Deviance:     1020
Residual Deviance: 64.5  AIC: 157
```

The deviance of 64.5 on 12 degrees of freedom seems to show that this model does not fit. Before we conclude that there is overdispersion, we need to eliminate other potential explanations. With about 100 eggs in each box, we have no problem with sparseness, but we do need to check for outliers and look at the model formula. A half-normal plot of the residuals is a good way to check for outliers:

```
> halfnorm(residuals(bmod))
```

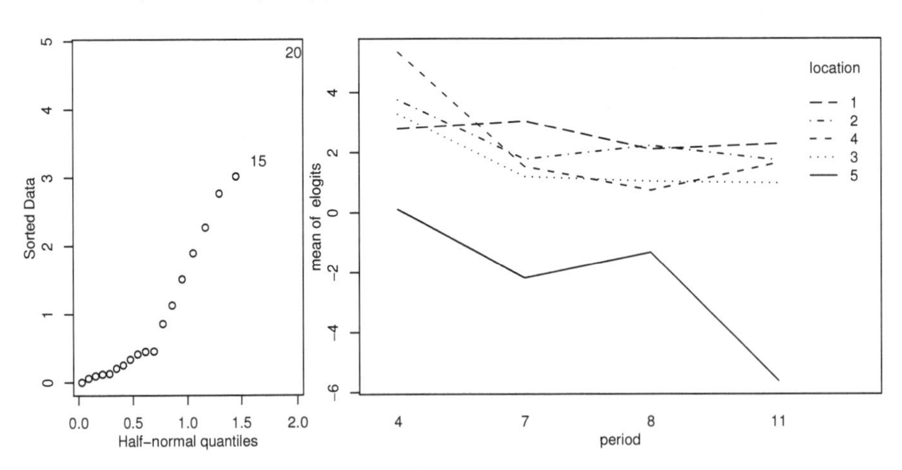

Figure 2.5 *Diagnostic plots for the trout egg model. A half-normal plot of the residuals is shown on the left and an interaction plot of the empirical logits is shown on the right.*

The half-normal plot is shown in the left panel of Figure 2.5. No single outlier is apparent. Perhaps one can discern a larger number of residuals which seem to follow a more dispersed distribution than the rest.

We can also check whether the predictors are correctly expressed by plotting the *empirical logits*. These are defined as:

$$\log\left(\frac{y+1/2}{m-y+1/2}\right)$$

The halves are added to prevent infinite values for groups consisting of all successes or failures. We now construct an interaction plot of the empirical logits:

```
> elogits <- log((troutegg$survive+0.5)/(troutegg$total-
  troutegg$survive+0.5))
> with(troutegg,interaction.plot(period,location,elogits))
```

Interaction plots are always difficult to interpret conclusively, but there is no obvious sign of large interactions. So there is no evidence that the linear model is inadequate. We do not have any outliers and the functional form of the model appears to be suitable, but the deviance is still larger than should be expected. Having eliminated these more obvious causes as the source of the problem, we may now put the blame on overdispersion. Possible reasons for the overdispersion include inhomogeneous trout eggs, variation in the experimental procedures or unknown variables affecting survival.

We can estimate the dispersion parameter as:

```
> (sigma2 <- sum(residuals(bmod,type="pearson")^2)/12)
[1] 5.3303
```

We see that this is substantially larger than one as it would be in the standard binomial GLM. We can now make F-tests on the predictors using:

```
> drop1(bmod,scale=sigma2,test="F")
Single term deletions
scale:  5.3303

         Df Deviance AIC F value   Pr(F)
<none>               64 157
location  4         914 308    39.5 8.1e-07
period    3         229 182    10.2  0.0013
Warning message:
F test assumes quasibinomial family in:
 drop1.glm(bmod, scale = sigma2, test = "F")
```

We see that both terms are clearly significant. It is necessary to specify the `scale` argument using the estimated value of σ^2. If this argument is omitted, the deviance will be used in the estimation of the dispersion parameter. For this particular dataset, it makes very little difference, but in some cases, using the deviance to estimate the dispersion gives inconsistent results. The warning message reminds us that the use of free dispersion parameter results in a model that is no longer a true binomial GLM, but rather what is known as a *quasi-binomial* GLM. More on such models may be found in Section 7.4.

No goodness of fit test is possible because we have a free dispersion parameter. We can use the dispersion parameter to scale up the estimates of the standard error as in:

```
> summary(bmod,dispersion=sigma2)
Coefficients:
            Estimate Std. Error z value Pr(>|z|)
(Intercept)    4.636      0.649    7.14  9.5e-13
location2     -0.417      0.568   -0.73    0.463
location3     -1.242      0.507   -2.45    0.014
location4     -0.951      0.528   -1.80    0.072
location5     -4.614      0.578   -7.99  1.4e-15
period7       -2.170      0.550   -3.94  8.1e-05
period8       -2.326      0.561   -4.15  3.4e-05
period11      -2.450      0.540   -4.53  5.8e-06
```

2.12 Matched Case-Control Studies

In a case-control study, we try to determine the effect of certain risk factors on the outcome. We understand that there are other confounding variables that may affect the outcome. One approach to dealing with these is to measure or record them, include them in the logistic regression model as appropriate and thereby control for their effect. But this method requires that we model these confounding variables with the correct functional form. This may be difficult. Also, making an appropriate adjustment is problematic when the distribution of the confounding variables is quite different in the cases and controls. So we might consider an alternative where the confounding variables are explicitly adjusted for in the design.

In a *matched case-control study*, we match each case (diseased person, defective object, success, etc.) with one or more controls that have the same or similar values of some set of potential confounding variables. For example, if we have a 56-year-old, Hispanic male case, we try to match him with some number of controls who are also 56-year-old Hispanic males. This group would be called a *matched set*. Obviously, the more confounding variables one specifies, the more difficult it will be to make the matches. Loosening the matching requirements, for example, accepting controls who are 50–60 years old might be necessary. Matching also gives us the possibility of adjusting for confounders that are difficult to measure. For example, suppose we suspect an environmental effect on the outcome. However, it is difficult to measure exposure, particularly when we may not know which substances are relevant. We could match subjects based on their place of residence or work. This would go some way to adjusting for the environmental effects.

Matched case-control studies also have some disadvantages apart from the difficulties of forming the matched sets. One loses the possibility of discovering the effects of the variables used to determine the matches. For example, if we match on sex, we will not be able to investigate a sex effect. Furthermore, the data will likely be far from a random sample of the population of interest. So although relative effects may be found, it may be difficult to generalize to the population.

Sometimes, cases are rare but controls are readily available. A $1:M$ design has M controls for each case. M is typically small and can even vary in size from matched set to matched set due to difficulties in finding matching controls and missing values.

Each additional control yields a diminished return in terms of increased efficiency in estimating risk factors — it is usually not worth exceeding $M = 5$.

For individual i in the jth matched set, we also observe a covariate vector x_{ij} which will include the risk factors of interest plus any other variables that we may wish to adjust for, but were unable for various reasons to include among the criteria used to match the sets. It is important that the decision to include a subject in the study be independent of the risk factors as in the unmatched case-control studies. Suppose we have n matched sets and that we take $i = 0$ to represent the case and $i = 1, \ldots, M$ to represent the controls. We propose a logistic regression model of the following form:

$$\text{logit}(p_j(x_{ij})) = \alpha_j + \beta^T x_{ij}$$

The α_j models the effect of the confounding variables in the jth matched set. Given a matched set j of $M+1$ subjects known to have one case and M controls, the conditional probability of the observed outcome, or, in other words, that subject $i = 0$ is the case and the rest are controls is:

$$\frac{\exp \beta^T x_{0j}}{\sum_{i=0}^{M} \exp \beta^T x_{ij}}$$

Notice that α_j cancels out in this expression. We may then form the conditional likelihood for the model by taking the product over all the matched sets:

$$L(\beta) = \prod_{j=1}^{n} \{1 + \sum_{i=1}^{M} \exp[\beta^T (x_{ij} - x_{0j})]\}^{-1}$$

We may now employ standard likelihood methods to make inference — see Breslow (1982) for details. The likelihood takes the same form as that used for the proportional hazards model used in survival analysis. This is convenient because we may use software developed for those models as we demonstrate below. Since the αs are not estimated, we cannot make predictions about individuals, but only make statements about the relative risks as measured by the βs. This same restriction also applies to the unmatched model, so this is nothing new.

In Le (1998), a matched case-control study is presented concerning the association between x-rays and childhood acute myeloid leukemia. The sets are matched on age, race and county of residence. For the most part, there is only one control for each case, but there are a few instances of two controls. We start with a look at the data:

```
> data(amlxray)
> head(amlxray)
    ID disease Sex downs age Mray MupRay MlowRay Fray Cray CnRay
1 7004       1   F    no   0   no     no      no   no   no     1
2 7004       0   F    no   0   no     no      no   no   no     1
3 7006       1   M    no   6   no     no      no   no  yes     3
4 7006       0   M    no   6   no     no      no   no  yes     2
5 7009       1   F    no   8   no     no      no   no   no     1
6 7009       0   F    no   8   no     no      no   no   no     1
```

Only the age is presented here as one of the matching variables. In the three sets shown here, we see that both subjects have the same age and the first is the case and the second is the control. The other variables are risk factors of interest.

Down syndrome is known to be a risk factor. There are only seven such subjects in the dataset:

```
> amlxray[amlxray$downs=="yes",1:4]
      ID disease Sex downs
7   7010       1   M   yes
17  7018       1   F   yes
78  7066       1   F   yes
88  7077       1   M   yes
173 7146       1   F   yes
196 7176       1   F   yes
210 7189       1   F   yes
```

We see that all seven subjects are cases. If we include this variable in the regression, its coefficient is infinite. Given this and the prior knowledge, it is simplest to exclude all these subjects and their associated matched subjects:

```
> (ii <- which(amlxray$downs=="yes"))
[1]   7  17  78  88 173 196 210
> ramlxray <- amlxray[-c(ii,ii+1),]
```

The variables `Mray`, `MupRay` and `MlowRay` record whether the mother has ever had an x-ray, ever had an upper body x-ray and ever had a lower body x-ray, respectively. These variables are closely associated, so we will pick just `Mray` for now and investigate the others more closely if indicated. We will also use `CnRay`, a four-level ordered factor grouping the number of x-rays that the child has received in preference to `Cray` which merely indicates whether the child has ever had an x-ray.

The `clogit` function fits a conditional logit model. Since the likelihood is identical with that from a proportional hazards model, it may be found in the `survival` package. The matched sets must designated by the `strata` function:

```
> library(survival)
> cmod <- clogit(disease ~ Sex+Mray+Fray+CnRay+strata(ID),ramlxray)
> summary(cmod)
          coef exp(coef) se(coef)      z      p
SexM     0.156      1.17    0.386  0.405 0.6900
Mrayyes  0.228      1.26    0.582  0.391 0.7000
Frayyes  0.693      2.00    0.351  1.974 0.0480
CnRay.L  1.941      6.96    0.621  3.127 0.0018
CnRay.Q -0.248      0.78    0.582 -0.426 0.6700
CnRay.C -0.580      0.56    0.591 -0.982 0.3300

        exp(coef) exp(-coef) lower .95 upper .95
SexM         1.17      0.855     0.549      2.49
Mrayyes      1.26      0.796     0.401      3.93
Frayyes      2.00      0.500     1.005      3.98
CnRay.L      6.96      0.144     2.063     23.51
CnRay.Q      0.78      1.281     0.249      2.44
CnRay.C      0.56      1.786     0.176      1.78

Rsquare= 0.089   (max possible= 0.499 )
Likelihood ratio test= 20.9  on 6 df,   p=0.00192
```

```
Wald test            = 14.5  on 6 df,   p=0.0246
Score (logrank) test = 18.6  on 6 df,   p=0.0049
```

The overall tests for significance of the predictors indicate that at least some of the variables are significant. We see that `Sex` and whether the mother had an x-ray are not significant. There seems little point in investigating the other x-ray variables associated with the mother. An x-ray on the father is marginally significant. However, the x-ray on the child has the clearest effect. Because this is an ordered factor, we have used linear, quadratic and cubic contrasts. Only the linear effect is significant.

The second table of coefficients gives us information helpful for interpreting the size of the effects. We see that the father having had an x-ray doubles the odds of the disease. The interpretation of the number of x-rays of the child is more difficult to interpret because of the coding. Since we have found only a linear effect, we convert `CnRay` to the numerical values 1–4 using `unclass`. We also drop the insignificant predictors:

```
> cmodr <- clogit(disease ~ Fray+unclass(CnRay)+strata(ID),ramlxray)
> summary(cmodr)
                 coef exp(coef) se(coef)    z       p
Frayyes         0.670      1.96    0.344 1.95 0.05100
unclass(CnRay) 0.814      2.26    0.237 3.44 0.00058

               exp(coef) exp(-coef) lower .95 upper .95
Frayyes             1.96      0.512     0.996      3.84
unclass(CnRay)      2.26      0.443     1.419      3.59
```

The codes for `Cnray` are 1 = none, 2 = 1 or 2 x-rays, 3 = 3 or 4 x-rays and 4 = 5 or more x-rays. We see that the odds of the disease increase by a factor of 2.26 as we move between adjacent categories. Notice that the father's x-ray variable is now just insignificant in this regression underlining its borderline status.

An incorrect analysis of this data ignores the matching structure and simply uses a binomial GLM:

```
> gmod <- glm(disease ~ Fray+unclass(CnRay),family=binomial,ramlxray)
> summary(gmod)
Coefficients:
               Estimate Std. Error z value Pr(>|z|)
(Intercept)      -1.162      0.301   -3.86  0.00011
Frayyes           0.500      0.308    1.63  0.10405
unclass(CnRay)    0.601      0.177    3.39  0.00071
```

The results are somewhat different.

Although we have found an effect due to x-rays of the child, we cannot conclude the effect is causal. After all, subjects only have x-rays when something is wrong, so it is quite possible that the x-rays are linked to some unknown causal factor.

Other examples of matched data may be found in Section 4.3.

Further Reading: See books by Collett (2003), Hosmer and Lemeshow (2000), Cox (1970), Harrell (2001), Menard (2002), Christensen (1997) and Kleinbaum and Klein (2002).

Exercises

1. The question concerns data from a case-control study of esophageal cancer in Ile-et-Vilaine, France. The data is distributed with R and may be obtained along with a description of the variables by:

```
> data(esoph)
> help(esoph)
```

   (a) Fit a binomial GLM with interactions between all three predictors. Use backward elimination to simplify the model as far as is reasonable.

   (b) All three factors are ordered and so special contrasts have been used appropriate for ordered factors involving linear, quadratic and cubic terms. Further simplification of the model is possible by eliminating some of these terms. Use the `unclass` function to convert some or all factors to a numerical representation and show how the model may be simplified.

   (c) Does your final model fit the data? Is the test you make accurate for this data?

   (d) Check for outliers in your final model.

   (e) What is the predicted effect of moving one category higher in alcohol consumption?

   (f) Compute a 95% confidence interval for this predicted effect.

   (g) Bearing in mind that this is a case-control study, what can be said about the predicted probability that a 25-year-old who does not smoke or drink will get esophageal cancer?

2. The dataset `wbca` comes from a study of breast cancer in Wisconsin. There are 681 cases of potentially cancerous tumors of which 238 are actually malignant. Determining whether a tumor is really malignant is traditionally determined by an invasive surgical procedure. The purpose of this study was to determine whether a new procedure called fine needle aspiration, which draws only a small sample of tissue, could be effective in determining tumor status.

   (a) Fit a binomial regression with `Class` as the response and the other nine variables as predictors. Report the residual deviance and associated degrees of freedom. Can this information be used to determine if this model fits the data? Explain.

   (b) Use AIC as the criterion to determine the best subset of variables. (Use the `step` function.)

   (c) Use the reduced model to predict the outcome for a new patient with predictor variables 1, 1, 3, 2, 1, 1, 4, 1, 1 (same order as above). Give a confidence interval for your prediction.

   (d) Suppose that a cancer is classified as benign if $p > 0.5$ and malignant if $p < 0.5$. Compute the number of errors of both types that will be made if this method is applied to the current data with the reduced model.

(e) Suppose we change the cutoff to 0.9 so that $p < 0.9$ is classified as malignant and $p > 0.9$ as benign. Compute the number of errors in this case. Discuss the issues in determining the cutoff.

(f) It is usually misleading to use the same data to fit a model and test its predictive ability. To investigate this, split the data into two parts — assign every third observation to a test set and the remaining two thirds of the data to a training set. Use the training set to determine the model and the test set to assess its predictive performance. Compare the outcome to the previously obtained results.

3. The National Institute of Diabetes and Digestive and Kidney Diseases conducted a study on 768 adult female Pima Indians living near Phoenix. The purpose of the study was to investigate factors related to diabetes. The data may be found in the the dataset `pima`.

   (a) Perform simple graphical and numerical summaries of the data. Can you find any obvious irregularities in the data? If you do, take appropriate steps to correct the problems.

   (b) Fit a model with the result of the diabetes test as the response and all the other variables as predictors. Can you tell whether this model fits the data?

   (c) What is the difference in the odds of testing positive for diabetes for a woman with a BMI at the first quartile compared with a woman at the third quartile, assuming that all other factors are held constant? Give a confidence interval for this difference.

   (d) Do women who test positive have higher diastolic blood pressures? Is the diastolic blood pressure significant in the regression model? Explain the distinction between the two questions and discuss why the answers are only apparently contradictory.

   (e) Perform diagnostics on the regression model, reporting any potential violations and any suggested improvements to the model.

   (f) Predict the outcome for a woman with predictor values 1, 99, 64, 22, 76, 27, 0.25, 25 (same order as in the dataset). Give a confidence interval for your prediction.

4. Aflatoxin B1 was fed to lab animals at various doses and the number responding with liver cancer recorded. The data may be found in the dataset `aflatoxin`.

   (a) Build a model to predict the occurrence of liver cancer. Compute the ED50 level.

   (b) Discuss the extrapolation properties of your chosen model for low doses.

5. A study was conducted to determine the effectiveness of a new teaching method in economics. The data may be found in the dataset `spector`. Write a report on how well the new method works.

6. Incubation temperature can affect the sex of turtles. An experiment was conducted with three independent replicates for each temperature and the number of male

and female turtles born was recorded and can be found in the `turtle` dataset. Check for evidence of overdispersion in a binomial model for the sex of the turtle.

7. The `infert` dataset from the `survival` package presents data from a study of infertility after spontaneous and induced abortion. Analyze and report on the factors related to infertility based on this data.

CHAPTER 3

Count Regression

When the response is a count (a positive integer), we can use a count regression model to explain this response in terms of the given predictors. Sometimes, the total count is bounded, in which case a binomial response regression should probably be used. In other cases, the counts might be sufficiently large that a normal approximation is justified so that a normal linear model may be used. We shall consider two distributions for counts. The Poisson and, less commonly, the negative binomial.

3.1 Poisson Regression

If Y is Poisson with mean $\mu > 0$, then:

$$P(Y = y) = \frac{e^{-\mu}\mu^y}{y!}, \qquad y = 0, 1, 2, \ldots$$

Now $EY = \text{var}\, Y = \mu$. The Poisson distribution arises naturally in several ways:

1. If the count is some number out of some possible total, then the response would be more appropriately modeled as a binomial. However, for small success probabilities and large totals, the Poisson is a good approximation and can be applied. For example, in modeling the incidence of rare forms of cancer, the number of people affected is a small proportion of the population in a given geographical area. A Poisson regression model can be used in preference to a binomial. If $\mu = np$ while $n \to \infty$, then $B(n, p)$ is well approximated by $Pois(\mu)$. Also, for small p, note that $\text{logit}(p) \approx \log p$, so that the use of the Poisson with a log link is comparable to the binomial with a logit link.
2. Suppose the probability of occurrence of an event in a given time interval is proportional to the length of that time interval and independent of the occurrence of other events. Then the number of events in any specified time interval will be Poisson distributed. Examples include modeling the number of incoming telephone calls to a service center or the number of earthquakes. However, in any real application, the assumptions are likely to be violated. For example, the rate of incoming telephone calls is likely to vary with the time of day while the timing of earthquakes are unlikely to be completely independent. Nevertheless, a good approximation may be sufficient.
3. Poisson distributions also arise naturally when the time between events is independent and identically exponentially distributed. We count the number of events in a given time period. This is effectively equivalent to the previous case, since the exponential distribution between events will result from the assumption of constant and independent probability of occurrence of an event in an interval.

If the count is the number falling into some level of a given category, then a multinomial response model or categorical data analysis should be used. For example, if we have counts of how many people have type O, A, B or AB blood and are interested in how that relates to race and gender, then a straight Poisson regression model will not be appropriate. We will see later that the Poisson distribution still comes into play in Chapter 5.

An important result concerning Poisson random variables is that their sum is also Poisson. Specifically, suppose that $Y_i \sim Pois(\mu_i)$ for $i = 1, 2, \ldots$ and are independent, then $\sum_i Y_i \sim Pois(\sum_i \mu_i)$. This is useful because sometimes we have access only to the aggregated data. If we assume the individual-level data is Poisson, then so is the summed data and Poisson regression can still be applied.

For 30 Galápagos Islands, we have a count of the number of plant species found on each island and the number that are endemic to that island. We also have five geographic variables for each island. The data was presented by Johnson and Raven (1973) and also appear in Weisberg (2005). We have filled in a few missing values that appeared in the original dataset for simplicity. We model the number of species using normal linear regression:

```
> data(gala)
> gala <- gala[,-2]
```

We throw out the Endemics variable (which falls in the second column of the dataframe) since we won't be using it in this analysis. We fit a linear regression and look at the residual vs. fitted plot:

```
> modl <- lm(Species ~ . , gala)
> plot(predict(modl),residuals(modl),xlab="Fitted",ylab="Residuals")
```

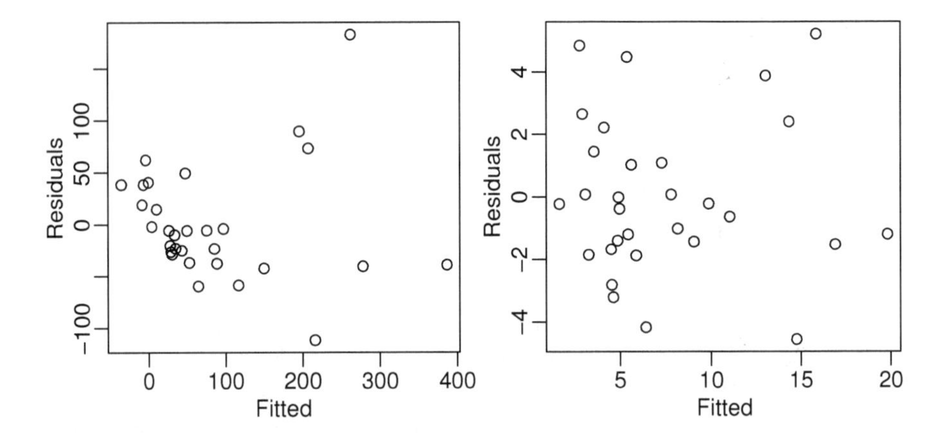

Figure 3.1 *Residual-fitted plots for the Galápagos dataset. The plot on the left is for a model with the original response while that on the right is for the square-root transformed response.*

We see clear evidence of nonconstant variance in left panel of Figure 3.1. Some experimentation (or the use of the Box–Cox method) reveals that a square-root transformation is best:

```
> modt <- lm(sqrt(Species) ~ . , gala)
> plot(predict(modt),residuals(modt),xlab="Fitted",ylab="Residuals")
```

We now see in the right panel of Figure 3.1 that the nonconstant variance problem has been cleared up. Let's take a look at the fit:

```
> summary(modt)
Coefficients:
             Estimate Std. Error t value Pr(>|t|)
(Intercept)  3.391924   0.871268    3.89  0.00069
Area        -0.001972   0.001020   -1.93  0.06508
Elevation    0.016478   0.002441    6.75  5.5e-07
Nearest      0.024933   0.047950    0.52  0.60784
Scruz       -0.013483   0.009798   -1.38  0.18151
Adjacent    -0.003367   0.000805   -4.18  0.00033

Residual standard error: 2.77 on 24 degrees of freedom
Multiple R-Squared: 0.783,     Adjusted R-squared: 0.737
F-statistic: 17.3 on 5 and 24 degrees of freedom,  p-value: 2.87e-07
```

We see a fairly good fit ($R^2 = 0.78$) considering the nature of the variables. However, we achieved this fit at the cost of transforming the response. This makes interpretation more difficult. Furthermore, some of the response values are quite small (single digits) which makes us question the validity of the normal approximation. This model may be adequate, but perhaps we can do better. We develop a Poisson regression model:

Suppose we have count responses Y_i that we wish to model in terms of a vector of predictors x_i. Now if $Y_i \sim Pois(\mu_i)$, we need some way to link the μ_i to the x_i. We use a linear combination of the x_i to form the linear predictor $\eta_i = x_i^T \beta$. Since we require that $\mu_i \geq 0$, we can ensure this by using a log link function, that is:

$$\log \mu_i = \eta_i = x_i^T \beta$$

So, as with the binomial regression models of the previous chapter, this models also has a linear predictor and a link function.

Now, the log-likelihood is:

$$l(\beta) = \sum_{i=1}^{n} (y_i x_i^T \beta - \exp(x_i^T \beta) - \log y_i!)$$

Differentiating with respect to β_j gives the MLE as the solution to:

$$\sum_{i=1}^{n} (y_i - \exp(x_i^T \hat{\beta})) x_{ij} = 0 \qquad \forall j$$

which can be more compactly written as:

$$X^T y = X^T \hat{\mu}$$

The normal equations for the least squares estimate of β in normal linear models take the same form when we set $\hat{\mu} = X\hat{\beta}$. The equations for β for a binomial regression with a logit link also take the same form. This would not be true for other link functions. The link function having this property is known as the *canonical link*.

However, there is no explicit formula for $\hat{\beta}$ for the Poisson (or binomial) regression and we must resort to numerical methods to find a solution. We fit the Poisson regression model to the Galápagos data:

```
> modp <- glm(Species ~ .,family=poisson,gala)
> summary(modp)
Deviance Residuals:
   Min      1Q  Median      3Q     Max
-8.275  -4.497  -0.944   1.917  10.185

Coefficients:
             Estimate Std. Error z value Pr(>|z|)
(Intercept)  3.1548079  0.0517495   60.96   < 2e-16
Area        -0.0005799  0.0000263  -22.07   < 2e-16
Elevation    0.0035406  0.0000874   40.51   < 2e-16
Nearest      0.0088256  0.0018213    4.85 0.0000013
Scruz       -0.0057094  0.0006256   -9.13   < 2e-16
Adjacent    -0.0006630  0.0000293  -22.61   < 2e-16

(Dispersion parameter for poisson family taken to be 1)

    Null deviance: 3510.73  on 29  degrees of freedom
Residual deviance:  716.85  on 24  degrees of freedom
AIC: 889.7

Number of Fisher Scoring iterations: 5
```

Using the same arguments as for binomial regression, we develop a deviance for the Poisson regression:

$$D = 2\sum_{i=1}^{n}(y_i \log(y_i/\hat{\mu}_i) - (y_i - \hat{\mu}_i))$$

This Poisson deviance is also known as the *G-statistic*.

The same asymptotic inference may be employed as for the binomial model. We can judge the goodness of fit of a proposed model by checking the deviance of the model against a χ^2 distribution with degrees of freedom equal to that of the model. We can compare nested models by taking the difference of the deviances and comparing to a χ^2 distribution with degrees of freedom equal to the difference in the number of parameters for the two models. We can test the significance of individual predictors and construct confidence intervals for β using the standard errors, $se(\hat{\beta})$, although, as before, it is better to use profile likelihood methods.

An alternative and perhaps better-known goodness of fit measure is the Pearson's X^2 statistic:

$$X^2 = \sum_{i=1}^{n} \frac{(y_i - \hat{\mu}_i)^2}{\hat{\mu}_i}$$

In this example, we see that the residual deviance is 717 on 24 degrees of freedom which indicates an ill-fitting model if the Poisson is the correct model for the response. We check the residuals to see if the large deviance can be explained by an outlier:

```
> halfnorm(residuals(modp))
```

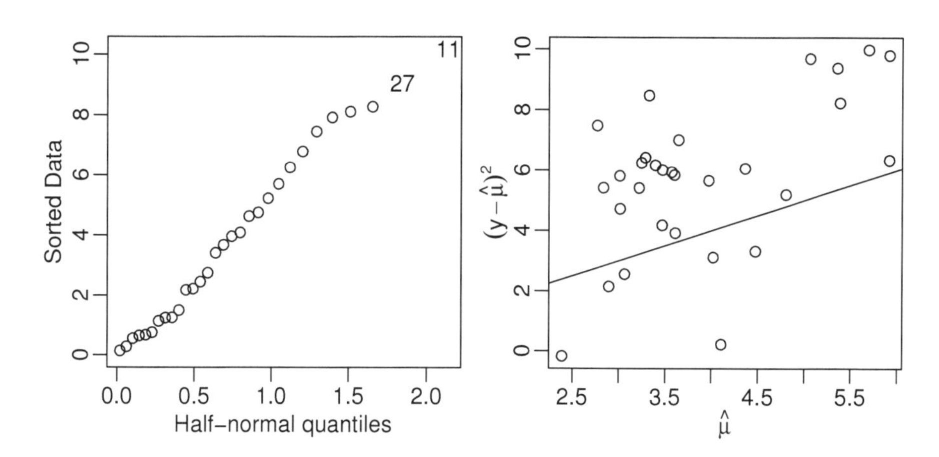

Figure 3.2 *Half-normal plot of the residuals of the Poisson model is shown on the left and the relationship between the mean and variance is shown on the right. A line representing mean equal to variance is also shown.*

The half-normal plot of the residuals shown in Figure 3.2 shows no outliers. It could be that the structural form of the model needs some improvement, but some experimentation with different forms for the predictors will reveal that there is little scope for improvement. Furthermore, the proportion of deviance explained by this model, 1-717/3510=0.796, is about the same as in the linear model above.

For a Poisson distribution, the mean is equal to the variance. Let's investigate this relationship for this model. It is difficult to estimate the variance for a given value of the mean, but $(y-\hat{\mu})^2$ does serve as a crude approximation. We plot this estimated variance against the mean, as seen in the second panel of Figure 3.2:

```
> plot(log(fitted(modp)),log((gala$Species-fitted(modp))^2),
  xlab=expression(hat(mu)),ylab=expression((y-hat(mu))^2))
> abline(0,1)
```

We see that the variance is proportional to, but larger than, the mean. When the variance assumption of the Poisson regression model is broken but the link function and choice of predictors is correct, the estimates of β are consistent, but the standard errors will be wrong. We cannot determine which predictors are statistically significant in the above model using the output we have.

The Poisson distribution has only one parameter and so is not very flexible for empirical fitting purposes. We can generalize by allowing ourselves a dispersion parameter. Over- or underdispersion can occur in Poisson models. For example, suppose the Poisson response Y has rate λ which is itself a random variable. The tendency to fail for a machine may vary from unit to unit even though they are the same model. We can model this by letting λ be gamma distributed with $E\lambda = \mu$ and $\text{var}\ \lambda = \mu/\phi$. Now Y is negative binomial with mean $EY = \mu$. The mean is the same as the Poisson, but the variance var $Y = \mu(1+\phi)/\phi$ which is not equal to μ. In this case, overdispersion would occur.

If we know the specific mechanism, as in the above example, we could model the response as a negative binomial or other more flexible distribution. If the mechanism is not known, we can introduce a dispersion parameter ϕ such that var $Y = \phi EY = \phi\mu$. $\phi = 1$ is the regular Poisson regression case, while $\phi > 1$ is overdispersion and $\phi < 1$ is underdispersion.

The dispersion parameter may be estimated using:

$$\hat{\phi} = \frac{X^2}{n-p} = \frac{\sum_i (y_i - \hat{\mu}_i)^2/\hat{\mu}_i}{n-p}$$

We estimate the dispersion parameter in our example by:

```
> (dp <- sum(residuals(modp,type="pearson")^2)/modp$df.res)
[1] 31.749
```

We can then adjust the standard errors and so forth in the summary as follows:

```
> summary(modp,dispersion=dp)
Coefficients:
             Estimate Std. Error z value Pr(>|z|)
(Intercept)  3.154808   0.291590   10.82  < 2e-16
Area        -0.000580   0.000148   -3.92  8.9e-05
Elevation    0.003541   0.000493    7.19  6.5e-13
Nearest      0.008826   0.010262    0.86     0.39
Scruz       -0.005709   0.003525   -1.62     0.11
Adjacent    -0.000663   0.000165   -4.01  6.0e-05

(Dispersion parameter for poisson family taken to be 31.749)

    Null deviance: 3510.73  on 29  degrees of freedom
Residual deviance:  716.85  on 24  degrees of freedom
AIC: 889.7
```

Notice that the estimation of the dispersion and the regression parameters is independent, so choosing a dispersion other than one has no effect on the regression parameter estimates. Notice also that there is some similarity in which variables are picked out as significant and which not when compared with the linear regression model.

When comparing Poisson models with overdispersion, an F-test rather than a χ^2 test should be used. As in normal linear models, the variance, or dispersion parameter in this case, needs to be estimated. This requires the use of the F-test. So to test the significance of each of the predictors relative to the full model, use:

```
> drop1(modp,test="F")
Single term deletions

Model:
Species ~ Area + Elevation + Nearest + Scruz + Adjacent
          Df Deviance  AIC F value   Pr(F)
<none>              717  890
```

```
Area       1     1204 1375   16.32 0.00048
Elevation  1     2390 2560   56.00   1e-07
Nearest    1      739  910    0.76 0.39336
Scruz      1      814  984    3.24 0.08444
Adjacent   1     1341 1512   20.91 0.00012
Warning message:
F test assumes quasipoisson family in: drop1.glm(modp, test = "F")
```

The z-statistics from the `summary()` are less reliable and so the F-test is preferred. In this example, there is little practical difference between the two.

3.2 Rate Models

The number of events observed may depend on a size variable that determines the number of opportunities for the events to occur. For example, if we record the number of burglaries reported in different cities, the observed number will depend on the number of households in these cities. In other cases, the size variable may be time. For example, if we record the number of customers served by a sales worker, we must take account of the differing amounts of time worked.

Sometimes, it is possible to analyze such data using a binomial response model. For the burglary example above, we might model the number of burglaries out of the number of households. However, if the proportional is small, the Poisson approximation to the binomial is effective. Furthermore, in some examples, the total number of potential cases may not be known exactly. The modeling of rare diseases illustrates this issue as we may know the number of cases but not have precise population data. Sometimes, the binomial model simply cannot be used. In the burglary example, some households may be affected more than once. In the customer service example, the size variable is not a count. An alternative approach is to model the ratio. However, there are often difficulties with normality and unequal variance when taking this approach, particularly if the counts are small.

In Purott and Reeder (1976), some data is presented from an experiment conducted to determine the effect of gamma radiation on the numbers of chromosomal abnormalities (`ca`) observed. The number (`cells`), in hundreds of cells exposed in each run, differs. The dose amount (`doseamt`) and the rate (`doserate`) at which the dose is applied are the predictors of interest. We may format the data for observation like this:

```
> data(dicentric)
> round(xtabs(ca/cells ~ doseamt+doserate, dicentric),2)
       doserate
doseamt  0.1 0.25  0.5    1  1.5    2  2.5    3    4
    1   0.05 0.05 0.07 0.07 0.06 0.07 0.07 0.07 0.07
    2.5 0.16 0.28 0.29 0.32 0.38 0.41 0.41 0.37 0.44
    5   0.48 0.82 0.90 0.88 1.23 1.32 1.34 1.24 1.43
```

We can plot the data as seen in the first panel of Figure 3.3:

```
> with(dicentric,interaction.plot(doseamt,doserate,ca/cells))
```

We might try modeling the rate directly. We see that the effect of the dose rate may be multiplicative, so we log this variable in the following model:

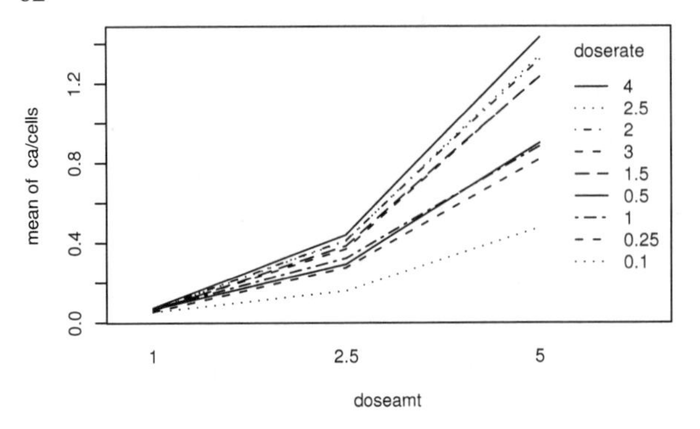

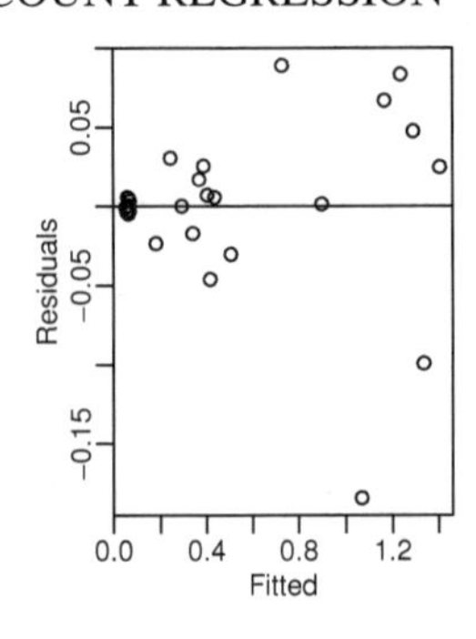

Figure 3.3 *Chromosomal abnormalities rate response is shown on the left and a residuals vs. fitted plot of a linear model fit to these data is shown on the right.*

```
> lmod <- lm(ca/cells ~ log(doserate)*factor(doseamt), dicentric)
> summary(lmod)$adj
[1] 0.98444
```

As can be seen from the adjusted R^2, this model fits well. However, a look at the diagnostics reveals a problem, as seen in the second panel of Figure 3.3:

```
> plot(residuals(lmod) ~ fitted(lmod),xlab="Fitted",ylab="Residuals")
> abline(h=0)
```

We might prefer an approach that directly models the count response. We need to use the log of the number of cells because we expect this to have a multiplicative effect on the response:

```
> dicentric$dosef <- factor(dicentric$doseamt)
> pmod <- glm(ca ~ log(cells)+log(doserate)*dosef,
  family=poisson,dicentric)
> summary(pmod)
Coefficients:
                        Estimate Std. Error z value Pr(>|z|)
(Intercept)              -2.7653     0.3812   -7.25    4e-13
log(cells)                1.0025     0.0514   19.52  < 2e-16
log(doserate)             0.0720     0.0355    2.03  0.04240
dosef2.5                  1.6298     0.1027   15.87  < 2e-16
dosef5                    2.7667     0.1229   22.52  < 2e-16
log(doserate):dosef2.5    0.1611     0.0484    3.33  0.00087
log(doserate):dosef5      0.1932     0.0430    4.49    7e-06

(Dispersion parameter for poisson family taken to be 1)

    Null deviance: 916.127  on 26  degrees of freedom
Residual deviance:  21.748  on 20  degrees of freedom
AIC: 211.2
```

We can relate this Poisson model with a log link back to a linear model for the ratio response:

$$\log(\texttt{ca/cells}) = X\beta$$

This can be rearranged as

$$\log \texttt{ca} = \log \texttt{cells} + X\beta$$

We are using log `cells` as a predictor. Checking above, we can see that the coefficient of 1.0025 is very close to one. This suggests fitting a model with the coefficient fixed as one. In this manner, we are modeling the rate of chromosomal abnormalities while still maintaining the count response for the Poisson model. This is known as a *rate model*. We fix the coefficient as one by using an *offset*. Such a term on the predictor side of the model equation has no parameter attached:

```
> rmod <- glm(ca ~ offset(log(cells))+log(doserate)*dosef,
  family=poisson,dicentric)
> summary(rmod)
Coefficients:
                        Estimate Std. Error z value  Pr(>|z|)
(Intercept)              -2.7467     0.0343  -80.16   < 2e-16
log(doserate)             0.0718     0.0352    2.04   0.04130
dosef2.5                  1.6254     0.0495   32.86   < 2e-16
dosef5                    2.7611     0.0435   63.49   < 2e-16
log(doserate):dosef2.5    0.1612     0.0483    3.34   0.00084
log(doserate):dosef5      0.1935     0.0424    4.56 0.0000051

(Dispersion parameter for poisson family taken to be 1)

    Null deviance: 4753.00  on 26  degrees of freedom
Residual deviance:   21.75  on 21  degrees of freedom
AIC: 209.2
```

Not surprisingly, the coefficients are only slightly different from the previous model. We see from the residual deviance that the model fits well. Previous analyses have posited a quadratic effect in dose; indeed, the observed coefficients speak against a purely linear effect. However, given that we have only three dose levels, we can hardly check whether quadratic is appropriate. Given the significant interaction effect, we can see that the effect of the dose rate is different depending on the overall dose. We can see that the combination of a high dose, delivered quickly, has a greater combined effect than the main effect estimates would suggest. More on the analysis of this data may be found in Frome and DuFrain (1986).

3.3 Negative Binomial

Given a series of independent trials, each with probability of success p, let Z be the number of trials until the kth success. Then:

$$P(Z = z) = \binom{z-1}{k-1} p^k (1-p)^{z-k} \qquad z = k, k+1, \ldots$$

The negative binomial can arise naturally in several ways. One can envision a system that can withstand k hits. The probability of a hit in a given time period is p. The negative binomial also arises from the generalization of the Poisson where the parameter λ is gamma distributed. The negative binomial also comes up as a limiting distribution for urn schemes that can be used to model contagion.

We get a more convenient parameterization if we let $Y = Z - k$ and $p = (1+\alpha)^{-1}$ so that:

$$P(Y = y) = \binom{y+k-1}{k-1} \frac{\alpha^y}{(1+\alpha)^{y+k}}, \qquad y = 0, 1, 2, \ldots$$

then $EY = \mu = k\alpha$ and $\text{var}\, Y = k\alpha + k\alpha^2 = \mu + \mu^2/k$.

The log-likelihood is then:

$$\sum_{i=1}^{n} \left(y_i \log \frac{\alpha}{1+\alpha} - k \log(1+\alpha) + \sum_{j=0}^{y_i - 1} log(j+k) - \log y_i! \right)$$

The most convenient way to link the mean response μ to a linear combination of the predictors X is:

$$\eta = x^T \beta = \log \frac{\alpha}{1+\alpha} = \log \frac{\mu}{\mu + k}$$

We can regard k as fixed and determined by the application or as an additional parameter to be estimated. More on regression models for negative binomial responses may found in Cameron and Trivedi (1998) and Lawless (1987).

Consider this example. ATT ran an experiment varying five factors relevant to a wave-soldering procedure for mounting components on printed circuit boards. The response variable, skips, is a count of how many solder skips appeared to a visual inspection. The data comes from Comizzoli, Landwehr, and Sinclair (1990). We start with a Poisson regression:

```
> data(solder)
> modp <- glm(skips ~ . , family=poisson, data=solder)
> deviance(modp)
[1] 1829
> df.residual(modp)
[1] 882
```

We see that the full model has a residual deviance of 1829 on 882 degrees of freedom. This is not a good fit. Perhaps including interaction terms will improve the fit:

```
> modp2  <- glm(skips ~ (Opening +Solder + Mask + PadType + Panel)^2 ,
  family=poisson, data=solder)
> deviance(modp2)
[1] 1068.8
> pchisq(deviance(modp2),df.residual(modp2),lower=FALSE)
[1] 1.1307e-10
```

The fit is improved but not enough to conclude that the model fits. We could try adding more interactions but that would make interpretation increasingly difficult. A check for outliers reveals no problem.

An alternative model for counts is the negative binomial. The functions for fitting come from the MASS package — see Venables and Ripley (2002) for more details. We can specify the link parameter k. Here we choose $k = 1$ to demonstrate the method, although there is no substantive motivation from this application to use this value:

```
> library(MASS)
> modn <- glm(skips ~ .,negative.binomial(1),solder)
> modn
Coefficients:
(Intercept)     OpeningM     OpeningS   SolderThin       MaskA3
    -1.6993       0.5085       1.9997       1.0489       0.6571
     MaskA6       MaskB3       MaskB6   PadTypeD6    PadTypeD7
     2.5265       1.2726       2.0803      -0.4612       0.0161
  PadTypeL4    PadTypeL6    PadTypeL7   PadTypeL8    PadTypeL9
     0.4688      -0.4711      -0.2949      -0.0849      -0.5213
  PadTypeW4    PadTypeW9        Panel
    -0.1425      -1.4836       0.1693

Degrees of Freedom: 899 Total (i.e. Null);  882 Residual
Null Deviance:      1740
Residual Deviance: 559  AIC: 3880
```

We could experiment with different values of k, but there is a more direct way of achieving this by allowing the parameter k to vary and be estimated using maximum likelihood in:

```
> modn <- glm.nb(skips ~ .,solder)
> summary(modn)
Coefficients:
            Estimate Std. Error z value Pr(>|z|)
(Intercept)  -1.4225     0.1427   -9.97  < 2e-16
OpeningM      0.5029     0.0798    6.31  2.9e-10
OpeningS      1.9132     0.0715   26.75  < 2e-16
SolderThin    0.9393     0.0536   17.52  < 2e-16
MaskA3        0.5898     0.0965    6.11  9.9e-10
MaskA6        2.2673     0.1018   22.27  < 2e-16
MaskB3        1.2110     0.0964   12.57  < 2e-16
MaskB6        1.9904     0.0922   21.58  < 2e-16
PadTypeD6    -0.4659     0.1124   -4.15  3.4e-05
PadTypeD7    -0.0331     0.1067   -0.31  0.75611
PadTypeL4     0.3827     0.1026    3.73  0.00019
PadTypeL6    -0.5784     0.1141   -5.07  4.0e-07
PadTypeL7    -0.3666     0.1109   -3.30  0.00095
PadTypeL8    -0.1589     0.1082   -1.47  0.14199
PadTypeL9    -0.5660     0.1139   -4.97  6.8e-07
PadTypeW4    -0.2004     0.1087   -1.84  0.06526
PadTypeW9    -1.5646     0.1362  -11.49  < 2e-16
Panel         0.1637     0.0314    5.21  1.8e-07

(Dispersion parameter for Negative Binomial(4.3972) family taken to be 1)

    Null deviance: 4043.3  on 899  degrees of freedom
Residual deviance: 1008.3  on 882  degrees of freedom
AIC: 3683

Number of Fisher Scoring iterations: 1
```

```
              Theta:  4.397
          Std. Err.:  0.495

 2 x log-likelihood:  -3645.309
```

We see that $\hat{k} = 4.397$ with a standard error of 0.495. We can compare negative binomial models using the usual inferential techniques.

Further Reading: See books by Cameron and Trivedi (1998) and Agresti (2002).

Exercises

1. The dataset `discoveries` lists the numbers of "great" inventions and scientific discoveries in each year from 1860 to 1959. Has the discovery rate remained constant over time?

2. The `salmonella` data was collected in a salmonella reverse mutagenicity assay. The predictor is the dose level of quinoline and the response is the numbers of revertant colonies of TA98 salmonella observed on each of three replicate plates. Show that a Poisson GLM is inadequate and that some overdispersion must be allowed for. Do not forget to check out other reasons for a high deviance.

3. The `ships` dataset found in the `MASS` package gives the number of damage incidents and aggregate months of service for different types of ships broken down by year of construction and period of operation. Develop a model for the rate of incidents, describing the effect of the important predictors.

4. The dataset `africa` gives information about the number of military coups in sub-Saharan Africa and various political and geographical information. Develop a simple but well-fitting model for the number of coups. Give an interpretation of the effect of the variables you include in your model on the response.

5. The `dvisits` data comes from the Australian Health Survey of 1977–78 and consist of 5190 single adults where young and old have been oversampled.

   (a) Build a Poisson regression model with `doctorco` as the response and `sex`, `age`, `agesq`, `income`, `levyplus`, `freepoor`, `freerepa`, `illness`, `actdays`, `hscore`, `chcond1` and `chcond2` as possible predictor variables. Considering the deviance of this model, does this model fit the data?

   (b) Plot the residuals and the fitted values — why are there lines of observations on the plot?

   (c) Use backward elimination with a critical p-value of 5% to reduce the model as much as possible. Report your model.

   (d) What sort of person would be predicted to visit the doctor the most under your selected model?

   (e) For the last person in the dataset, compute the predicted probability distribution for their visits to the doctor, i.e., give the probability they visit 0, 1, 2, etc. times.

(f) Fit a comparable (Gaussian) linear model and graphically compare the fits. Describe how they differ.

6. Components are attached to an electronic circuit card assembly by a wave-soldering process. The soldering process involves baking and preheating the circuit card and then passing it through a solder wave by conveyor. Defects arise during the process. The design is 2^{7-3} with three replicates. The data is presented in the dataset `wavesolder`.

   Assuming that the replicates are independent, analyze the data. Write a report on the analysis that summarizes the substantive conclusions and includes the highlights of your analysis.

7. The dataset `esdcomp` was recorded on 44 doctors working in an emergency service at a hospital to study the factors affecting the number of complaints received. Build a model for the number of complaints received and write a report on your conclusions.

CHAPTER 4

Contingency Tables

A contingency table is used to show cross-classified categorical data on two or more variables. The variables can be *nominal* or *ordinal*. A nominal variable has categories with no natural ordering; for example, consider the automotive companies Ford, General Motors and Toyota. An ordering could be imposed using some criterion like sales, but there is nothing inherent in the categories that makes any particular ordering obvious. An ordinal variable does have a natural default ordering. For example, a disease might be recorded as absent, mild or severe. The five-point Likert scale ranging through strongly disagree, disagree, neutral, agree and strongly agree is another example.

An *interval scale* is an ordinal variable that has categories with a distance measure. This is often the result of continuous data that has been discretized into intervals. For example, age groups 0–18, 18–34, 34–55 and 55+ might be used to record age information. If the intervals are relatively wide, then methods for ordinal data can be used where the additional information about the intervals may be useful in the modeling. If the intervals are quite narrow, then we could replace interval response with the midpoint of the interval and then use continuous data methods. One could argue that all so-called continuous data is of this form, because such data cannot be measured with arbitrary precision. Height might be given to the nearest centimeter, for example.

4.1 Two-by-Two Tables

The data shown in Table 4.1 were collected as part of a quality improvement study at a semiconductor factory. A sample of wafers was drawn and cross-classified according to whether a particle was found on the die that produced the wafer and whether the wafer was good or bad. More details on the study may be found in Hall (1994). The data might have arisen under several possible sampling schemes:

Quality	No Particles	Particles	Total
Good	320	14	334
Bad	80	36	116
Total	400	50	450

Table 4.1 *Study of the relationship between wafer quality and the presence of particles on the wafer.*

1. We observed the manufacturing process for a certain period of time and observed 450 wafers. The data were then cross-classified. We could use a Poisson model.
2. We decided to sample 450 wafers. The data were then cross-classified. We could use a multinomial model.
3. We selected 400 wafers without particles and 50 wafers with particles and then recorded the good or bad outcome. We could use a binomial model.
4. We selected 400 wafers without particles and 50 wafers with particles that also included, by design, 334 good wafers and 116 bad ones. We could use hypergeometric model.

The first three sampling schemes are all plausible. The fourth scheme seems less likely in this example, but we include it for completeness. Such a scheme is more attractive when one level of each variable is relatively rare and we choose to over-sample both levels to ensure some representation.

The main question of interest concerning these data is whether the presence of particles on the wafer affects the quality outcome. We shall see that all four sampling schemes lead to exactly the same conclusion. First, let's set up the data in a convenient form for analysis:

```
> y <- c(320,14,80,36)
> particle <- gl(2,1,4,labels=c("no","yes"))
> quality <- gl(2,2,labels=c("good","bad"))
> wafer <- data.frame(y,particle,quality)
> wafer
    y particle quality
1 320       no    good
2  14      yes    good
3  80       no     bad
4  36      yes     bad
```

We will need the data in this form with one observation per line for our model fitting, but usually we prefer to observe it table form:

```
> (ov <- xtabs(y ~ quality+particle))
       particle
quality no  yes
   good 320  14
   bad   80  36
```

Poisson Model

Suppose we assume that the process is observed for some period of time and we count the number of occurrences of the possible outcomes. It would be natural to view these outcomes occurring at different rates and that we could form Poisson model for these rates. Suppose we fit an additive model:

```
> modl <- glm(y ~ particle+quality, poisson)
> summary(modl)
Coefficients:
            Estimate Std. Error z value Pr(>|z|)
(Intercept)   5.6934     0.0572   99.54   <2e-16
particleyes  -2.0794     0.1500  -13.86   <2e-16
```

```
qualitybad   -1.0576     0.1078   -9.81   <2e-16

(Dispersion parameter for poisson family taken to be 1)

    Null deviance: 474.10  on 3  degrees of freedom
Residual deviance:  54.03  on 1  degrees of freedom
```

The null model, which suggests all four outcomes occur at the same rate, does not fit because the deviance of 474.1 is very large for three degrees of freedom. The additive model, with a deviance of 54.03 is clearly an improvement over this. We might also want to test the significance of the individual predictors. We could use the z-values, but it is better to use the likelihood ratio test based on the differences in the deviance (not that it matters much for this particular dataset):

```
> drop1(modl,test="Chi")
Single term deletions

Model:
y ~ particle + quality
         Df Deviance AIC LRT Pr(Chi)
<none>            54  84
particle  1      364 392 310  <2e-16
quality   1      164 192 110  <2e-16
```

We see that both predictors are significant relative to the full model. By examining the coefficients, we see that wafers without particles occur at a significantly higher rate than wafers with particles. Similarly, we see that good-quality wafers occur at a significantly higher rate than bad-quality wafers.

The model coefficients are closely related to the marginal totals in the table. The maximum likelihood estimates satisfy:

$$X^T y = X^T \hat{\mu}$$

where the $X^T y$ is, in this example:

```
> (t(model.matrix(modl)) %*% y)[,]
(Intercept) particleyes  qualitybad
        450          50         116
```

So we see that the fitted values, $\hat{\mu}$, are a function of marginal totals. This fact is exploited in an alternative fitting method known as *iterative proportional fitting*. The `glm` function in R, however, uses Fisher scoring, described in Section 6.2. In any case, the log-likelihood (ignoring any terms not involving μ) is:

$$\log L = \sum_i y_i \log \mu_i$$

which is maximized to obtain the fit.

The analysis so far has told us nothing about the relationship between the presence of particles and the quality of the wafer. The additive model posits:

$$\log \mu = \gamma + \alpha_i + \beta_j$$

where α represents the particle effect and β represents the quality outcome and $i, j = 1, 2$. γ is the intercept term. Due to the log link, the predicted rate for the response in any cell in the table is formed from the product of the rates for the corresponding levels of the two predictors. There is no interaction term and so good- or bad-quality outcomes occur independently of whether a particle was found on the wafer. This model has a deviance of 54.03 on one degree of freedom and so does not fit the data.

The addition of an interaction term would saturate the model and so would have zero deviance and degrees of freedom. So an hypothesis comparing the models with and without interaction would use a test statistic of 54.03 on one degree of freedom. The hypothesis of no interaction would be rejected.

Multinomial Model

Suppose we assume that the total sample size was fixed at 450 and that the frequency of the four possible outcomes was recorded. In these circumstances, it is natural to use a multinomial distribution to model the response. Let y_{ij} be the observed response in cell (i, j) and let p_{ij} be the probability that an observation falls in that cell and let n be the sample size. The probability of the observed response under the multinomial is then:

$$\frac{n!}{\prod_i \prod_j y_{ij}} \prod_i \prod_j p_{ij}^{y_{ij}}$$

Now the p_{ij} will be linked to the predictor information according to the model we choose. To estimate the parameters, we would maximize the log-likelihood:

$$\log L = \sum_i \sum_j y_{ij} \log p_{ij}$$

where terms not involving p_{ij} are ignored. Notice that this takes essentially the same form as for the Poisson model above.

The main hypothesis of interest is whether the quality and presence of a particle on the wafer are independent. Let p_i for $i = 1,2$ be the probabilities of the two quality outcomes and p_j for $j = 1,2$ be the probability of the two particle categories. Let p_{ij} be the probability of a particular joint outcome. Under independence, $p_{ij} = p_i p_j$. Using the fact that probabilities must sum to one, the maximum likelihood estimates are:

$$\hat{p}_i = \sum_j y_{ij}/n \qquad \text{and} \qquad \hat{p}_j = \sum_i y_{ij}/n$$

We can compute these for the wafer data as, respectively:

```
> (pp <- prop.table( xtabs(y ~ particle)))
particle
     no     yes
0.88889 0.11111
> (qp <- prop.table( xtabs(y ~ quality)))
quality
   good     bad
0.74222 0.25778
```

The fitted values are then $\hat{\mu}_{ij} = np_i p_j = \sum_i y_{ij} \sum_j y_{ij}/n$ or:

```
> (fv <- outer(qp,pp)*450)
       particle
quality     no    yes
   good 296.89 37.111
```

```
  bad   103.11 12.889
```

To test the fit, we compare this model against the saturated model, for which $\hat{\mu}_{ij} = y_{ij}$. So the deviance is:

$$2\sum_i \sum_j y_{ij} \log(y_{ij}/\mu_{ij})$$

which computes to:

```
> 2*sum(ov*log(ov/fv))
[1] 54.03
```

which is the same deviance we observed in the Poisson model. So we see that the test for independence in the multinomial model coincides with the test for no interaction in the Poisson model. The latter test is easier to execute in R, so we shall usually take that approach.

This connection between the Poisson and multinomial is no surprise due to the following result. Let $Y_1, \ldots, Y_k$ be independent Poisson random variables with means $\lambda_1, \ldots \lambda_k$, then the joint distribution of $Y_1, \ldots, Y_k | \sum_i Y_i = n$ is multinomial with probabilities $p_j = \lambda_j / \sum_i \lambda_i$.

One alternative to the deviance is the Pearson X^2 statistic:

$$X^2 = \sum_{i,j} \frac{(y_{ij} - \hat{\mu}_{ij})^2}{\hat{\mu}_{ij}}$$

which takes the value:

```
> sum( (ov-fv)^2/fv)
[1] 62.812
```

Yates' continuity correction subtracts 0.5 from $y_{ij} - \hat{\mu}_{ij}$ when this value is positive and adds 0.5 when it is negative. This gives superior results for small samples. This correction is implemented in:

```
> prop.test(ov)

        2-sample test for equality of proportions with
        continuity correction

data:  ov
X-squared = 60.124, df = 1, p-value = 8.907e-15
```

The deviance-based test is preferred to the Pearson's X^2.

Binomial

It would also be natural to view the presence of the particle as affecting the quality of wafer. We would view the quality as the response and the particle status as a predictor. We might fix the number of wafers with no particles at 400 and the number with particles as 50 and then observe the outcome. We could then use a binomial model for the response for both groups. Let's see what happens:

```
> (m <- matrix(y,nrow=2))
     [,1] [,2]
[1,]  320   80
[2,]   14   36
> modb <- glm(m ~ 1, family=binomial)
```

```
> deviance(modb)
[1] 54.03
```

We fit the null model which suggests that the probability of the response is the same in both the particle and no particle group. This hypothesis of *homogeneity* corresponds exactly to the test of independence and the deviance is exactly the same.

For larger contingency tables, where there are more than two rows (or columns), we can use a multinomial model for each row. This model is more accurately called a *product* multinomial model to distinguish it from the unrestricted multinomial model introduced above.

Hypergeometric

The remaining case is where both marginal totals are fixed. This situation is rather less common in practice, but does suggest a more accurate test for independence. This sampling scheme can arise when classifying objects into one of two types when the true proportions of each type are known in advance. For example, suppose you are given 10 true or false statements and told that 5 are true and 5 are false. You are asked to sort the statements into true and false. We can generate a two-by-two table of the correct classification against the observed classification generated. Under the hypergeometric distribution and the assumption of independence, the probability of the observed table is:

$$\frac{(y_{11}+y_{12})!(y_{11}+y_{21})!(y_{12}+y_{22})!(y_{21}+y_{22})!}{y_{11}!y_{12}!y_{21}!y_{22}!n!}$$

If we fix any number in the table, say y_{11}, the remaining three numbers are completely determined because the row and column totals are known. There is a limited number of values which y_{11} can possibly take and we can compute the probability of all these outcomes. Specifically, we can compute the total probability of all outcomes more extreme than the one observed. This method is called *Fisher's exact test.* We may execute it as follows:

```
> fisher.test(ov)

        Fisher's Exact Test for Count Data

data:  ov
p-value = 2.955e-13
alternative hypothesis: true odds ratio is not equal to 1
95 percent confidence interval:
  5.0906 21.5441
sample estimates:
odds ratio
    10.213
```

Notice that the odds ratio, which is $(y_{11}y_{22})/(y_{12}y_{21})$, takes the value:

```
> (320*36)/(14*80)
[1] 10.286
```

and is a measure of the association for which an exact confidence interval may be calculated as we see in the output.

Fisher's test is attractive because the null distribution for the deviance and Pearson's χ^2 test statistics is only approximately χ^2 distributed. This approximation is particularly suspect for tables with small counts making an exact method valuable. The Fisher test becomes more difficult to compute for larger tables and some approximations may be necessary. However, for larger tables, the χ^2 approximation will tend to be very accurate.

4.2 Larger Two-Way Tables

Snee (1974) presents data on 592 students cross-classified by hair and eye color.

```
> data(haireye)
> haireye
    y   eye  hair
1    5 green BLACK
2   29 green BROWN
..etc..
16   7 brown BLOND
```

The data is more conveniently displayed using:

```
> (ct <- xtabs(y ~ hair + eye, haireye))
       eye
hair     green hazel blue brown
  BLACK    5    15    20    68
  BROWN   29    54    84   119
  RED     14    14    17    26
  BLOND   16    10    94     7
```

We can execute the usual Pearson's χ^2 test for independence as:

```
> summary(ct)
Call: xtabs(formula = y ~ hair + eye, data = haireye)
Number of cases in table: 592
Number of factors: 2
Test for independence of all factors:
        Chisq = 138, df = 9, p-value = 2.3e-25
```

where we see that hair and eye color are clearly not independent.

One option for displaying contingency table data is the *dotchart*:

```
> dotchart(ct)
```

which may be seen in the first panel of Figure 4.1. The mosaic plot, described in Hartigan and Kleiner (1981), divides the plot region according to the frequency of each level in a recursive manner:

```
> mosaicplot(ct,color=TRUE,main=NULL,las=1)
```

In the plot shown in the second panel of Figure 4.1, the area is first divided according to the frequency of hair color. Within each hair color, the area is then divided according to the frequency of eye color. A different plot could be constructed by reversing the order of `hair` and `eye` in the `xtabs` command above. We can now readily see the frequency of various outcomes. We see, for example, that brown hair and brown

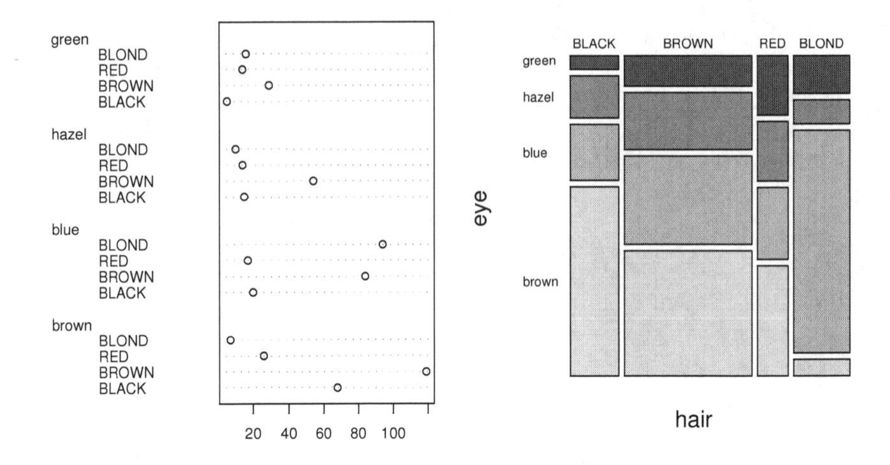

Figure 4.1 *Dotchart and Mosaic Plot*

eyes is the most common combination while green eyes and black hair is the least common.

Now we fit the Poisson GLM:

```
> modc <- glm(y ~ hair+eye,family=poisson,haireye)
> summary(modc)
Coefficients:
            Estimate Std. Error z value Pr(>|z|)
(Intercept)    2.458      0.152   16.14   <2e-16
hairBROWN      0.974      0.113    8.62   <2e-16
hairRED       -0.419      0.153   -2.75    0.006
hairBLOND      0.162      0.131    1.24    0.216
eyehazel       0.374      0.162    2.30    0.021
eyeblue        1.212      0.142    8.51   <2e-16
eyebrown       1.235      0.142    8.69   <2e-16

(Dispersion parameter for poisson family taken to be 1)

    Null deviance: 453.31  on 15  degrees of freedom
Residual deviance: 146.44  on  9  degrees of freedom
AIC: 241.0
```

We see that most of the levels of hair and eye color show up as significantly different from the reference levels of black hair and green eyes. But this merely indicates that there are higher numbers of people with some hair colors than others and some eye colors than others. We already know this. We are more interested in the relationship between hair and eye color. The deviance of 146.44 on nine degrees freedom shows that they are clearly dependent. This does not tell us how they are dependent. To study this, we can use a kind of residual analysis for contingency tables called *correspondence analysis*.

Compute the Pearson residuals r_P and write them in the matrix form R_{ij}, where $i = 1,\ldots,r$ and $j = 1,\ldots,c$, according to the structure of the data. Perform the singular value decomposition:

$$R_{r\times c} = U_{r\times w} D_{w\times w} V^T_{w\times c}$$

where r is the number of rows, c is the number of columns and $w = \min(r,c)$. U and V are called the right and left singular vectors, respectively. D is a diagonal matrix with sorted elements d_i, called *singular values*. Another way of writing this is:

$$R_{ij} = \sum_{k=1}^{w} U_{ik} d_k V_{jk}$$

As with eigendecompositions, it is not uncommon for the first few singular values to be much larger than the rest. Suppose that the first two dominate so that:

$$R_{ij} \approx U_{i1} d_1 V_{j1} + U_{i2} d_2 V_{j2}$$

We usually absorb the ds into U and V for plotting purposes so that we can assess the relative contribution of the components. Thus:

$$\begin{aligned} R_{ij} &\approx (U_{i1}\sqrt{d_1}) \times (V_{j1}\sqrt{d_1}) + (U_{i2}\sqrt{d_2}) \times (V_{j2}\sqrt{d_2}) \\ &\equiv U_{i1}V_{j1} + U_{i2}V_{j2} \end{aligned}$$

where in the latter expression we have redefined the Us and Vs to include the $\sqrt{d}$.

The two-dimensional correspondence plot displays U_{i2} against U_{i1} and V_{j2} against V_{j1} on the same graph. So the points on the plot will either represent a row level (U) or a column level (V). We compute the plot for the hair and eye color data:

```
> z <- xtabs(residuals(modc,type="pearson")~hair+eye,haireye)
> svdz <- svd(z,2,2)
> leftsv <- svdz$u %*% diag(sqrt(svdz$d[1:2]))
> rightsv <- svdz$v %*% diag(sqrt(svdz$d[1:2]))
> ll <- 1.1*max(abs(rightsv),abs(leftsv))
> plot(rbind(leftsv,rightsv),asp=1,xlim=c(-ll,ll),ylim=c(-ll,ll),
  xlab="SV1",ylab="SV2",type="n")
> abline(h=0,v=0)
> text(leftsv,dimnames(z)[[1]])
> text(rightsv,dimnames(z)[[2]])
```

The plot is shown in Figure 4.2. The correspondence analysis plot can be interpreted in light of the following observations:

- $\sum d_i^2$ = Pearson's X^2 is called the *inertia*. When $r = c$, d_i^2 are the eigenvalues of R.
- Look for large values of $|U_i|$ indicating that the row i profile is different. For example, the point for blonds in Figure 4.2 is far from the origin indicating that the distribution of eye colors within this group of people is not typical. In contrast, we see that the point for people with brown hair is close to the origin, indicating an eye color distribution that is close to the overall average. The same type of observation is true for the columns, $|V_j|$. Points distant from the origin mean that the level associated with the column j profile is different in some way.
- If row and column levels appear close together on the plot and far from the origin, we can see that there will be a large positive residual associated with this

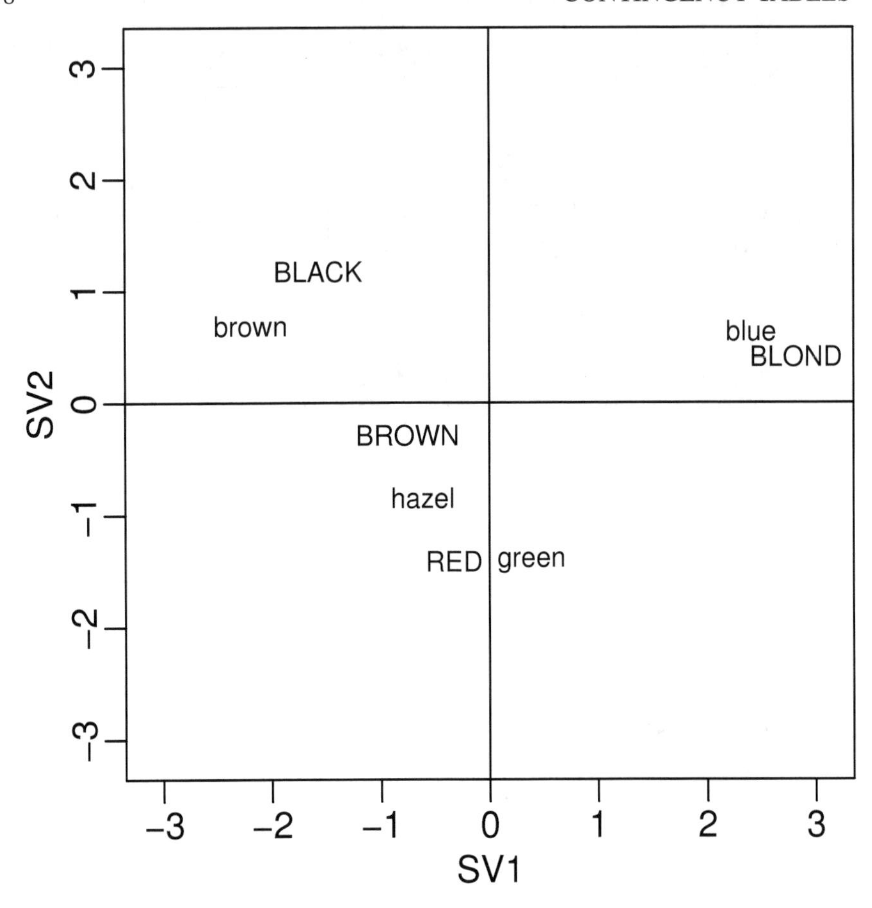

Figure 4.2 *Correspondence analysis for hair-eye combinations. Hair colors are given in upper-case letters and eye colors are given in lower-case letters.*

particular combination indicating a strong positive association. For example, we see that blue eyes and blond hair occur close together on the plot and far from the origin indicating a strong association. On the other hand, if the two points are situated diametrically apart on either side of the origin, we may expect a large negative residual indicating a strong negative association. For example, there are relatively fewer people with blond hair and brown eyes than would be expected under independence.

- If points representing two rows or two column levels are close together, this indicates that the two levels will have a similar pattern of association. In some cases, one might consider combining the two levels. For example, people with hazel or green eyes have similar hair color distributions and we might choose to combine these two categories.

- Because the distance between points is of interest, it is important that the plot is scaled so that the visual distance is proportionately correct. This does not happen automatically, because the default behavior of plots is to fill the plot region out to the specified aspect ratio.

There are several competing ways to construct contingency tables. See Venables and Ripley (2002) who provide the function `corresp` in the MASS package. See also Blasius and Greenacre (1998) for a survey of methods for visualizing categorical data.

4.3 Matched Pairs

In the typical two-way contingency tables, we display accumulated information about two categorical measures on the same object. In matched pairs, we observe one measure on two matched objects.

In Stuart (1955), data on the vision of a sample of women is presented. The left and right eye performance is graded into four categories:

```
> data(eyegrade)
> (ct <- xtabs(y ~ right+left, eyegrade))
        left
right    best second third worst
  best   1520  266    124    66
  second  234 1512    432    78
  third   117  362   1772   205
  worst    36   82    179   492
```

If we check for independence:

```
> summary(ct)
Call: xtabs(formula = y ~ right + left, data = eyegrade)
Number of cases in table: 7477
Number of factors: 2
Test for independence of all factors:
        Chisq = 8097, df = 9, p-value = 0
```

We are not surprised to find strong evidence of dependence. A more interesting hypothesis for such matched pair data is symmetry. Is $p_{ij} = p_{ji}$? We can fit such a model by defining a factor where the levels represent the symmetric pairs for the off-diagonal elements. There is only one observation for each level down the diagonal:

```
> (symfac <- factor(apply(eyegrade[,2:3],1,
  function(x) paste(sort(x),collapse="-"))))
 [1] best-best     best-second   best-third    best-worst
 [5] best-second   second-second second-third  second-worst
 [9] best-third    second-third  third-third   third-worst
[13] best-worst    second-worst  third-worst   worst-worst
10 Levels: best-best best-second best-third ... worst-worst
```

We now fit this model:

```
> mods <- glm(y ~ symfac, eyegrade, family=poisson)
> c(deviance(mods),df.residual(mods))
```

```
[1] 19.249  6.000
> pchisq(deviance(mods),df.residual(mods),lower=F)
[1] 0.0037629
```

Here, we see evidence of a lack of symmetry. It is worth checking the residuals:

```
> round(xtabs(residuals(mods) ~ right+left, eyegrade),3)
        left
right     best   second third  worst
  best    0.000  1.001  0.317  2.008
  second -1.023  0.000  1.732 -0.225
  third  -0.320 -1.783  0.000  0.928
  worst  -2.219  0.223 -0.949  0.000
```

We see that the residuals above the diagonal are mostly positive, while they are mostly negative below the diagonal. So there are generally more poor left, good right eye combinations than the reverse. Furthermore, we can compute the marginals:

```
> margin.table(ct,1)
right
  best second  third  worst
  1976   2256   2456    789
> margin.table(ct,2)
left
  best second  third  worst
  1907   2222   2507    841
```

We see that there are somewhat more poor left eyes and good right eyes, so perhaps marginal homogeneity does not hold here. The assumption of symmetry implies marginal homogeneity (the reverse is not necessarily true). We may observe data where there is a difference in the frequencies of the levels of the rows and columns, but still be interested in symmetry. Suppose we set:

$$p_{ij} = \alpha_i \beta_j \gamma_{ij}$$

where $\gamma_{ij} = \gamma_{ji}$. This will allow for some symmetry while allowing for different marginals. This is the *quasi-symmetry* model. Now:

$$\log EY_{ij} = \log np_{ij} = \log n + \log \alpha_i + \log \beta_j + \log \gamma_{ij}$$

So we can fit this model using:

```
> modq <- glm(y ~ right+left+symfac, eyegrade, family=poisson)
> pchisq(deviance(modq),df.residual(modq),lower=F)
[1] 0.06375
```

We see that this model does fit. It can be shown that marginal homogeneity together with quasi-symmetry implies symmetry. One can test for marginal homogeneity by comparing the symmetry and quasi-symmetry models:

```
> anova(mods,modq,test="Chi")
Analysis of Deviance Table

Model 1: y ~ symfac
Model 2: y ~ right + left + symfac
  Resid. Df Resid. Dev Df Deviance P(>|Chi|)
```

```
1         6       19.25
2         3        7.27  3     11.98       0.01
```

So we find evidence of a lack of marginal homogeneity. This test is only appropriate if quasi-symmetry already holds.

When we examine the data here, we do see that many people do have symmetric vision. These entries lie down the diagonal. We might ask whether there is independence between left and right eyes among those people whose vision is not symmetric. This is the *quasi-independence* hypothesis and we can test it by omitting the data from the diagonal:

```
> modqi <- glm(y ~ right+left, eyegrade, family=poisson,
  subset=-c(1,6,11,16))
> pchisq(deviance(modqi),df.residual(modqi),lower=F)
[1] 4.4118e-41
```

This model does not fit. This is not surprising since we can see that the entries adjacent to the diagonal are larger than those further away. The difference in vision between the two eyes is likely to be smaller than expected under independence.

4.4 Three-Way Contingency Tables

In Appleton, French, and Vanderpump (1996), a 20-year follow-up study on the effects of smoking is presented. In the period 1972–74, a larger study, which also considered other issues, categorized women into smokers and nonsmokers and according to their age group. In the follow-up, the researchers recorded whether the subjects were dead or still alive. Only smokers or women who had never smoked are presented here. Relatively few smokers quit and these women have been excluded from the data. The cause of death is not reported here. Here is the data:

```
> data(femsmoke)
> femsmoke
    y smoker dead   age
1   2    yes  yes 18-24
2   1     no  yes 18-24
3   3    yes  yes 25-34
....
28  0     no   no   75+
```

We can combine the data over age groups to produce:

```
> (ct <- xtabs(y ~ smoker+dead,femsmoke))
      dead
smoker yes no
   yes 139 443
   no  230 502
```

We can compute the proportions of dead and alive for smokers and nonsmokers:

```
> prop.table(ct,1)
      dead
smoker yes     no
   yes 0.23883 0.76117
   no  0.31421 0.68579
```

We see that 76% of smokers have survived for 20 years while only 69% of nonsmokers have survived. Thus smoking appears to have a beneficial effect on longevity. We can check the significance of this difference:

```
> summary(ct)
Call: xtabs(formula = y ~ smoker + dead, data = femsmoke)
Number of cases in table: 1314
Number of factors: 2
Test for independence of all factors:
        Chisq = 9.1, df = 1, p-value = 0.0025
```

So the difference cannot be reasonably ascribed to chance variation. However, if we consider the relationship within a given age group, say 55–64:

```
> (cta <- xtabs(y ~ smoker+dead,femsmoke, subset=(age=="55-64")))
      dead
smoker yes no
   yes 51  64
   no  40  81
> prop.table(cta,1)
      dead
smoker yes     no
   yes 0.44348 0.55652
   no  0.33058 0.66942
```

We see that 56% of the smokers have survived compared to 67% of the nonsmokers. This advantage to nonsmokers holds throughout all the age groups. Thus the *marginal* association where we add over the age groups is different from the *conditional* association observed within age groups. Data where this effect is observed are an example of *Simpson's paradox*. The paradox is named after Simpson (1951), but dates back to Yule (1903).

Let's see why the effect occurs here:

```
> prop.table(xtabs(y ~ smoker+age, femsmoke),2)
      age
smoker 18-24   25-34   35-44   45-54   55-64   65-74   75+
   yes 0.47009 0.44128 0.47391 0.62500 0.48729 0.21818 0.16883
   no  0.52991 0.55872 0.52609 0.37500 0.51271 0.78182 0.83117
```

We see that smokers are more concentrated in the younger age groups and younger people are more likely to live for another 20 years. This explains why the marginal table gave an apparent advantage to smokers which is, in fact, illusory because once we control for age, we see that smoking has a negative effect on longevity.

It is interesting to note that the dependence in the 55–64 age group is not statistically significant:

```
> fisher.test(cta)

        Fisher's Exact Test for Count Data

data:  cta
p-value = 0.08304
alternative hypothesis: true odds ratio is not equal to 1
```

```
95 percent confidence interval:
 0.92031 2.83340
sample estimates:
odds ratio
    1.6103
```

However, this is just a subset of the data. Suppose we compute the odds ratios in all the age groups:

```
> ct3 <- xtabs(y ~ smoker+dead+age,femsmoke)
> apply(ct3, 3, function(x) (x[1,1]*x[2,2])/(x[1,2]*x[2,1]))
  18-24   25-34   35-44   45-54   55-64   65-74     75+
2.30189 0.75372 2.40000 1.44175 1.61367 1.14851     NaN
```

We see that there is some variation in the odds ratio, but they are all greater than one with the exception of the 25–34 age group. We could test for independence in each 2×2 table, but it is better to use a combined test. The Mantel–Haenszel test is designed to test independence in 2×2 tables across K strata. It only makes sense to use this test if the relationship is similar in each stratum. For this data, the observed odds ratios do not vary greatly, so the use of the test is justified.

Let the entries in the $2 \times 2 \times K$ table be y_{ijk}. If we assume a hypergeometric distribution in each 2×2 table, then y_{11k} is sufficient for each table given that we assume that the marginal totals for each table carry no information. The Mantel–Haenszel statistic is:

$$\frac{(|\sum_k y_{11k} - \sum_k E y_{11k}| - 1/2)^2}{\sum_k \text{var } y_{11k}}$$

where the expectation and variance are computed under the null hypothesis of independence in each stratum. The statistic is approximately χ_1^2 distributed under the null, although it is possible to make an exact calculation for smaller datasets. The statistic as stated above is due to Mantel and Haenszel (1959), but a version without the half continuity correction was published by Cochran (1954). For this reason, it is sometimes known as the Cochran–Mantel–Haenszel statistic.

We compute the statistic for the data here:

```
> mantelhaen.test(ct3,exact=TRUE)

        Exact conditional test of independence in 2 x 2 x k
        tables

data:  ct3
S = 139, p-value = 0.01591
alternative hypothesis: true common odds ratio is not equal to 1
95 percent confidence interval:
 1.0689 2.2034
sample estimates:
common odds ratio
           1.5303
```

We used the exact method in preference to the approximation. We see that a statistically significant association is revealed once we combine the information across strata.

Now let's consider a linear models approach to investigating how three factors interact. Let p_{ijk} be the probability that an observation falls into the (i, j, k) cell. Let p_i be the marginal probability that the observation falls into the ith cell of the first variable, p_j be the marginal probability that the observation falls into the jth cell of the second variable and p_k be the marginal probability that the observation falls into the kth cell of the third variable.

Mutual Independence: If all three variables are independent, then:

$$p_{ijk} = p_i p_j p_k$$

Now $EY_{ijk} = np_{ijk}$ so:

$$\log EY_{ijk} = \log n + \log p_i + \log p_j + \log p_k$$

So the main effects-only model corresponds to mutual independence. The coding we use will determine exactly how the parameters relate to the margin totals of the table although typically we will not be especially interested in these. Since independence is the simplest possibility, this model is the null model in an investigation of this type. The model $\log EY_{ijk} = \mu$ would suggest that all the cells have equal probability. It is rare that such a model would have any interest so the model above makes for a more appropriate null.

We can test for independence using the Pearson's χ^2 test:

```
> summary(ct3)
Call: xtabs(formula = y ~ smoker + dead + age, data = femsmoke)
Number of cases in table: 1314
Number of factors: 3
Test for independence of all factors:
        Chisq = 791, df = 19, p-value = 2.1e-155
```

We can also fit the appropriate linear model:

```
> modi <- glm(y ~ smoker + dead + age, femsmoke, family=poisson)
> c(deviance(modi),df.residual(modi))
[1] 735  19
```

Although the statistics for the two tests are somewhat different, in either case, we see a very large value for the degrees of freedom. We conclude that this model does not fit the data.

We can show that the coefficients of this model correspond to the marginal proportions. For example, consider the smoker factor:

```
> (coefsmoke <- exp(c(0,coef(modi)[2])))
        smokerno
  1.0000  1.2577
> coefsmoke/sum(coefsmoke)
        smokerno
 0.44292  0.55708
```

We see that these are just the marginal proportions for the smokers and nonsmokers in the data:

```
> prop.table(xtabs(y ~ smoker, femsmoke))
smoker
```

```
    yes      no
0.44292 0.55708
```

This just serves to emphasize the point that the main effects of the model just convey information that we already know and is not the main interest of the study.

Joint Independence: Let p_{ij} be the (marginal) probability that the observation falls into a $(i, j, \cdot)$ cell where any value of the third variable is acceptable. Now suppose that the first and second variable are dependent, but jointly independent of the third. Then:

$$p_{ijk} = p_{ij}p_k$$

We can represent this as:

$$\log EY_{ijk} = \log n + \log p_{ij} + \log p_k$$

Using the hierarchy principle, we would also include the main effects corresponding to the interaction term $\log p_{ij}$. So the log-linear model with just one interaction term corresponds to joint independence. The specific interaction term tells us which pair of variables is dependent. For example, we fit a model that says age is jointly independent of smoking and life status:

```
> modj <- glm(y ~ smoker*dead + age, femsmoke, family=poisson)
> c(deviance(modj),df.residual(modj))
[1] 725.8  18.0
```

Although this represents a small improvement over the mutual independence model, the deviance is still very high for the degrees of freedom and it is clear that this model does not fit the data. There are two other joint independence models that have the two other interaction terms. These models also fit badly.

Conditional Independence: Let $p_{ij|k}$ be the probability that an observation falls in cell $(i, j, \cdot)$ given that we know the third variable takes the value k. Now suppose we assert that the first and second variables are independent given the value of the third variable, then:

$$p_{ij|k} = p_{i|k}p_{j|k}$$

which leads to:

$$p_{ijk} = p_{ik}p_{jk}/p_k$$

This results in the model:

$$\log EY_{ijk} = \log n + \log p_{ik} + \log p_{jk} - \log p_k$$

Again, using the hierarchy principle, we would also include the main effects corresponding to the interaction terms and we would have model with main effects and two interaction terms. The minus for the $\log p_k$ term is irrelevant. The nature of the conditional independence can be determined by observing which of one of the three possible two-way interactions does not appear in the model.

The most plausible conditional independence model for our data is:

```
> modc <- glm(y ~ smoker*age + age*dead, femsmoke, family=poisson)
> c(deviance(modc),df.residual(modc))
[1] 8.327 7.000
```

We see that the deviance is only slightly larger than the degrees of freedom indicating a fairly good fit. This indicates that smoking is independent of life status given age. However, bear in mind that we do have some zeroes and other small numbers in the table and so there is some doubt as to the accuracy of the χ^2 approximation here. It is generally better to compare models rather than assess the goodness of fit.

Uniform Association: We might consider a model with all two-way interactions:

$$\log EY_{ijk} = \log n + \log p_i + \log p_j + \log p_k + \log p_{ij} + \log p_{ik} + \log p_{jk}$$

The model has no three-way interaction and so it is not saturated. There is no simple interpretation in terms of independence. Consider our example:

```
> modu <- glm(y ~ (smoker+age+dead)^2, femsmoke, family=poisson)
```

Now we compute the fitted values and determine the odds ratios for each age group based on these fitted values:

```
> ctf <- xtabs(fitted(modu) ~ smoker+dead+age,femsmoke)
> apply(ctf, 3, function(x) (x[1,1]*x[2,2])/(x[1,2]*x[2,1]))
 18-24  25-34  35-44  45-54  55-64  65-74    75+
1.5333 1.5333 1.5333 1.5333 1.5333 1.5333 1.5333
```

We see that the odds ratio is the same for every age group. Thus the uniform association model asserts that for every level of one variable, we have the same association for the other two variables.

The information may also be extracted from the coefficients of the fit. Consider the odds ratio for smoking and life status for a given age group:

$$(EY_{11k}EY_{22k})/(EY_{12k}EY_{21k})$$

This will be precisely the coefficient for the smoking and life-status term. We extract this:

```
> exp(coef(modu)['smokerno:deadno'])
smokerno:deadno
         1.5333
```

We see that this is exactly the odds ratio we found above. The other interaction terms may be interpreted similarly.

Model Selection: Log-linear models are hierarchical, so it makes sense to start with the most complex model and see how far it can be reduced. We can use analysis of deviance to compare models. We start with the saturated model:

```
> modsat <- glm(y ~ smoker*age*dead, femsmoke, family=poisson)
> drop1(modsat,test="Chi")
Single term deletions

Model:
y ~ smoker * age * dead
                Df Deviance   AIC   LRT Pr(Chi)
<none>             3.0e-10 190.2
smoker:age:dead  6      2.4 180.6   2.4    0.88
```

We see that the three-way interaction term may be dropped. Now we consider dropping the two-way terms:

```
> drop1(modu,test="Chi")
Single term deletions

Model:
y ~ (smoker + age + dead)^2
            Df Deviance AIC LRT Pr(Chi)
<none>                2 181
smoker:age   6       93 259  90  <2e-16
smoker:dead  1        8 185   6   0.015
age:dead     6      632 798 630  <2e-16
```

Two of the interaction terms are strongly significant, but the `smoker:dead` term is only just statistically significant. This term corresponds to the test for conditional independence of smoking and life status given age group. We see that the conditional independence does not hold. This tests the same hypothesis as the Mantel–Haenszel test above. In this case the p-values for the two tests are very similar.

Binomial Model: For some three-way tables, it may be reasonable to regard one variable as the response and the other two as predictors. In this example, we could view life status as the response. Since this variable has only two levels, we can model it using a binomial GLM. For more than two levels, a multinomial model would be required.

We construct a binomial response model:

```
> ybin <- matrix(femsmoke$y,ncol=2)
> modbin <- glm(ybin ~ smoker*age, femsmoke[1:14,], family=binomial)
```

This model is saturated, so we investigate a simplification:

```
> drop1(modbin,test="Chi")
Single term deletions

Model:
ybin ~ smoker * age
           Df Deviance  AIC  LRT Pr(Chi)
<none>         5.3e-10 75.0
smoker:age  6      2.4 65.4  2.4    0.88
```

We see that the interaction term may be dropped, but now we check if we may drop further terms:

```
> modbinr <- glm(ybin ~ smoker+age, femsmoke[1:14,], family=binomial)
> drop1(modbinr,test="Chi")
Single term deletions

Model:
ybin ~ smoker + age
       Df Deviance AIC LRT Pr(Chi)
<none>           2  65
smoker  1        8  69   6   0.015
age     6      632 683 630  <2e-16
```

We see that both main effect terms are significant, so no further simplification is possible. This model is effectively equivalent to the uniform association model above. Check the deviances:

```
> deviance(modu)
[1] 2.3809
> deviance(modbinr)
[1] 2.3809
```

We see that they are identical. We can extract the same odds ratio from the parameter estimates as above:

```
> exp(-coef(modbinr)[2])
smokerno
  1.5333
```

The change in sign is simply due to which outcome is considered a success in the binomial GLM. So we can identify the binomial GLM with a corresponding Poisson GLM and the numbers we will obtain will be identical. We would likely prefer the binomial analysis where one factor can clearly be identified as the response and we would prefer the Poisson GLM approach when the relationship between the variables is more symmetric. However, there is one important difference between the two approaches. The null model for the binomial GLM:

```
> modbinull <- glm(ybin ~ 1, femsmoke[1:14,], family=binomial)
> deviance(modbinull)
[1] 641.5
```

is associated with this two-way interaction model for the Poisson GLM:

```
> modj <- glm(y ~ smoker*age + dead, femsmoke, family=poisson)
> deviance(modj)
[1] 641.5
```

So the binomial model implicitly assumes an association between `smoker` and `age`. In this particular dataset, there are more younger smokers than older ones, so the association is present. However, what if there was no association? One could argue that the Poisson GLM approach would be superior because it would allow us to drop this term and achieve a simpler model. On the other hand, one could argue that if the relationship between the response and the two predictors is the main subject of interest, then we lose little by conditioning out the marginal combined effect of age and smoking status, whether it is significant or not.

Correspondence Analysis: We cannot directly apply the correspondence analysis method described above for two-way tables. However, we could combine two of the factors into a single factor by considering all possible combinations of the two level. To make the choice of which two levels to combine, we would pick the pair whose association is least interesting to us. We could apply this to the smoking dataset here, but because there are only two levels of smoking and life status, the plot is not very interesting.

4.5 Ordinal Variables

Some variables have a natural order. One can use the methods for nominal variables described earlier in this chapter, but more information can be extracted by taking advantage of the structure of the data. Sometimes one might identify a particular ordinal variable as the response. In such cases, the methods of Section 5.3 can be used.

However, sometimes one is simply interested in modeling the association between ordinal variables. Here the use of *scores* can be helpful.

Consider a two-way table where both variables are ordinal. We may assign scores u_i and v_j to the rows and columns such that $u_1 \le u_2 \le \cdots \le u_I$ and $v_1 \le v_2 \le \cdots \le v_J$. The assignment of scores requires some judgment. If you have no particular preference, even spacing allows for the simplest interpretation. If you have an interval scale, for example, 0–10 years old, 10–20 years old, 20–40 years old and so on, midpoints are often used. It is a good idea to check that the inference is robust to the assignment of scores by trying some alternative choices. If your qualitative conclusions are changed, this is an indication that you cannot make any strong finding.

Now fit the *linear-by-linear association* model:

$$\log EY_{ij} = \log \mu_{ij} = \log np_{ij} = \log n + \alpha_i + \beta_j + \gamma u_i v_j$$

So $\gamma = 0$ means independence while γ represents the amount of association and can be positive or negative. γ is rather like an (unscaled) correlation coefficient. Consider underlying (latent) continuous variables which are discretized by the cutpoints u_i and v_j. We can then identify γ with the correlation coefficient of the latent variables

Consider an example drawn from a subset of the 1996 American National Election Study (Rosenstone, Kinder, and Miller (1997)). Considering just the data on party affiliation and level of education, we can construct a two-way table:

```
> data(nes96)
> xtabs( ~ PID + educ, nes96)
          educ
PID         MS HSdrop HS Coll CCdeg BAdeg MAdeg
  strDem     5 19     59 38   17    40    22
  weakDem    4 10     49 36   17    41    23
  indDem     1  4     28 15   13    27    20
  indind     0  3     12  9    3     6     4
  indRep     2  7     23 16    8    22    16
  weakRep    0  5     35 40   15    38    17
  strRep     1  4     42 33   17    53    25
```

Both variables are ordinal in this example. We need to convert this to a dataframe with one count per line to enable model fitting.

```
> (partyed <- as.data.frame.table(xtabs( ~ PID + educ, nes96)))
       PID   educ Freq
1   strDem     MS    5
2  weakDem     MS    4
3   indDem     MS    1
...etc....
```

If we fit a nominal-by-nominal model, we find no evidence against independence:

```
> nomod <- glm(Freq ~ PID + educ, partyed, family= poisson)
> pchisq(deviance(nomod),df.residual(nomod),lower=F)
[1] 0.26961
```

However, we can take advantage of the ordinal structure of both variables and define some scores. As there seems to be no strong reason to the contrary, we assign evenly spaced scores: one to seven for both `PID` and `educ`:

```
> partyed$oPID <- unclass(partyed$PID)
> partyed$oeduc <- unclass(partyed$educ)
```

Now fit the linear-by-linear association model and compare to the independence model:

```
> ormod <- glm(Freq ~ PID + educ + I(oPID*oeduc), partyed,
  family= poisson)
> anova(nomod,ormod,test="Chi")
Analysis of Deviance Table

Model 1: Freq ~ PID + educ
Model 2: Freq ~ PID + educ + I(oPID * oeduc)
  Resid. Df Resid. Dev Df Deviance P(>|Chi|)
1        36       40.7
2        35       30.6  1     10.2    0.0014
```

We see that there is some evidence of an association. So we see that using the ordinal information gives us more power to detect an association. We can examine $\hat{\gamma}$:

```
> summary(ormod)$coef['I(oPID * oeduc)',]
  Estimate Std. Error    z value    Pr(>|z|)
 0.0287446  0.0090617  3.1720850  0.0015135
```

We see that $\hat{\gamma}$ is 0.0287. The *p*-value here can also be used to test the significance of the association although, as a Wald test, it is less reliable than the likelihood ratio test we used first. We see that $\hat{\gamma}$ is positive, which, given the way that we have assigned the scores, mean that a higher level of education is associated with a greater probability of tending to the Republican end of the spectrum.

Just to check the robustness of the assignment of the scores, it is worth trying some different choices. For example, suppose we choose scores so that there is more of a distinction between Democrats and Independents as well as Independents and Republicans. Our assignment of scores for `apid` below achieves this. Another idea might be that people who complete high school or less are not different; that those who go to college, but do not get a BA degree are not different and that those who get a BA or higher are not different. My assignment of scores in `aedu` achieves this:

```
> apid <- c(1,2,5,6,7,10,11)
> aedu <- c(1,1,1,2,2,3,3)
> ormoda <- glm(Freq ~ PID + educ + I(apid[oPID]*aedu[oeduc]),
  partyed, family= poisson)
> anova(nomod,ormoda,test="Chi")
Analysis of Deviance Table

Model 1: Freq ~ PID + educ
Model 2: Freq ~ PID + educ + I(apid[oPID] * aedu[oeduc])
  Resid. Df Resid. Dev Df Deviance P(>|Chi|)
1        36       40.7
2        35       30.9  1      9.8    0.0017
```

The numerical outcome is slightly different, but the result is still significant. Some experimentation with other plausible choices indicates that we can be fairly confident about the association here.

The association parameter may be interpreted in terms of log-odds. For example, consider the log-odds ratio for adjacent entries in both rows and columns:

$$\log \frac{\mu_{ij}\mu_{i+1,j+1}}{\mu_{i,j+1}\mu_{i+1,j}} = \gamma(u_{i+1} - u_i)(v_{j+1} - v_j)$$

For evenly spaced scores, these log-odds ratios will all be equal. For our example, where the scores are spaced one apart, the log-odds ratio is γ. To illustrate this point, consider the fitted values under the linear-by-linear association model:

```
> round(xtabs(predict(ormod,type="response") ~ PID + educ, partyed),2)
          educ
PID        MS    HSdrop HS    Coll  CCdeg BAdeg MAdeg
  strDem    3.58 13.36  59.22 41.34 18.34 42.46 21.71
  weakDem   2.92 11.22  51.20 36.78 16.80 40.02 21.06
  indDem    1.59  6.27  29.45 21.78 10.23 25.09 13.59
  indind    0.49  2.00   9.65  7.34  3.55  8.96  5.00
  indRep    1.12  4.71  23.41 18.33  9.13 23.70 13.60
  weakRep   1.61  6.95  35.59 28.68 14.69 39.28 23.19
  strRep    1.69  7.49  39.48 32.74 17.26 47.49 28.85
```

Now compute log-odds ratio for, say, the lower two-by-two table:

```
> log(39.28*28.85/(47.49*23.19))
[1] 0.028585
```

We see this is, but for rounding, equal to $\hat{\gamma}$.

It is always worth examining the residuals to check if there is more structure than the model suggests. We use the raw response residuals (the unscaled difference between observed and expected) because we would like to see effects which are large in an absolute sense.

```
> round(xtabs(residuals(ormod,type="response") ~ PID + educ, partyed),2)
          educ
PID        MS    HSdrop HS    Coll  CCdeg BAdeg MAdeg
  strDem    1.42  5.64  -0.22 -3.34 -1.34 -2.46  0.29
  weakDem   1.08 -1.22  -2.20 -0.78  0.20  0.98  1.94
  indDem   -0.59 -2.27  -1.45 -6.78  2.77  1.91  6.41
  indind   -0.49  1.00   2.35  1.66 -0.55 -2.96 -1.00
  indRep    0.88  2.29  -0.41 -2.33 -1.13 -1.70  2.40
  weakRep  -1.61 -1.95  -0.59 11.32  0.31 -1.28 -6.19
  strRep   -0.69 -3.49   2.52  0.26 -0.26  5.51 -3.85
```

We do see some indications of remaining structure. For example, we see many more weak Republicans with some college than expected while fewer Republicans with master's degrees or higher. There may not be a monotone relationship between party affiliation and educational level.

To investigate this effect, we might consider an ordinal-by-nominal model where we now treat education as a nominal variable. This is called a *column effects* model because the columns (which are the education levels here) are not assigned scores and we will estimate their effect instead. A *row effects* model is effectively the same model except with the roles of the variables reversed. The model takes the form:

$$\log EY_{ij} = \log \mu_{ij} = \log np_{ij} = \log n + \alpha_i + \beta_j + u_i\gamma_j$$

where the γ_j are called the column effects. Equality of the γ_js corresponds to the hypothesis of independence. We fit this model for our data:

```
> cmod <- glm(Freq ~ PID + educ + educ:oPID, partyed, family= poisson)
```

We can compare this to the independence model:

```
> anova(nomod,cmod,test="Chi")
Analysis of Deviance Table

Model 1: Freq ~ PID + educ
Model 2: Freq ~ PID + educ + educ:oPID
  Resid. Df Resid. Dev Df Deviance P(>|Chi|)
1        36       40.7
2        30       22.8  6      18.0    0.0063
```

We find that the column-effects model is preferred. Now examine the fitted coefficients, looking at just the interaction terms as the main effects have no particular interest:

```
> summary(cmod)$coef[14:19,]
                  Estimate Std. Error   z value Pr(>|z|)
educMS:oPID     -0.3122169   0.154051 -2.026710 0.042692
educHSdrop:oPID -0.1944513   0.077228 -2.517891 0.011806
educHS:oPID     -0.0553470   0.048196 -1.148384 0.250810
educColl:oPID    0.0044605   0.050603  0.088147 0.929760
educCCdeg:oPID  -0.0086994   0.060667 -0.143395 0.885978
educBAdeg:oPID   0.0345539   0.048782  0.708330 0.478740
```

The last coefficient, `educMAdeg:oPID`, is not identifiable and so this may be taken as zero. If there was really a monotone trend in the effect of educational level on party affiliation, we would expect these coefficients to be monotone. However, we can see that they are not. However, if we compare this to the linear-by-linear association model:

```
> anova(ormod,cmod,test="Chi")
Analysis of Deviance Table

Model 1: Freq ~ PID + educ + I(oPID * oeduc)
Model 2: Freq ~ PID + educ + educ:oPID
  Resid. Df Resid. Dev Df Deviance P(>|Chi|)
1        35      30.57
2        30      22.76  5      7.81       0.17
```

We see that the simpler linear-by-linear association is preferred to the more complex column-effects model. Nevertheless, if the linear-by-linear association were a good fit, we would expect the observed column-effect coefficients to be roughly evenly spaced. Looking at these coefficients, we observe that for high school and above, the coefficients are not significantly different from zero while for the lowest two categories, there is some difference. This suggests an alternate assignment of scores for education:

```
> aedu <- c(1,1,2,2,2,2,2)
> ormodb <- glm(Freq ~ PID + educ + I(oPID*aedu[oeduc]),
  partyed, family= poisson)
```

```
> deviance(ormodb)
[1] 28.451
> deviance(ormod)
[1] 30.568
```

We see that the deviance of this model is even lower than our original model. This gives credence to the view that whether a person finishes high school or not is the determining factor in party affiliation. However, since we used the data itself to assign the scores and come up with this hypothesis, we would be tempting fate to then use the data again to test this hypothesis.

The use of scores can be helpful in reducing the complexity of models for categorical data with ordinal variables. It is especially useful in higher dimensional tables where a reduction in the number of parameters is particularly welcome. The use of scores can also sharpen our ability to detect associations.

Further Reading: See books by Agresti (2002), Bishop, Fienberg, and Holland (1975), Haberman (1977), Le (1998), Leonard (2000), Powers and Xie (2000), Santner and Duffy (1989) and Simonoff (2003).

Exercises

1. The dataset `parstum` contains cross-classified data on marijuana usage by college students as it relates to the alcohol and drug usage of the parents. Analyze the data as if both factors were nominal. Redo the analysis treating both factors as ordinal. Contrast the results.
2. The dataset `melanoma` gives data on a sample of patients suffering from melanoma (skin cancer) cross-classified by the type of cancer and the location on the body. Determine whether the type and location are independent. Examine the residuals to determine whether any dependence can be ascribed to particular type/location combinations.
3. Data on social mobility of men in the UK may be found in `cmob`. A sample of men aged 45–64 was drawn from the 1971 census and 1981 census and the social class of the man was recorded at each timepoint. The classes are I = professional, II = semiprofessional, IIIN = skilled nonmanual, IIIM = skilled manual, IV = semiskilled, V = unskilled.
   (a) Check for symmetry, quasi-symmetry, marginal homogeneity and quasi-independence.
   (b) Develop a score-based model. Find some good-fitting scores.
4. The dataset `death` contains data on murder cases in Florida in 1977. The data is cross-classified by the race (black or white) of the victim, of the defendant and whether the death penalty was given.
   (a) Consider the frequency with which the death penalty is applied to black and white defendants, both marginally and conditionally, with respect to the race of the victim. Is this an example of Simpson's paradox? Are the observed differences in the frequency of application of the death penalty statistically significant?

(b) Determine the most appropriate dependence model between the variables.

(c) Fit a binomial regression with death penalty as the response and show the relationship to your model in the previous question.

5. The dataset `sexfun` comes from a questionnaire from 91 couples in the Tucson, Arizona, area. Subjects answered the question "Sex is fun for me and my partner". The possible answers were "never or occasionally", "fairly often", "very often" and "almost always".

   (a) Check for symmetry, quasi-symmetry, marginal homogeneity and quasi-independence.

   (b) Develop a score-based model. Find some good-fitting scores.

6. The dataset `suicide` contains one year of suicide data from the United Kingdom cross-classified by sex, age and method.

   (a) Determine the most appropriate dependence model between the variables.

   (b) Collapse the sex and age of the subject into a single six-level factor containing all combinations of sex and age. Conduct a correspondence analysis and give an interpretation of the plot.

   (c) Repeat the correspondence analysis separately for males and females. Does this analysis reveal anything new compared to the combined analysis in the previous question?

7. A student newspaper conducted a survey of student opinions about the Vietnam War in May 1967. Responses were classified by sex, year in the program and one of four opinions. The survey was voluntary. The data may be found in the dataset `uncviet`.

   (a) Conduct an analysis of the patterns of dependence in the data assuming that all variables are nominal.

   (b) Assign scores to the year and opinion and fit an appropriate model. Interpret the trends in opinion over the years. Check the sensitivity of your conclusions to the assignment of the scores.

8. The dataset `HairEyeColor` contains the same data analyzed in this chapter as `haireye`. Repeat the analysis in the text for each sex and make a comparison of the conclusions.

9. A sample of psychiatry patients were cross-classified by their diagnosis and whether a drug treatment was prescribed. The data may be found in `drugpsy`. Is the chance that drugs will be prescribed constant across diagnoses?

10. The `UCBAdmissions` dataset presents data on applicants to graduate school at Berkeley for the six largest departments in 1973 classified by admission and sex.

    (a) Show that this provides an example of Simpson's paradox.

    (b) Determine the most appropriate dependence model between the variables.

    (c) Fit a binomial regression with admissions status as the response and show the relationship to your model in the previous question.

CHAPTER 5

Multinomial Data

The multinomial distribution is an extension of the binomial to the situation where the response can take more than two values. Let Y_i be a random variable that takes one of a finite number of values, $1,2,\ldots,J$. Let $p_{ij} = P(Y_i = j)$ so $\sum_{j=1}^{J} p_{ij} = 1$. As with binary data (the case where $J = 2$), we may encounter both grouped and ungrouped data. Let Y_{ij} be the number of observations falling into category j for group or individual i and let $n_i = \sum_j Y_{ij}$. For ungrouped data, $n_i = 1$ and one and only one of $Y_{i1},\ldots,Y_{iJ}$ is equal to one and the rest are zero. The Y_{ij}, conditional on the total n_i, follow a *multinomial* distribution:

$$P(Y_{i1} = y_{i1},\ldots,Y_{iJ} = y_{iJ}) = \frac{n_i}{y_{i1}!\cdots y_{iJ}!} p_{i1}^{y_{i1}} \cdots p_{iJ}^{y_{iJ}}$$

We must also distinguish between *nominal* multinomial data where there is no natural order to the categories and *ordinal* multinomial data where there is an order. The *multinomial logit* model is intended for nominal data. It can be used for ordinal data, but the information about order will not be used.

5.1 Multinomial Logit Model

As with the binomial logit model, we must find a way to link the probabilities p_{ij} to the predictors x_i, while ensuring that the probabilities are restricted between zero and one. We can use a similar idea:

$$\eta_{ij} = x_i^T \beta_j = \log \frac{p_{ij}}{p_{i1}}, \qquad j = 2,\ldots,J$$

We must obey the constraint that $\sum_{j=1}^{J} p_{ij} = 1$, so it is convenient to declare one of the categories as the *baseline*, say, $j = 1$. So we set $p_{i1} = 1 - \sum_{j=2}^{J} p_{ij}$ and have:

$$p_{ij} = \frac{\exp(\eta_{ij})}{1 + \sum_{j=2}^{J} \exp(\eta_{ij})}$$

Note that $\eta_{i1} = 0$. We may estimate the parameters of this model using maximum likelihood and then use the standard methods of inference.

Consider an example drawn from a subset of the 1996 American National Election Study (Rosenstone, Kinder, and Miller (1997)). For simplicity, we consider only the age, education level and income group of the respondents. Our response will be party identification of the respondent: Democrat, Independent or Republican. The original data involved more than three categories; we collapse this to three, again for simplicity of the presentation.

```
> data(nes96)
> sPID <- nes96$PID
> levels(sPID) <- c("Democrat","Democrat","Independent","Independent",
  "Independent","Republican","Republican")
> summary(sPID)
   Democrat Independent  Republican
        380         239         325
> inca <- c(1.5,4,6,8,9.5,10.5,11.5,12.5,13.5,14.5,16,18.5,21,23.5,
  27.5,32.5,37.5,42.5,47.5,55,67.5,82.5,97.5,115)
> nincome <- inca[unclass(nes96$income)]
> summary(nincome)
   Min. 1st Qu.  Median    Mean 3rd Qu.    Max.
    1.5    23.5    37.5    46.6    67.5   115.0
> table(nes96$educ)
    MS HSdrop     HS   Coll  CCdeg  BAdeg  MAdeg
    13     52    248    187     90    227    127
```

The income variable in the original data was an ordered factor with income ranges. We have converted this to a numeric variable by taking the midpoint of each range.

Let's start with a graphical look at the relationship between the predictors and the response. The response is at the individual level and so we need to group the data just to get a sense of how the party identification is associated with the predictors. We cut the age and income predictors into seven levels and used the approximate midpoints of the ranges to label the groups:

```
> matplot(prop.table(table(nes96$educ,sPID),1),type="l",
  xlab="Education",ylab="Proportion",lty=c(1,2,5))
> cutinc <- cut(nincome,7)
> il <- c(8,26,42,58,74,90,107)
> matplot(il,prop.table(table(cutinc,sPID),1),lty=c(1,2,5),
  type="l",ylab="Proportion",xlab="Income")
> cutage <- cut(nes96$age,7)
> al <- c(24,34,44,54,65,75,85)
> matplot(al,prop.table(table(cutage,sPID),1),lty=c(1,2,5),
  type="l",ylab="Proportion",xlab="Age")
```

The plots are shown in Figure 5.1. We see that proportion of Democrats falls with educational status, reaching a plateau for the college educated. We see the proportion of Republicans rising with educational level and reaching a similar plateau. As income increases, we observe an increase in the proportion of Republicans and Independents and a decrease in the proportion of Democrats. The relationship of party to age is not clear. This is cross-sectional rather than longitudinal data, so we cannot say anything about what might happen to an individual with, for example, increasing income. We can only expect to make conclusions about the relative probability of party affiliations for different individuals with different incomes.

We might ask whether the trends we see in the observed proportions are statistically significant. We need to model the data to answer this question. We fit a multinomial logit model. The `multinom` function is part of the `nnet` package described in Venables and Ripley (2002):

```
> library(nnet)
```

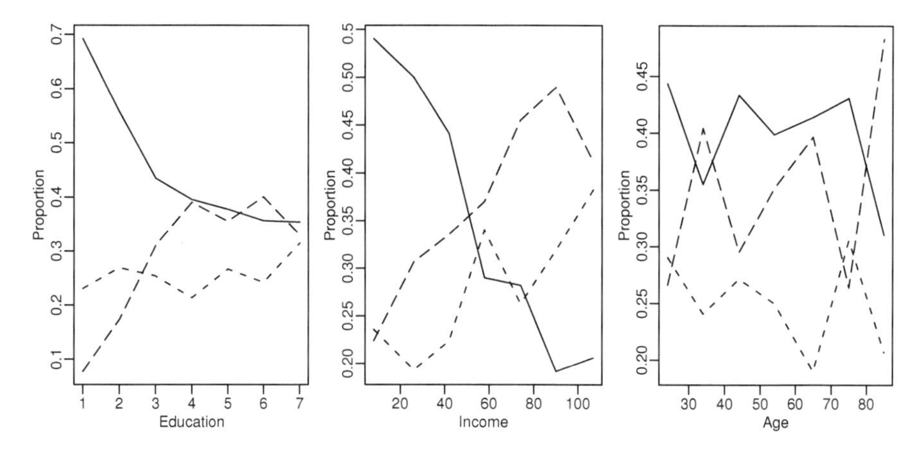

Figure 5.1 *Relationship between party affiliation and education, age and income. Democrats are shown with solid line, Republicans with a dashed line and Independents with a dotted line. Education is categorized into seven levels described in the text. Income is in thousands of dollars.*

```
> mmod <- multinom(sPID ~ age + educ + nincome, nes96)
# weights:  30 (18 variable)
initial  value 1037.090001
iter  10 value 990.568608
iter  20 value 984.319052
final  value 984.166272
converged
```

The program uses the optimization method from the neural net trainer in `nnet` to compute the maximum likelihood, but there is no deeper connection to neural networks.

We can select which variables to include in the model based the AIC criterion using a stepwise search method (output edited to show only the decision information):

```
> mmodi <- step(mmod)
          Df    AIC
- educ     6 1996.5
- age     16 2003.6
<none>    18 2004.3
- nincome 16 2045.9

          Df    AIC
- age      4 1993.4
<none>     6 1996.5
- nincome  4 2048.9

          Df    AIC
<none>     4 1993.4
- nincome  2 2045.3
```

At the first stage of the search, we see that omitting education would be the best option to reduce the AIC criterion. At the next step, age is removed resulting in a model with only income.

We can also use the standard likelihood methods to derive a test to compare nested models. For example, we can fit a model without education and then compare the deviances:

```
> mmode <- multinom(sPID ~ age +  nincome, nes96)
> deviance(mmode) - deviance(mmod)
[1] 16.206
> pchisq(16.206,mmod$edf-mmode$edf,lower=F)
[1] 0.18198
```

We see that education is not significant relative to the full model. This may seem somewhat surprising given the plot in Figure 5.1, but the large differences between proportions of Democrats and Republicans occur for groups with low education which represent only a small number of people.

We can obtain predicted values for specified values of income. For example, suppose we pick the midpoints of the income groups we selected for the earlier plot:

```
> predict(mmodi,data.frame(nincome=il),type="probs")
  Democrat Independent Republican
1  0.55663     0.19552    0.24786
2  0.48049     0.22546    0.29405
3  0.41343     0.25094    0.33564
4  0.34939     0.27432    0.37629
5  0.29033     0.29486    0.41481
6  0.23758     0.31211    0.45031
7  0.18917     0.32668    0.48415
```

We see that the probability of being Republican or Independent increases with income. The default form just gives the most probable category:

```
> predict(mmodi,data.frame(nincome=il))
[1] Democrat   Democrat   Democrat   Republican Republican
[6] Republican Republican
```

We may also examine the coefficients to gain an understanding of the relationship between the predictor and the response:

```
> summary(mmodi)
Coefficients:
            (Intercept)  nincome
Independent    -1.17493 0.016087
Republican     -0.95036 0.017665

Std. Errors:
            (Intercept)   nincome
Independent     0.15361 0.0028497
Republican      0.14169 0.0026525

Residual Deviance: 1985.4
AIC: 1993.4
```

The intercept terms model the probabilities of the party identification for an income of zero. We can see the relationship from this calculation:

```
> cc <- c(0,-1.17493,-0.95036)
> exp(cc)/sum(exp(cc))
[1] 0.58982 0.18216 0.22802
> predict(mmodi,data.frame(nincome=0),type="probs")
   Democrat Independent  Republican
    0.58982     0.18216     0.22802
```

The slope terms represent the log-odds of moving from the baseline category of Democrat to Independent and Republican, respectively, for a unit change of $1000 in income. We can see more explicitly what this means by predicting probabilities for incomes $1000 apart and then computing the log-odds:

```
> (pp <- predict(mmodi,data.frame(nincome=c(0,1)),type="probs"))
  Democrat Independent Republican
1  0.58982     0.18216    0.22802
2  0.58571     0.18382    0.23047
> log(pp[1,1]*pp[2,2]/(pp[1,2]*pp[2,1]))
[1] 0.016087
> log(pp[1,1]*pp[2,3]/(pp[1,3]*pp[2,1]))
[1] 0.017665
```

Log-odds can be difficult to interpret particularly with many predictors and interactions. Sometimes, computing predicted probabilities for a selected range of predictors can provide better intuition.

It is possible to fit a multinomial logit model using a Poisson GLM. Recall that independent Poisson variates conditional on their total are multinomial as described in Section 3.1. We can exploit this fact by declaring a factor that has a level for each multinomial observation in the data; we call this the *response factor*. We then treat the individual components of the multinomial response as Poisson responses. For ungrouped data, such as the current example, this means that one response will be one and the rest zero.

We set up these variables, also illustrating what happens with the first four individuals:

```
> sPID[1:4]
[1] Republican Democrat   Democrat   Democrat
Levels: Democrat Independent Republican
> cm <- diag(3)[unclass(sPID),]
> cm[1:4,]
     [,1] [,2] [,3]
[1,]    0    0    1
[2,]    1    0    0
[3,]    1    0    0
[4,]    1    0    0
> y <- as.numeric(t(cm))
> resp.factor <- gl(944,3)
```

The three Poisson responses correspond to the different affiliations so we need to label which is which:

```
> cat.factor <- gl(3,1,3*944,labels=c("D","I","R"))
```

We also need to replicate the predictor:

```
> rnincome <- rep(nincome,each=3)
```

Now examine the form of the reorganized data:

```
> head(data.frame(y,resp.factor,cat.factor,rnincome))
  y resp.factor cat.factor rnincome
1 0           1          D      1.5
2 0           1          I      1.5
3 1           1          R      1.5
4 1           2          D      1.5
5 0           2          I      1.5
6 0           2          R      1.5
```

As with the contingency table models, the null model has only main effects:

```
> nullmod <- glm(y ~ resp.factor + cat.factor, family=poisson)
```

The effect of income is modeled with an interaction with party affiliation:

```
> glmod <- glm(y ~ resp.factor + cat.factor + cat.factor:rnincome,
  family=poisson)
```

We find that the deviance is the same as the multinomial model above:

```
> deviance(glmod)
[1] 1985.4
> deviance(mmodi)
[1] 1985.4
```

The coefficients also correspond:

```
> coef(glmod)[c(1,945:949)]
         (Intercept)          cat.factorI          cat.factorR
          -0.5119613           -1.1749375           -0.9503621
cat.factorD:rnincome cat.factorI:rnincome cat.factorR:rnincome
          -0.0176645           -0.0015777                   NA
> coef(mmodi)
            (Intercept)  nincome
Independent    -1.17493 0.016087
Republican     -0.95036 0.017665
```

The parameterization is slightly different for the Poisson GLM. Because only two interaction parameters are identifiable, the last one, being inestimable, is not estimated. This has the effect of making Republicans the reference level rather than Democrats as in the multinomial model. We see that the sign of the Republican-Democrat contrast is reversed and that we may obtain the Independent-Democrat contrast from the Poisson GLM by computing:

```
> 0.016087-0.017665
[1] -0.001578
```

So we may obtain the same results using the Poisson GLM, but the `multinom` function is more transparent. However, the point is that the multinomial logit can be viewed as a GLM-type model, which allows us to apply all the common methodology developed for GLMs.

5.2 Hierarchical or Nested Responses

Consider the following data collected by Lowe, Roberts, and Lloyd (1971) by way of McCullagh and Nelder (1989) concerning live births with deformations of the central nervous system in south Wales:

```
> data(cns)
> cns
            Area NoCNS An Sp Other Water      Work
1        Cardiff  4091  5  9     5   110 NonManual
2        Newport  1515  1  7     0   100 NonManual
3        Swansea  2394  9  5     0    95 NonManual
4     GlamorganE  3163  9 14     3    42 NonManual
5     GlamorganW  1979  5 10     1    39 NonManual
6     GlamorganC  4838 11 12     2   161 NonManual
7      MonmouthV  2362  6  8     4    83 NonManual
8  MonmouthOther  1604  3  6     0   122 NonManual
9        Cardiff  9424 31 33    14   110    Manual
10       Newport  4610  3 15     6   100    Manual
11       Swansea  5526 19 30     4    95    Manual
12    GlamorganE 13217 55 71    19    42    Manual
13    GlamorganW  8195 30 44    10    39    Manual
14    GlamorganC  7803 25 28    12   161    Manual
15     MonmouthV  9962 36 37    13    83    Manual
16 MonmouthOther  3172  8 13     3   122    Manual
```

`NoCNS` indicates no central nervous system(CNS) malformation. `An` denotes anencephalus while `Sp` denotes spina bifida and `Other` represents other malformations. `Water` is water hardness and the subjects are categorized by the type of work performed by the parents. We might consider a multinomial response with four categories. However, we can see that most births suffer no malformation and so this category dominates the other three. It is better to consider this as a hierarchical response as depicted in Figure 5.2. Now consider the multinomial likelihood for the ith observation which is proportional to:

$$p_{i1}^{y_{i1}} p_{i2}^{y_{i2}} p_{i3}^{y_{i3}} p_{i4}^{y_{i4}}$$

Define $p_{ic} = p_{i2} + p_{i3} + p_{i4}$ which is probability of a birth with some kind of CNS malformation. We can then write the likelihood as:

$$p_{i1}^{y_{i1}} p_{ic}^{y_{i2}+y_{i3}+y_{i4}} \times \left(\frac{p_{i2}}{p_{ic}}\right)^{y_{i2}} \left(\frac{p_{i3}}{p_{ic}}\right)^{y_{i3}} \left(\frac{p_{i4}}{p_{ic}}\right)^{y_{i4}}$$

The first part of the product is now a binomial likelihood for a `CNS` vs. `NoCNS` response. The second part of the product is now a multinomial likelihood for the three `CNS` categories conditional of the presence of `CNS`. For example, p_{i2}/p_{ic} is the conditional probability of an anencephalus birth given that a malformation has occurred for the ith observation. We can now separately develop a binomial model for whether malformation occurs and a multinomial model for the type of malformation.

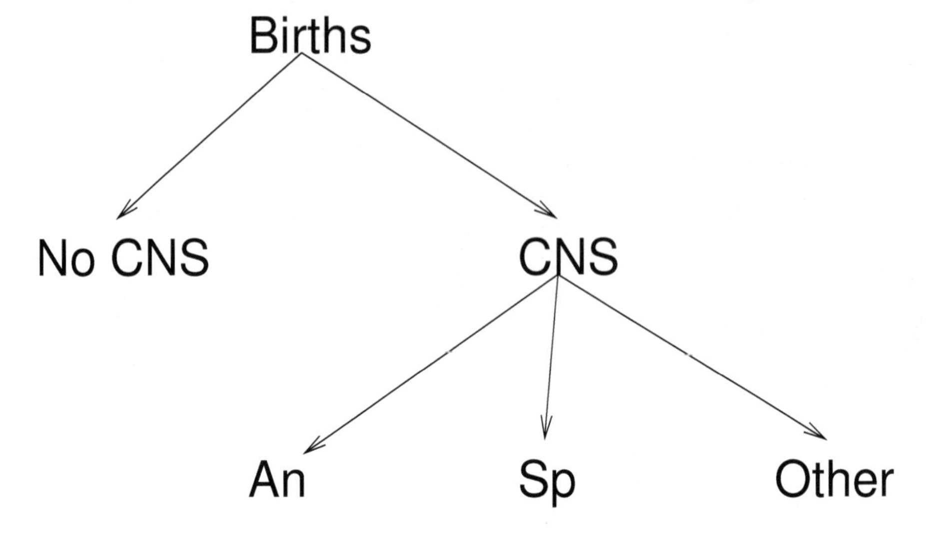

Figure 5.2 *Hierarchical response for birth types.*

We start with the binomial model. First we accumulate the number of CNS births and plot the data with the response on the logit scale as shown in the first panel of Figure 5.3:

```
> cns$CNS <- cns$An+cns$Sp+cns$Other
> plot(log(CNS/NoCNS) ~ Water, cns, pch=as.character(Work))
```

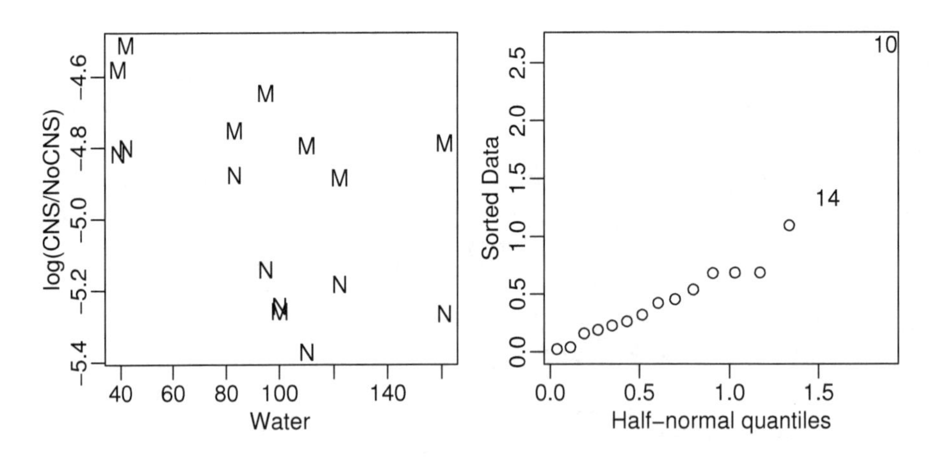

Figure 5.3 *The first plot shows the empirical logits for the proportion of CNS births related to water hardness and profession (M=Manual, N=Nonmanual). The second is a half-normal plot of the residuals of the chosen model.*

We observe that the proportion of CNS births falls with increasing water hardness and is higher for manual workers. We observe one observation (manual, Newport) that may be an outlier. Notice that the `Area` is confounded with the `Water` hardness, so we cannot put both these predictors in our model. We try them both and compare:

```
> binmodw <- glm(cbind(CNS,NoCNS) ~ Water + Work, cns, family=binomial)
> binmoda <- glm(cbind(CNS,NoCNS) ~ Area + Work, cns, family=binomial)
> anova(binmodw,binmoda,test="Chi")
Analysis of Deviance Table

Model 1: cbind(CNS, NoCNS) ~ Water + Work
Model 2: cbind(CNS, NoCNS) ~ Area + Work
  Resid. Df Resid. Dev Df Deviance P(>|Chi|)
1        13      12.36
2         7       3.08  6     9.29      0.16
```

One can view this test as a check for linear trend in the effect of water hardness. We find that the simpler model using `Water` is acceptable. A check for an interaction effect revealed nothing significant although a look at the residuals is worthwhile:

```
> halfnorm(residuals(binmodw))
```

In the second plot of Figure 5.3, we see an outlier corresponding to Newport manual workers. This case deserves closer examination. Finally, a look at the chosen model:

```
> summary(binmodw)
Coefficients:
               Estimate Std. Error z value Pr(>|z|)
(Intercept)   -4.432580   0.089789  -49.37  < 2e-16
Water         -0.003264   0.000968   -3.37  0.00075
WorkNonManual -0.339058   0.097094   -3.49  0.00048

(Dispersion parameter for binomial family taken to be 1)

    Null deviance: 41.047  on 15  degrees of freedom
Residual deviance: 12.363  on 13  degrees of freedom
AIC: 102.5
```

The residual deviance is close to the degrees of freedom indicating a reasonable fit to the data. We see that since:

```
> exp(-0.339058)
[1] 0.71244
```

births to nonmanual workers have a 29% lower chance of CNS malformation. Water hardness ranges from about 40 to 160. So a difference of 120 would decrease the odds of CNS malformation by about 32%.

Now consider a multinomial model for the three malformation types conditional on a malformation having occurred. As this data is grouped, in contrast to the `nes96` example, it is most convenient to present the response as a matrix:

```
> cmmod <- multinom(cbind(An,Sp,Other) ~ Water + Work, cns)
```

We find that neither predictor has much effect:

```
> nmod <- step(cmmod)
        Df    AIC
- Water  4 1381.1
```

```
- Work    4 1381.2
<none>    6 1383.5

       Df    AIC
- Work  2 1378.5
<none>  4 1381.1
```

which leaves us with a null final model:

```
> nmod
Coefficients:
      (Intercept)
Sp        0.28963
Other    -0.98083

Residual Deviance: 1374.5
```

The fitted proportions are:

```
> cc <- c(0,0.28963,-0.98083)
> names(cc) <- c("An","Sp","Other")
> exp(cc)/sum(exp(cc))
     An      Sp   Other
0.36888 0.49279 0.13833
```

So we find that water hardness and parents' profession are related to the probability of a malformed birth, but that they have no effect on the type of malformation.

Observe that if we fit a multinomial logit model to all four categories:

```
> multinom(cbind(NoCNS,An,Sp,Other) ~ Water + Work, cns)
Coefficients:
      (Intercept)       Water WorkNonManual
An        -5.4551 -0.00290884      -0.36388
Sp        -5.0710 -0.00432305      -0.24359
Other     -6.5947 -0.00051358      -0.64219

Residual Deviance: 9391
AIC: 9409
```

We find that both `Water` and `Work` are significant, but that the fact that they do not distinguish the type of malformation is not easily discovered from this model.

5.3 Ordinal Multinomial Responses

Suppose we have J ordered categories and that for individual i, with ordinal response Y_i, $p_{ij} = P(Y_i = j)$ for $j = 1,\ldots,J$. With an ordered response, it is often easier to work with the cumulative probabilities, $\gamma_{ij} = P(Y_i \le j)$. The cumulative probabilities are increasing and invariant to combining adjacent categories. Furthermore, $\gamma_{iJ} = 1$, so we need only model $J-1$ probabilities.

As usual, we must link the γs to the covariates x. We will consider three possibilities which all take the form:

$$g(\gamma_{ij}) = \theta_j - x_i^T \beta$$

We will consider three possibilities for the link function g: the logit, the probit and the complementary log-log. Notice that we have explicitly specified the intercepts, θ_j, so that the vector x_i does not include an intercept. Furthermore, β does not depend on j so that we assume that the predictors have a uniform effect on the response categories in a sense that we will shortly make clear.

Suppose that Z_i is some unobserved continuous variable that might be thought of as the real underlying latent response. We only observe a discretized version of Z_i in the form of Y_i where $Y_i = j$ is observed if $\theta_{j-1} < Z_i \leq \theta_j$. Further suppose that $Z_i - \beta^T x_i$ has distribution F then:

$$P(Y_i \leq j) = P(Z_i \leq \theta_j) = P(Z_i - \beta^T x_i \leq \theta_j - \beta^T x_i) = F(\theta_j - \beta^T x_i)$$

Now if, for example, F follows the logistic distribution, where $F(x) = e^x/(1+e^x)$, then:

$$\gamma_{ij} = \frac{\exp(\theta_j - \beta^T x_i)}{1+\exp(\theta_j - \beta^T x_i)}$$

and so we would have a logit model for the cumulative probabilities γ_{ij}. Choosing the normal distribution for the latent variable leads to a probit model while the choice of an extreme value distribution leads to the complementary log-log. This *latent* variable explanation for the model is displayed in Figure 5.4. Notice that if $\beta > 0$, as x_i

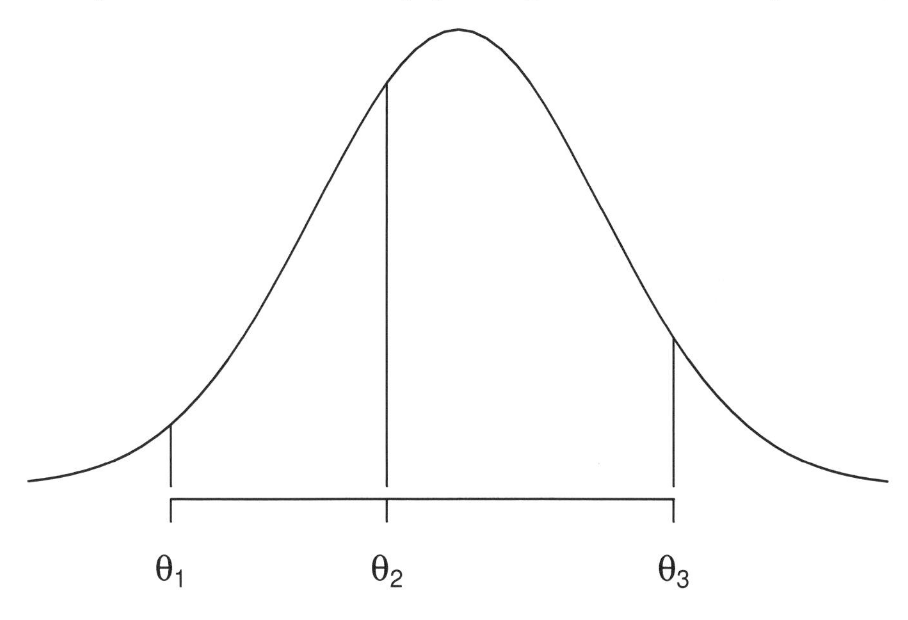

Figure 5.4 *Latent variable view of an ordered multinomial response. Here, four discrete response can occur, depending on the position of Z relative to the cutpoints θ_j. As x changes, the cutpoints will move together to change the relative probabilities of the four responses.*

increases, $P(Y_i = J)$ will also increase. This explains the use of the minus sign in the definition of the model because it allows for the more intuitive interpretation of the sign of β.

Proportional Odds Model: Let $\gamma_j(x_i) = P(Y_i \leq j|x_i)$ then the *proportional odds* model, which uses the logit link, is:

$$\log \frac{\gamma_j(x_i)}{1-\gamma_j(x_i)} = \theta_j - \beta^T x_i, \quad j = 1,\ldots,J-1$$

It is so called because the relative odds for $y \leq j$ comparing x_1 and x_2 are:

$$\left(\frac{\gamma_j(x_1)}{1-\gamma_j(x_1)}\right) / (\frac{\gamma_j(x_2)}{1-\gamma_j(x_2)}) = \exp(-\beta^T(x_1 - x_2))$$

This does not depend on j. Of course, the assumption of proportional odds does need to be checked for a given dataset.

Returning to the `nes96` dataset, suppose we assume that Independents fall somewhere between Democrats and Republicans. We would then have an ordered multinomial response. We can then fit this using the `polr` function from the MASS library described in Venables and Ripley (2002):

```
> library(MASS)
> pomod <- polr(sPID ~ age + educ + nincome, nes96)
```

The deviance and number of parameters for this model are:

```
> c(deviance(pomod),pomod$edf)
[1] 1984.2   10.0
```

which can be compared to the corresponding multinomial logit model:

```
> c(deviance(mmod),mmod$edf)
[1] 1968.3   18.0
```

The proportional odds model uses fewer parameters, but does not fit quite as well. Typically, the output from the proportional odds model is easier to interpret. We may use an AIC-based variable selection method:

```
> pomodi <- step(pomod)
Start:  AIC= 2004.2
 sPID ~ age + educ + nincome


          Df  AIC
- educ     6 2003
<none>       2004
- age      1 2004
- nincome  1 2039

Step:  AIC= 2002.8
 sPID ~ age + nincome


          Df  AIC
- age      1 2001
<none>       2003
- nincome  1 2047
```

```
Step:  AIC= 2001.4
 sPID ~ nincome

          Df  AIC
<none>        2001
- nincome  1 2045
```

Thus we finish with a model including just income as we did with the earlier multinomial model. We could also use a likelihood ratio test to compare the models:

```
> deviance(pomodi)-deviance(pomod)
[1] 11.151
> pchisq(11.151,pomod$edf-pomodi$edf,lower=F)
[1] 0.13217
```

We see that the simplification to just income is justifiable. We can check the proportional odds assumption by computing the observed odds proportions with respect to, in this case, income levels. We have computed the log-odds difference between γ_1 and γ_2:

```
> pim <- prop.table(table(nincome,sPID),1)
> logit(pim[,1])-logit(pim[,1]+pim[,2])
     1.5        4        6        8      9.5     10.5     11.5
-0.90079 -2.06142 -0.75769 -1.00330 -2.30259 -0.30830 -0.79851
    12.5     13.5     14.5       16     18.5       21     23.5
-1.89712 -1.25276 -1.17865 -0.41285 -0.35424 -1.51413 -1.65345
    27.5     32.5     37.5     42.5     47.5       55     67.5
-0.74678 -0.52252 -0.92326 -1.02962 -0.82198 -1.42760 -1.18261
    82.5     97.5      115
-0.98676 -1.48292 -1.70660
```

It is questionable whether these can be considered sufficiently constant, but at least there is no trend. Now consider the interpretation of the fitted coefficients:

```
> summary(pomodi)
Coefficients:
           Value Std. Error t value
nincome 0.013120  0.0019708  6.6572

Intercepts:
                         Value  Std. Error t value
Democrat|Independent      0.209  0.112       1.863
Independent|Republican   1.292  0.120      10.753

Residual Deviance: 1995.36
AIC: 2001.36
```

We can say that the odds of moving from Democrat to Independent/Republican category (or from Democrat/Independent to Republican) increase by a factor of $\exp(0.013120) = 1.0132$ as income increases by one unit ($1000). Notice that the log-odds are similar to those obtained in the multinomial logit model. The intercepts correspond to the θ_j. So for an income of $0, the predicted probability of being a Democrat is:

```
> ilogit(0.209)
[1] 0.55206
```

while that of being an Independent is:

```
> ilogit(1.292)-ilogit(0.209)
[1] 0.23242
```

with the remainder being Republicans. We can compute predicted values:

```
> predict(pomodi,data.frame(nincome=il,row.names=il),
  type="probs")
    Democrat Independent Republican
8    0.52602     0.24011    0.23387
26   0.46705     0.25415    0.27880
42   0.41535     0.26176    0.32290
58   0.36544     0.26418    0.37038
74   0.31827     0.26122    0.42051
90   0.27455     0.25311    0.47234
107  0.23242     0.23954    0.52804
```

Notice how the probability of being a Democrat uniformly decreases with income while that for being a Republican uniformly increases as income increases, but that the middle category of Independent increases then decreases. This type of behavior can be expected from the latent variable representation of the model.

We can illustrate the latent variable interpretation of proportional odds by computing the cutpoints for incomes of $0, $50,000 and $100,000:

```
> x <- seq(-4,4,by=0.05)
> plot(x,dlogis(x),type="l")
> abline(v=c(0.209,1.292))
> abline(v=c(0.209,1.292)-50*0.013120,lty=2)
> abline(v=c(0.209,1.292)-100*0.013120,lty=5)
```

The plot is shown in Figure 5.5.

Ordered Probit Model: If the latent variable Z_i has a standard normal distribution, then:

$$\Phi^{-1}(\gamma_j(x_i)) = \theta_j - \beta^T x_i \quad j = 1,\ldots,J-1$$

Applying this model to the `nes96` data, we find:

```
> opmod <- polr(sPID ~ nincome, method="probit")
> summary(opmod)
Coefficients:
           Value Std. Error t value
nincome 0.008182  0.0012078  6.7745

Intercepts:
                         Value  Std. Error t value
Democrat|Independent      0.128  0.069       1.851
Independent|Republican    0.798  0.072      11.040

Residual Deviance: 1994.89
AIC: 2000.89
```

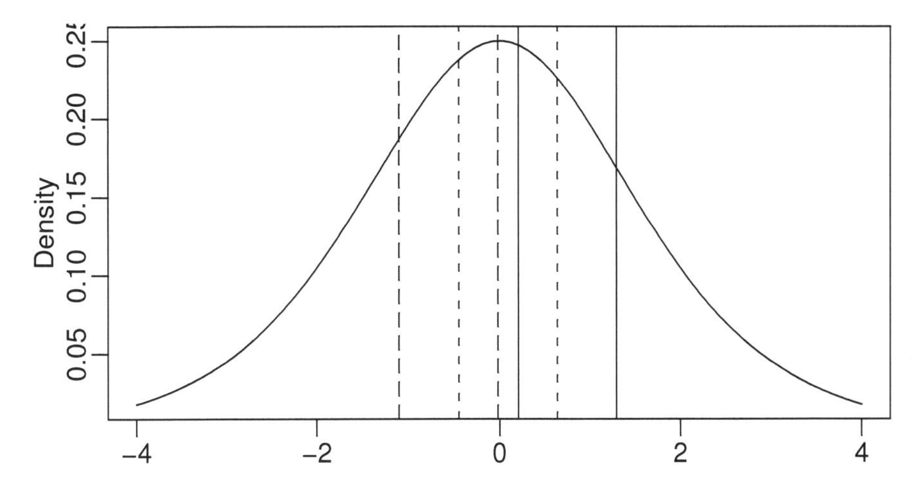

Figure 5.5 *Solid lines represent an income of $0, dotted lines are for $50,000 and dashed lines are for $100,000. Probability of being a Democrat is given by the area lying to the left of the leftmost of each pair of lines, while the probability of being a Republican is given by the area to the right of the rightmost of the pair. Independents are represented by the area in between.*

The deviance is similar to the logit version of this model, but the coefficients appear to be different. However, if we compute the same predictions:

```
> dems <- pnorm(0.128-il*0.008182)
> demind <- pnorm(0.798-il*0.008182)
> cbind(dems,demind-dems,1-demind)
        dems
[1,] 0.52494 0.24315 0.23192
[2,] 0.46624 0.25458 0.27918
[3,] 0.41463 0.26058 0.32479
[4,] 0.36446 0.26236 0.37318
[5,] 0.31651 0.25982 0.42366
[6,] 0.27147 0.25310 0.47543
[7,] 0.22739 0.24173 0.53088
```

We see that the predicted values are very similar to those seen for the logit. If the coefficients are appropriately rescaled, they are also very similar.

Proportional Hazards Model: A concept of *hazard* was developed in insurance applications. When issuing a life insurance policy, the insurer is interested in the probability that the person will die during the term of the policy given that they are alive now. This is not the same as the unconditional probability of death during the same time period. In other words, for example, we want to know the chance that a 55-year-old man will die in the next year, given that he is alive and aged 55. The unconditional probability that a man will die aged 55 is not particular useful for the purposes of insurance.

Suppose we use the complementary log-log in place of the logit above, that is:

$$\log(-\log(1-\gamma_j(x_i))) = \theta_j + \beta^T x_i$$

Then the *hazard* of category j is the probability of falling in category j given that your category is greater than j:

$$\text{Hazard}(j) = P(Y_i = j | Y_i \geq j) = \frac{P(Y_i = j)}{P(Y_i \geq j)} = \frac{p_j}{1-\gamma_{i,j-1}} = \frac{\gamma_{ij} - \gamma_{i,j-1}}{1-\gamma_{i,j-1}}$$

The corresponding latent variable distribution is the extreme value:

$$F(x) = 1 - \exp(-\exp(x))$$

The extreme value distribution is not symmetric like the logistic and normal and so there seems little justification for applying it to the `nes96` data, but the command is:

```
> polr(sPID ~ nincome, method="cloglog")
```

Generalization: The proportional hazards and odds models can be generalized by allowing β to vary that is

$$\log \frac{\gamma_j(x)}{1-\gamma_j(x)} = \theta_j - \beta_j^T x \quad j = 1, \ldots, k-1$$

but this loses the proportionality property.

Further Reading: For more on the analysis of ordered categorical data see the books by Agresti (1984), Clogg and Shihadeh (1994), Powers and Xie (2000) and Simonoff (2003).

Exercises

1. This `hsb` data was collected as a subset of the High School and Beyond study conducted by the National Education Longitudinal Studies program of the National Center for Education Statistics. The variables are gender; race; socioeconomic status; school type; chosen high school program type; scores on reading, writing, math, science, and social studies. We want to determine which factors are related to the choice of the type of program — academic, vocational, or general — that the students pursue in high school. The response is multinomial with three levels.

   (a) Fit a trinomial response model with the other relevant variables as predictors (untransformed).

   (b) Use backward elimination to reduce the model to one where all predictors are statistically significant. Give an interpretation of the resulting model.

   (c) For the student with id 99, compute the predicted probabilities of the three possible choices.

2. Data were collected from 39 students in a University of Chicago MBA class and may be found in the dataset `happy`.

   (a) Build a model for the level of happiness as a function of the other variables.

   (b) Interpret the parameters of your chosen model.

   (c) Predict the happiness distribution for subject whose parents earn $30,000 a year, who is lonely, not sexually active and has no job.

3. A student newspaper conducted a survey of student opinions about the Vietnam War in May 1967. Responses were classified by sex, year in the program and one of four opinions. The survey was voluntary. The data may be found in the dataset `uncviet`.

   (a) Treat the opinion as the response and the sex and year as predictors. Build a proportional odds model, giving an interpretation to the estimates.

   (b) If you completed the analysis of this same dataset as a question in the previous chapter, compare and contrast the results of the two analyses.

4. The `pneumo` data gives the number of coal miners classified by radiological examination into one of three categories of pneumonoconiosis and by the number of years spent working at the coal face divided into eight categories.

   (a) Treating the pneumonoconiosis status as response variable as nominal, build a model for predicting the frequency of the three outcomes in terms of length of service and use it to predict the outcome for a miner with 25 years of service.

   (b) Repeat the analysis with the pneumonoconiosis status being treated as ordinal.

   (c) Now treat the response variable as hierarchical with top level indicating whether the miner has the disease and the second level indicating, given they have the disease, whether they have a moderate or severe case.

   (d) Compare the three analyses.

5. The `debt` data arise from a large postal survey on the psychology of debt. The frequency of credit card use is a three-level factor ranging from never, through occasionally to regularly. Build a model for predicting credit card use as a function of the other variables. Write a report describing the nature of the effect of the predictors on the response.

6. The National Youth Survey collected a sample of 11–17 year-olds with 117 boys and 120 girls, asking questions about marijuana usage. The data may be found in `potuse`. This data is actually longitudinal — the same boys and girls are followed for five years. However, for the purposes of this question, imagine that the data is cross-sectional, that is, a different sample of boys and girls are sampled each year. Build a model for the different levels of marijuana usage, describing the trend over time and the difference between the sexes.

CHAPTER 6

Generalized Linear Models

In previous chapters, we have seen how to model a binomial or Poisson response. Multinomial response models can often be recast as Poisson responses and the standard linear model with a normal (Gaussian) response is already familiar. Although these models each have their distinctive characteristics, we observe some common features in all of them that we can abstract to form the *generalized linear model* (GLM). By developing a theory and constructing general methods for GLMs, we are able to tackle a wider range of data with different types of response variables. GLMs were introduced by Nelder and Wedderburn (1972) while McCullagh and Nelder (1989) provides a book-length treatment.

6.1 GLM Definition

A GLM is defined by specifying two components. The response should be a member of the exponential family distribution and the link function describes how the mean of the response and a linear combination of the predictors are related.

Exponential family: In a GLM the distribution of Y is from the exponential family of distributions which take the general form:

$$f(y|\theta,\phi) = \exp\left[\frac{y\theta - b(\theta)}{a(\phi)} + c(y,\phi)\right]$$

The θ is called the *canonical parameter* and represents the location while ϕ is called the *dispersion parameter* and represents the scale. We may define various members of the family by specifying the functions a, b, and c. The most commonly used examples are:

1. Normal or Gaussian:

$$f(y|\theta,\phi) = \frac{1}{\sqrt{2\pi}\sigma}\exp\left[-\frac{(y-\mu)^2}{2\sigma^2}\right]$$

$$= \exp\left[\frac{y\mu - \mu^2/2}{\sigma^2} - \frac{1}{2}\left(\frac{y^2}{\sigma^2} + \log(2\pi\sigma^2)\right)\right]$$

So we can write $\theta = \mu$, $\phi = \sigma^2$, $a(\phi) = \phi$, $b(\theta) = \theta^2/2$ and $c(y,\phi) = -(y^2/\phi + \log(2\pi\phi))/2$.

2. Poisson:

$$f(y|\theta,\phi) = e^{-\mu}\mu^y/y!$$

$$= \exp(y\log\mu - \mu - \log y!)$$

So we can write $\theta = \log(\mu)$, $\phi \equiv 1$, $a(\phi) = 1$, $b(\theta) = \exp(\theta)$ and $c(y,\phi) = \quad \log y!$.

3. Binomial:

$$f(y|\theta,\phi) = \binom{n}{y}\mu^y(1-\mu)^{n-y}$$

$$= \exp\left(y\log\mu + (n-y)\log(1-\mu) + \log\binom{n}{y}\right)$$

$$= \exp\left(y\log\frac{\mu}{1-\mu} + n\log(1-\mu) + \log\binom{n}{y}\right)$$

So we see that $\theta = \log\frac{\mu}{1-\mu}$, $b(\theta) = -n\log(1-\mu) = n\log(1+\exp\theta)$ and $c(y,\phi) = \log\binom{n}{y}$.

The gamma and inverse Gaussian are other lesser-used members of the exponential family that are covered in Chapter 7. Notice that in the normal density, the ϕ parameter is free (as it is also for the gamma density) while for the Poisson and binomial it is fixed at one. This is because the Poisson and binomial are one parameter families while the normal and gamma have two parameters. In fact, some authors reserve the term *exponential family* distribution for cases where ϕ is not used, while using the term *exponential dispersion family* for cases where it is. This has important consequences for the analysis.

Some other densities, such as the negative binomial and the Weibull distribution, are not members of the exponential family, but they are sufficiently close that the GLM can be fit with some modifications. It is also possible to fit distributions that are not in the exponential family using the GLM-style approach, but there are some additional complications.

The exponential family distributions have mean and variance:

$$\begin{aligned} EY &= \mu = b'(\theta) \\ \text{var}\, Y &= b''(\theta)a(\phi) \end{aligned}$$

The mean is a function of θ only while the variance is a product of functions of the location and the scale. $b''(\theta)$ is called the *variance function* and describes how the variance relates to the mean.

In the Gaussian case, $b''(\theta) = 1$ and so the variance is independent of the mean. For other distributions, this is not true, making the Gaussian case exceptional. We can introduce weights by setting:

$$a(\phi) = \phi/w$$

where w is a known weight that varies between observations.

Link function: Let us suppose we may express the effect of the predictors on the response through a *linear predictor:*

$$\eta = \beta_0 + \beta_1 x_1 + \cdots + \beta_p x_p = x^T\beta$$

The link function, g, describes how the mean response, $EY = \mu$, is linked to the covariates through the linear predictor:

$$\eta = g(\mu)$$

In principle, any monotone continuous and differentiable function will do, but there are some convenient and common choices for the standard GLMs.

In the Gaussian linear model, the identity link, $\eta = \mu$ is the obvious selection, but another choice would give $y = g^{-1}(x^T\beta) + \varepsilon$. This does not correspond directly to a transform on the response: $g(y) = x^T\beta + \varepsilon$ as, for example, in a Box–Cox type transformation. In a GLM, the link function is assumed known whereas in a *single index model*, g is estimated.

For the Poisson GLM, the mean μ must be positive so $\eta = \mu$ will not work conveniently since η can be negative. The standard choice is $\mu = e^{\eta}$ so that $\eta = \log \mu$ which ensures $\mu > 0$. This log link means that additive effects of x lead to multiplicative effects on μ.

For the binomial GLM, let p be the probability of success and let this be our μ if we define the response as the proportion rather than the count. This requires that $0 \le p \le 1$. There are several commonly used ways to ensure this: the logistic, probit and complementary log-log links. These are discussed in detail in Chapter 2.

The *canonical link* has g such that $\eta = g(\mu) = \theta$, the canonical parameter of the exponential family distribution. This means that $g(b'(\theta)) = \theta$. The canonical links for the common GLMs are shown in Table 6.1. If a canonical link is used, X^TY is

Family	Link	Variance Function
Normal	$\eta = \mu$	1
Poisson	$\eta = \log \mu$	μ
Binomial	$\eta = \log(\mu/(1-\mu))$	$\mu(1-\mu)$
Gamma	$\eta = \mu^{-1}$	μ^2
Inverse Gaussian	$\eta = \mu^{-2}$	μ^3

Table 6.1 *Canonical links for GLMs.*

sufficient for β. The canonical link is mathematically and computationally convenient and is often the natural choice of link. However, one is not required to use the canonical link and sometimes context may compel another choice.

6.2 Fitting a GLM

The parameters, β, of a GLM can be estimated using maximum likelihood. The log-likelihood for single observation, where $a_i(\phi) = \phi/w_i$, is:

$$\log L(\theta_i, \phi; y_i) = w_i[\frac{y_i\theta_i - b(\theta_i)}{\phi}] + c(y_i, \phi)$$

So for independent observations, the log-likelihood will be $\sum_i \log L(\theta_i, \phi; y_i)$. Sometimes we can maximize this analytically and find an exact solution for the MLE $\hat{\beta}$, but the Gaussian GLM is the only common case where this is possible. Typically, we must use numerical optimization. By applying the Newton–Raphson method with Fisher scoring, McCullagh and Nelder (1989) show that the optimization is equivalent to iteratively reweighted least squares (IRWLS).

The procedure can be understood intuitively by analogy to the procedure for the Gaussian linear model $Y = X\beta + \varepsilon$. Suppose var $Y \propto f(\hat{\eta})$ where $\hat{y} = \hat{\eta} = X\hat{\beta}$. We would use weights w_i where $w_i^{-1} = f(\hat{\eta})$. Since the weights are a function of $\hat{\beta}$ an iterative fitting procedure would be needed. We might set the weights all equal to one, estimate $\hat{\beta}$, use this to recompute the weights, reestimate $\hat{\beta}$, and so on until convergence.

We can use a similar idea to fit a GLM. Roughly speaking, we want to regress $g(y)$ on X with weights inversely proportional to var $g(y)$. However, $g(y)$ might not make sense in some cases — for example, in the binomial GLM. So we linearize $g(y)$ as follows: Let $\eta = g(\mu)$ and $\mu = EY$. Now do a one-step expansion:

$$\begin{aligned} g(y) &\approx g(\mu) + (y-\mu)g'(\mu) \\ &= \eta + (y-\mu)\frac{d\eta}{d\mu} \\ &\equiv z \end{aligned}$$

and

$$\widehat{\text{var}}\, z = \left(\frac{d\eta}{d\mu}\right)^2 V(\hat{\mu}) = \frac{1}{w}$$

So the IRWLS procedure would be:

1. Set initial estimates $\hat{\eta}_0$ and $\hat{\mu}_0$.
2. Form the "adjusted dependent variable" $z_0 = \hat{\eta}_0 + (y - \hat{\mu}_0)\frac{d\eta}{d\mu}|_{\hat{\eta}_0}$.
3. Form the weights $w_0^{-1} = \left(\frac{d\eta}{d\mu}\right)^2|_{\hat{\eta}_0} V(\hat{\mu}_0)$.
4. Reestimate β to get $\hat{\eta}_1$.
5. Iterate steps 2–3–4 until convergence.

Notice that the fitting procedure uses only $\eta = g(\mu)$ and $V(\mu)$, but requires no further knowledge of the distribution of y. This point will be important later in Section 7.4. Estimates of variance may be obtained from:

$$\hat{\text{var}}(\hat{\beta}) = (X^T W X)^{-1}\hat{\phi}$$

which is comparable to the form used in weighted least squares with the exception that the weights are now a function of the response for a GLM.

Let's implement the procedure explicitly to understand how the fitting algorithm works. We use the Bliss data from Section 2.7 to illustrate this. Here is the fit we are trying to match:

```
> data(bliss)
> modl <- glm(cbind(dead,alive) ~ conc, family=binomial, bliss)
> summary(modl)$coef
            Estimate Std. Error z value    Pr(>|z|)
(Intercept)  -2.3238    0.41789 -5.5608 2.6854e-08
conc          1.1619    0.18142  6.4046 1.5077e-10
```

For a binomial response, we have:

$$\eta = \log\frac{\mu}{1-\mu} \qquad \frac{d\eta}{d\mu} = \frac{1}{\mu(1-\mu)} \qquad V(\mu) = \mu(1-\mu)/n \qquad w = n\mu(1-\mu)$$

where the variance is computed with the understanding that y is the proportion not the count. We use y for our initial guess for $\hat{\mu}$ which works here because none of the observed proportions are zero or one:

```
> y <- bliss$dead/30;  mu <- y
> eta <- logit(mu)
> z <- eta + (y-mu)/(mu*(1-mu))
> w <- 30*mu*(1-mu)
> lmod <- lm(z ~ conc, weights=w, bliss)
> coef(lmod)
(Intercept)        conc
    -2.3025      1.1536
```

It is interesting how close these initial estimates are to the converged values given above. This is not uncommon. Even so, to get a more precise result, iteration is necessary. We do five iterations here:

```
> for(i in 1:5){
+ eta <- lmod$fit
+ mu <- ilogit(eta)
+ z <- eta + (y-mu)/(mu*(1-mu))
+ w <- 30*mu*(1-mu)
+ lmod <- lm(z ~ bliss$conc, weights=w)
+ cat(i,coef(lmod),"\n")
+ }
1 -2.3237 1.1618
2 -2.3238 1.1619
3 -2.3238 1.1619
4 -2.3238 1.1619
5 -2.3238 1.1619
```

We can see that convergence is fast in this case. The *Fisher scoring iterations* referred to in the output record the number of iterations. In most cases, the convergence is rapid. If there is a failure to converge, this is often a sign of some problem with the model specification or unusual feature of the data. An example of such a problem with the estimation may be seen in Section 2.8. A look at the final (weighted) linear model reveals that:

```
> summary(lmod)
Coefficients:
            Estimate Std. Error t value Pr(>|t|)
(Intercept)  -2.3238     0.1462   -15.9  0.00054
conc          1.1619     0.0635    18.3  0.00036

Residual standard error: 0.35 on 3 degrees of freedom
```

The standard errors are not correct and can be computed (rather inefficiently) as follows:

```
> xm <- model.matrix(lmod)
> wm <- diag(w)
> sqrt(diag(solve(t(xm) %*% wm %*% xm)))
[1] 0.41787 0.18141
```

Now $\widehat{\text{var}}(\hat{\beta}) = (X^T WX)^{-1}$ because $\phi = 1$ for the binomial model but in the Gaussian linear model $\widehat{\text{var}}(\hat{\beta}) = (X^T WX)^{-1}\hat{\sigma}^2$. To get the correct standard errors from the `lm` fit, we need to scale out the $\hat{\sigma}$ as follows:

```
> summary(lmod)$coef[,2]/summary(lmod)$sigma
(Intercept)        conc
    0.41789     0.18142
```

These calculations are shown for illustration purposes only and are done more efficiently and reliably by the `glm` function.

6.3 Hypothesis Tests

When considering the choice of model for some data, we should define the range of possibilities. The *null* model is the smallest model we will entertain while the *full* or *saturated* model is the most complex.

The null model represents the situation where there is no relation between the predictors and the response. Usually this means we fit a common mean μ for all y, that is, one parameter only. For the Gaussian GLM, this is the model $y = \mu + \varepsilon$. For some contingency table models, there will be additional parameters that represent row or column totals or other such constraints. In these cases, the null model will have more than one parameter.

In the saturated model, the data is explained exactly. Typically, we need to use n parameters for n data points. This can often be achieved by fitting a sufficiently high-order polynomial or by treating the numerical values of quantitative predictors as codes, thereby changing them into qualitative predictors. If enough interactions are included, the model will be saturated. This model tells us no more than the data itself and is usually uninformative.

A statistical model describes how we partition the data into systematic structure and random variation. The null model represents one extreme where the data is represented entirely as random variation, while the saturated or full model represents the data as being entirely systematic.

The full model does give us a measure of how well any model could possibly fit and so we might consider the difference between the log-likelihood for the full model, $l(y, \phi|y)$, and that for the model under consideration, $l(\hat{\mu}, \phi|y)$, expressed as a likelihood ratio statistic:

$$2(l(y,\phi|y) - l(\hat{\mu},\phi|y))$$

Provided that the observations are independent and for an exponential family distribution, when $a_i(\phi) = \phi/w_i$, this simplifies to:

$$\sum_i 2w_i(y_i(\tilde{\theta}_i - \hat{\theta}_i) - b(\tilde{\theta}_i) + b(\hat{\theta}_i))/\phi$$

where $\tilde{\theta}$ are the estimates under the full (saturated) model and $\hat{\theta}$ are the estimates under the model of interest. The above can be written simply as $D(y,\hat{\mu})/\phi$ where $D(y,\hat{\mu})$ is called the `deviance` and $D(y,\hat{\mu})/\phi$ is called the `scaled deviance`. Deviances for the common GLMs are shown in Table 6.2.

GLM	Deviance
Gaussian	$\sum_i (y_i - \hat{\mu}_i)^2$
Poisson	$2\sum_i [y_i \log(y_i/\hat{\mu}_i) - (y_i - \hat{\mu}_i)]$
Binomial	$2\sum_i [y_i \log(y_i/\hat{\mu}_i) + (m - y_i)\log((m - y_i)/(m - \hat{\mu}_i))]$
Gamma	$2\sum_i [-\log(y_i/\hat{\mu}_i) + (y_i - \hat{\mu}_i)/\hat{\mu}_i]$
Inverse Gaussian	$\sum_i (y_i - \hat{\mu}_i)^2/(\hat{\mu}_i^2 y_i)$

Table 6.2 *For the binomial $y_i \sim B(m, p_i)$ and $\mu_i = mp_i$ that is μ is the count and not proportion in this formula. For the Poisson, the deviance is known as the G-statistic. The second term $\sum_i (y_i - \hat{\mu}_i)$ is usually zero if an intercept term is used in the model.*

Pearson's X^2 statistic:

$$X^2 = \sum_i \frac{(y_i - \hat{\mu}_i)^2}{V(\hat{\mu}_i)}$$

where $V(\hat{\mu}) = \text{var}\ (\hat{\mu})$ is an alternative measure of discrepancy that is sometimes used in place of the deviance.

There are two main types of hypothesis test we shall employ. The goodness of fit test simply asks whether the current model fits the data. The other type of test compares two nested models where the smaller model represents a linear restriction on the parameters of the larger model. The goodness of fit test can be viewed as model comparison test if we identify the smaller model with the model of interest and the larger model with the full or saturated model.

For the goodness of fit test, we use the fact that, under certain conditions, provided the model is correct, the scaled Deviance and the Pearson's X^2 statistic are both asymptotically χ^2 with degrees of freedom equal to the number of identifiable parameters. For GLMs such as the Gaussian, we usually do not know the value of the dispersion parameter, ϕ, and so this test cannot be used. For the binomial and the Poisson, $\phi = 1$, and so the test is practical. However, the accuracy of the asymptotic approximation is dubious for smaller datasets. For a binary, that is a 0-1 response, the approximation is worthless.

For comparing a larger model, Ω, to a smaller nested model, ω the difference in the scaled deviances, $D_\omega - D_\Omega$ is asymptotically χ^2 with degrees of freedom equal to the difference in the number of identifiable parameters in the two models. For the Gaussian model and other models where the dispersion ϕ is usually not known, this test cannot be directly used. However, if we insert an estimate of ϕ we may compute an F-statistic of the form:

$$\frac{(D_\omega - D_\Omega)/(df_\omega - df_\Omega)}{\hat{\phi}}$$

where $\hat{\phi} = X^2/(n-p)$ is a good estimate of the dispersion. For the Gaussian model, $\hat{\phi} = RSS_\Omega/df_\Omega$, and the resulting F-statistic has an exact F distribution for the null. For other GLMs with free dispersion parameters, the statistic is only approximately F distributed.

For every GLM except the Gaussian, an approximate null distribution must be used whose accuracy may be in doubt particularly for smaller samples. However, the approximation is better when comparing models than for the goodness of fit statistic.

Let's consider the possible tests on the Bliss insect data:

```
> summary(modl)
Coefficients:
            Estimate Std. Error z value Pr(>|z|)
(Intercept)   -2.324      0.418   -5.56  2.7e-08
conc           1.162      0.181    6.40  1.5e-10

(Dispersion parameter for binomial family taken to be 1)

    Null deviance: 64.76327  on 4  degrees of freedom
Residual deviance:  0.37875  on 3  degrees of freedom
```

We are able to make a goodness of fit test by examining the size of the residual deviance compared to its degrees of freedom:

```
> 1-pchisq(deviance(modl),df.residual(modl))
[1] 0.9446
```

where we see the p-value is large indicating no evidence of a lack of fit. As with lack of fit tests for Gaussian linear models, this outcome does not mean that this model is correct or that no better models exist. We can also quickly see that the null model would be inadequate for the data since the null deviance of 64.7 is very large for four degrees of freedom.

We can also test for the significance of the linear concentration term by comparing the current model to the null model:

```
> anova(modl,test="Chi")
Analysis of Deviance Table
Model: binomial, link: logit

Terms added sequentially (first to last)
     Df Deviance Resid. Df Resid. Dev P(>|Chi|)
NULL                    4       64.8
conc  1     64.4        3        0.4     1e-15
```

We see that the concentration term is clearly significant. We can also fit and test a more complex model:

```
> modl2 <- glm(cbind(dead,alive) ~ conc+I(conc^2), family=binomial,bliss)
> anova(modl,modl2,test="Chi")
  Resid. Df Resid. Dev Df Deviance P(>|Chi|)
1         3      0.379
2         2      0.195  1     0.183     0.669
```

We can see that there is no need for a quadratic term in the model. The same information could be extracted with:

```
> anova(modl2,test="Chi")
```

We may also take a Wald test approach. We may use the standard error of the parameter estimates to construct a z-statistic of the form $\hat{\beta}/se(\hat{\beta})$. This has an asymptotically normal null distribution. For the Bliss data, for the concentration term, we have $z = 1.162/0.181 = 6.40$. Thus the (approximate) p-value for the Wald test of the concentration parameter being equal to zero is 1.5e^{-10} and thus we clearly reject the null here. Remember that this is again only an approximate test except in the special case of the Gaussian GLM where the z-statistic is the t-statistic and has an exact t-distribution. The difference of deviances test is preferred to the Wald test due, in part, to the problem noted by Hauck and Donner (1977).

6.4 GLM Diagnostics

As with standard linear models, it is important to check the adequacy of the assumptions that support the GLM. The diagnostic methods for GLMs mirror those used for Gaussian linear models. However, some adaptations are necessary and, depending on the type of GLM, not all diagnostic methods will be applicable.

Residuals: Residuals represent the difference between the data and the model and are essential to explore the adequacy of the model. In the Gaussian case, the residuals are $\hat{\varepsilon} = y - \hat{\mu}$. These are called response residuals for GLMs, but since the variance of the response is not constant for most GLMs, some modification is necessary. We would like residuals for GLMs to be defined such that they can be used in a similar way as in the Gaussian linear model.

The *Pearson residual* is comparable to the standardized residuals used for linear models and is defined as:

$$r_P = \frac{y - \hat{\mu}}{\sqrt{V(\hat{\mu})}}$$

where $V(\mu) \equiv b''(\theta)$. These are just a rescaling of $y - \hat{\mu}$. Notice that $\sum r_P^2 = X^2$ and hence the name. Pearson residuals can be skewed for nonnormal responses.

The *deviance residuals* are defined by analogy to Pearson residuals. The Pearson residual was r_P such that $\sum r_P^2 = X^2$, so we set the deviance residual as r_D such that $\sum r_D^2 = \text{Deviance} = \sum d_i$. Thus:

$$r_D = sign(y - \hat{\mu})\sqrt{d_i}$$

For example, in the Poisson:

$$r_D = sign(y - \hat{\mu})[2(y \log y/\hat{\mu} - y + \hat{\mu})]^{1/2}$$

Let's examine the types of residuals available to us using the Bliss data. We can obtain the deviance residuals as:

```
> residuals(modl)
[1] -0.451015  0.359696  0.000000  0.064302 -0.204493
```

These are the default choice of residuals. The Pearson residuals are:

```
> residuals(modl,"pearson")
        1         2         3         4         5
-0.432523  0.364373  0.000000  0.064147 -0.208107
```

which are just slightly different from the deviance residuals. The response residuals are:

```
> residuals(modl,"response")
```

```
         1          2          3          4          5
-0.0225051  0.0283435  0.0000000  0.0049898 -0.0108282
```

which is just the response minus the fitted value:

```
> bliss$dead/30 - fitted(modl)
         1          2          3          4          5
-0.0225051  0.0283435  0.0000000  0.0049898 -0.0108282
```

Finally, the so-called working residuals are:

```
> residuals(modl,"working")
        1         2         3         4         5
-0.277088  0.156141  0.000000  0.027488 -0.133320
> modl$residuals
        1         2         3         4         5
-0.277088  0.156141  0.000000  0.027488 -0.133320
```

Note that it is important to use the `residuals()` function to get the deviance residuals which are most likely what is needed for diagnostic purposes. Using `$residuals` gives the working residuals which is not usually needed for diagnostics. We can now identify the working residuals as a by-product of the IRWLS fitting procedure:

```
> residuals(lmod)
          1           2           3           4           5
-2.7709e-01  1.5614e-01 -3.8463e-16  2.7488e-02 -1.3332e-01
```

Leverage and influence: For a linear model, $\hat{y} = Hy$, where H is the *hat matrix* that projects the data onto the fitted values. The leverages h_i are given by the diagonal of H and represent the potential of the point to influence the fit. They are solely a function of X and whether they are in fact influential will also depend on y. Leverages are somewhat different for GLMs. The IRWLS algorithm used to fit the GLM uses weights, w. These weights are just part of the IRWLS algorithm and are not user assigned. However, these do affect the leverage. We form a matrix $W = \text{diag}(w)$ and the hat matrix is:

$$H = W^{1/2}X(X^TWX)^{-1}X^TW^{1/2}$$

We extract the diagonal elements of H to get the leverages h_i. A large value of h_i indicates that the fit may be sensitive to the response at case i. Large leverages typically mean that the predictor values are unusual in some way. One important difference from the linear model case is that the leverages are no longer just a function of X and now depend on the response through the weights W. The leverages may be calculated as:

```
> influence(modl)$hat
      1       2       3       4       5
0.42550 0.41331 0.32238 0.41331 0.42550
```

As in the linear model case, we might choose to studentize the residuals as follows:

$$r_{SD} = \frac{r_D}{\sqrt{\hat{\phi}(1-h_i)}}$$

or compute jackknife residuals representing the difference between the observed response for case i and that predicted from the data with case i excluded, scaled appropriately. These are expensive to compute exactly and so an approximation due to

Williams (1987) can be used:

$$sign(y-\hat{\mu})\sqrt{(1-h_i)r_{SD}^2+h_i r_{SP}^2}$$

where $r_{SP}=r_P/\sqrt{1-h_i}$. These may be computed as:

```
> rstudent(modl)
        1         2         3         4         5
-0.584786  0.472135  0.000000  0.083866 -0.271835
```

Outliers may be detected by observing particularly large jackknife residuals.

Leverage only measures the potential to affect the fit whereas measures of influence more directly assess the effect of each case on the fit. We can examine the change in the fit from omitting a case by looking at the changes in the coefficients:

```
> influence(modl)$coef
  (Intercept)       conc
1  -0.2140015  0.0806635
2   0.1556719 -0.0470873
3   0.0000000  0.0000000
4  -0.0058417  0.0084177
5   0.0492639 -0.0365734
```

Alternatively, we can examine the Cook statistics:

$$D_i=\frac{(\hat{\beta}_{(i)}-\hat{\beta})^T(X^TWX)(\hat{\beta}_{(i)}-\hat{\beta})}{p\hat{\phi}}$$

which may be calculated as:

```
> cooks.distance(modl)
        1         2         3         4         5
0.1205927 0.0797100 0.0000000 0.0024704 0.0279174
```

We can see that the biggest change would occur by omitting the first observation. However, since this is a very small dataset with just five observations, we would not contemplate dropping cases. In any event, we see that the change in the coefficients would not qualitatively change the conclusion.

Model diagnostics: We may divide diagnostic methods into two types. Some methods are designed to detect single cases or small groups of cases that do not fit the pattern of the rest of the data. Outlier detection is an example of this. Other methods are designed to check the assumptions of the model. These methods can be subdivided into those that check the structural form of the model, such as the choice and transformation of the predictors, and those that check the stochastic part of the model, such as the nature of the variance about the mean response. Here, we focus on methods for checking the assumptions of the model.

For linear models, the plot of residuals against fitted values is probably the single most valuable graphic. For GLMs, we must decide on the appropriate scale for the fitted values. Usually, it is better to plot the linear predictors $\hat{\eta}$ rather than the predicted responses $\hat{\mu}$. We revisit the model for Galápagos data first presented in Section 3.1. Consider first a plot using $\hat{\mu}$ presented in the first panel of Figure 6.1:

```
> data(gala)
```

```
> gala <- gala[,-2]
> modp <- glm(Species ~ .,family=poisson,gala)
> plot(residuals(modp) ~ predict(modp,type="response"),
  xlab=expression(hat(mu)),ylab="Deviance residuals")
```

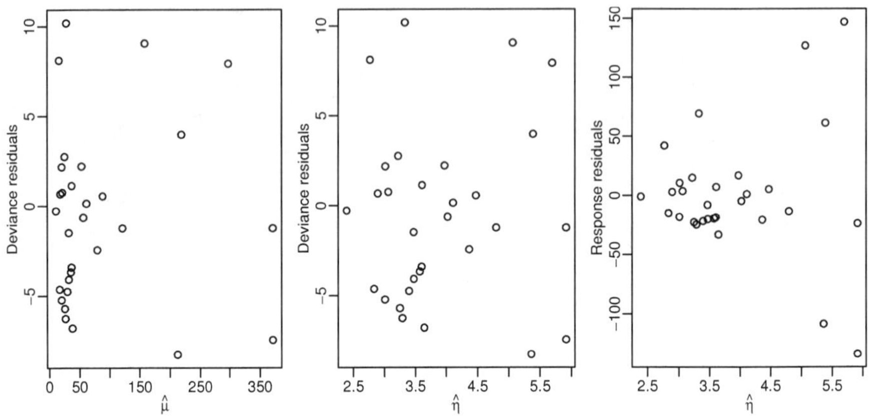

Figure 6.1 *Residual vs. fitted plots for the Galápagos model. The first uses fitted values in the scale of the response while the second uses fitted values in the scale of the linear predictor. The third plot uses response residuals while the first two use deviance residuals.*

There are just a few islands with a large predicted number of species while most predicted response values are small. This makes it difficult to see the relationship between the residuals and the fitted values because most of the points are compressed on the left of the display. Now we try plotting $\hat{\eta}$:

```
> plot(residuals(modp) ~ predict(modp,type="link"),
  xlab=expression(hat(eta)),ylab="Deviance residuals")
```

Now the points, shown in the second panel of Figure 6.1, are more evenly spaced in the horizontal direction. We are looking for two main features in such a plot. Is there any nonlinear relationship between the predicted values and the residuals? If so, this would be an indication of a lack of fit that might be rectified by a change in the model. For a linear model, we might consider a transformation of the response, but this is usually impractical for a GLM since it would change the assumed distribution of the response. We might also consider a change to the link function, but often this is undesirable since there a few choices of link function that lead to easily interpretable models. It is best if a change in the choice of predictors or transformations on these predictors can be made since this involves the least disruption to the GLM. For this particular plot, there is no evidence of nonlinearity.

The variance of the residuals with respect to the fitted values should also be inspected. The assumptions of the GLM would require constant variance in the plot and, in this case, this appears to be the case. A violation of this assumption would prompt a change in the model. We might consider a change in the variance function $V(\mu)$, but this would involve abandoning the Poisson GLM since this specifies a particular form for the variance function. We would need to use a quasi-likelihood GLM described in Section 7.4. Alternatively, we could employ a different GLM for a count response such as the negative binomial. Finally, we might use weights if we could identify some feature of the data that would suggest a suitable choice.

For all GLMs but the Gaussian, we have a nonconstant variance function. However, by using deviance residuals, we have already scaled out the variance function and so, provided the variance function is correct, we do expect to see constant variance in the plot. If we use response residuals, that is $y - \hat{\mu}$, as seen in the third panel of Figure 6.1:

```
> plot(residuals(modp,type="response") ~ predict(modp,type="link"),
  xlab=expression(hat(eta)),ylab="Response residuals")
```

We see a pattern of increasing variation consistent with the Poisson.

In some cases, plots of the residuals are not particularly helpful. For a binary response, the residual can only take two possible values for given predicted response. This is the most extreme situation, but similar discreteness can occur for binomial responses with small group sizes and Poisson responses that are small. Plots of residuals in these cases tend to show curved lines of points corresponding to the limited number of observed responses. Such artifacts can obscure the main purpose of the plot. Difficulties arise for binomial data where the covariate classes have very different sizes. Points on plots may represent just a few or a large number of individuals.

Investigating the nature of the relationship between the predictors and the response is another primary objective of diagnostic plots. Even before a model is fit to the data, we might simply plot the response against the predictors. For the Galápagos data, consider a plot of the number of species against the area of the island shown in the first panel of Figure 6.2:

```
> plot(Species ~ Area, gala)
```

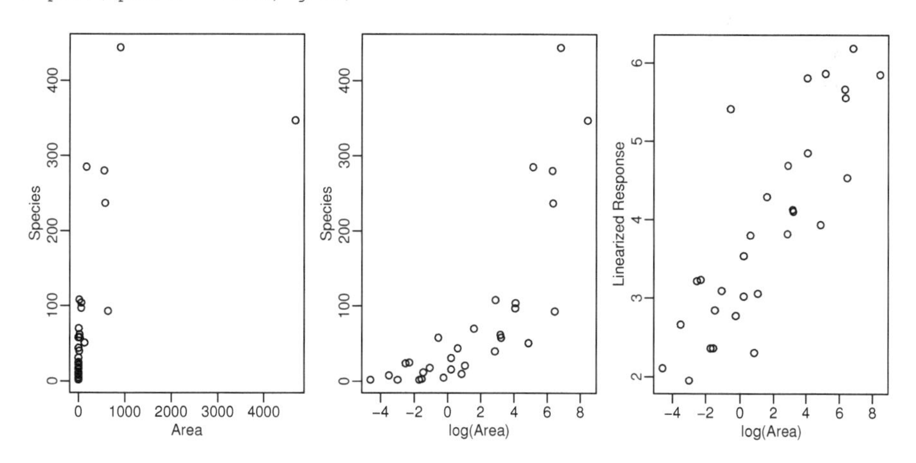

Figure 6.2 *Plots of the number of species against area for the Galápagos data. The first plot clearly shows a need for transformation, the second shows the advantage of using logged area, while the third shows the value of using the linearized response.*

We see that both variables have skewed distributions. We start with a log transformation on the predictor as seen in the second panel of Figure 6.2:

```
> plot(Species ~ log(Area), gala)
```

We see a curvilinear relationship between the predictor and the response. However, the default Poisson GLM uses a log link which we need to take into account. To allow for the choice of link function, we can plot the linearized response:

$$z = \eta + (y - \mu)\frac{d\eta}{d\mu}$$

as we see in the third panel of Figure 6.2:

```
> mu <- predict(modp,type="response")
> z <- predict(modp)+(gala$Species-mu)/mu
> plot(z ~ log(Area), gala,ylab="Linearized Response")
```

We now see a linear relationship suggesting that no further transformation of area is necessary. Notice that we used the current model in the computation of z. Some might prefer to use an initial guess here to avoid presuming the choice of model. For this dataset, we find that a log transformation of all the predictors is helpful:

```
> modpl <- glm(Species ~ log(Area) + log(Elevation) + log(Nearest) +
  log(Scruz+0.1) + log(Adjacent), family=poisson, gala)
> c(deviance(modp),deviance(modpl))
[1] 716.85 359.12
```

We see that this results in a substantial reduction in the deviance.

The disadvantage of simply examining the raw relationship between the response and the other predictors is that it fails to take into account the effect of the other predictors. Partial residual plots are used for linear models to make allowance for the effect of the other predictors while focusing on the relationship of interest. These can be adapted for use in GLMs by plotting $z - \hat{\eta} + \beta_j x_j$ versus x_j. The interpretation is the same as in the linear model case. We compute the partial residual plot for the (now logged) area, as shown in the first panel of Figure 6.3:

```
> mu <- predict(modpl,type="response")
> u <- (gala$Species-mu)/mu + coef(modpl)[2]*log(gala$Area)
> plot(u ~ log(Area), gala,ylab="Partial Residual")
> abline(0,coef(modpl)[2])
```

In this plot, we see no reason for concern. There is no nonlinearity indicating a need to transform nor are there any obvious outliers or influential points. Partial residuals can also be obtained from `residuals(., type="partial")` although an offset will be necessary if you want the regression line displayed correctly on the plot.

One can search for good transformations of the predictors in nongraphical ways. Polynomials terms or spline functions of the predictors can be experimented with, but generalized additive models, described in Chapter 12, offer a more direct way to discover some good transformations.

The link function is a fundamental assumption of the GLM. Quite often the choice of link function is set by the characteristics of the response, such as positivity, or by ease of interpretation, as with logit link for binomial GLMs. It is often difficult to contemplate alternatives. Nevertheless, it is worth checking to see whether the link assumption is not grossly wrong. Before doing this, it is important to eliminate other simpler violations of the assumptions that are more easily rectified such as outliers

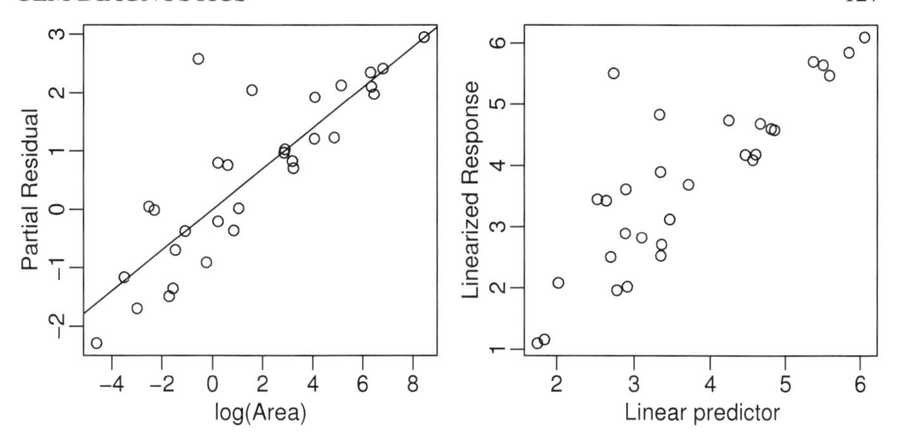

Figure 6.3 *A partial residual plot for log(Area) is shown on the left while a diagnostic for the link function is shown on the right.*

or transformations of the predictors. After these concerns have been eliminated, one can check the link assumption by making a plot of the linearized response z against linear predictor $\hat{\eta}$. An example of this is shown in the second panel of Figure 6.3:

```
> z <- predict(modpl)+(gala$Species-mu)/mu
> plot(z ~ predict(modpl), xlab="Linear predictor",
  ylab="Linearized Response")
```

In this case, we see no indication of a problem.

An alternative approach to checking the link function is to propose a family of link functions of which the current choice is a member. A range of links can then be fit and compared to the current choice. The approach is analogous to the Box–Cox method used for linear models. Alternative choices are easier to explore within the quasi-likelihood framework described in Section 7.4.

Unusual Points

We have already described the raw material of residuals, leverage and influence measures that can be used to check for points that do not fit the model or influence the fit unduly. Let's now see how to use graphical methods to examine these quantities.

The Q-Q plot of the residuals is the standard way to check the normality assumption on the errors typically made for a linear model. For a GLM, we do not expect the residuals to be normally distributed, but we are still interested in detecting outliers. For this purpose, it is better to use a half-normal plot that compares the sorted absolute residuals and the quantiles of the half-normal distribution:

$$\Phi^{-1}\left(\frac{n+i}{2n+1}\right) \qquad i=1,\ldots,n$$

The residuals are not expected to be normally distributed, so we are not looking for an approximate straight line. We only seek outliers which may be identified as points off the trend. A half-normal plot is better for this purpose because in a sense the resolution of the plot is doubled by having all the points in one tail.

Since we are more specifically interested in outliers, we should plot the jackknife residuals. An example for the Galápagos model is shown in the first panel of Figure 6.4:

```
> halfnorm(rstudent(modpl))
```

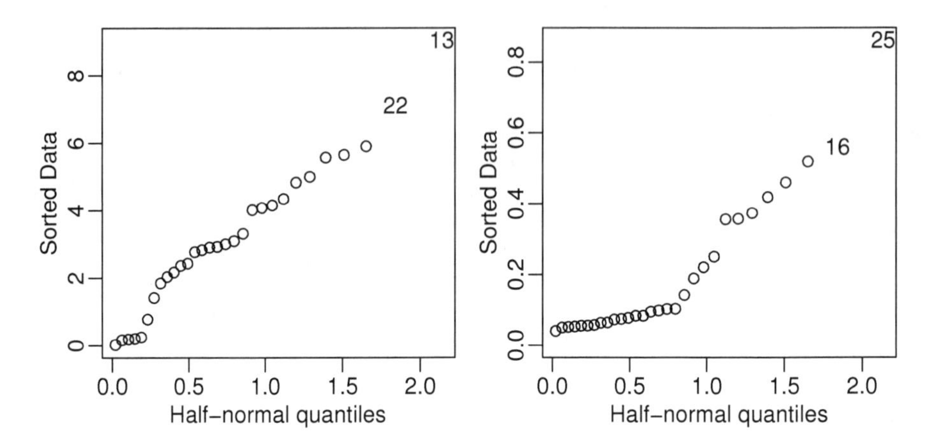

Figure 6.4 *Half-normal plots of the jackknife residuals on the left and the leverages on the right.*

We see no sign of outliers in the plot. The half-normal plot is also useful for positive-valued diagnostics such as the leverages and the Cook statistics. A look at the leverages is shown in the second panel of Figure 6.4:

```
> gali <- influence(modpl)
> halfnorm(gali$hat)
```

There is some indication that case 25, Santa Cruz island, may have some leverage. The predictor `Scruz` is the distance from Santa Cruz island which is zero for this case. This posed a problem for making the log transformation and explains why we added 0.1 to this variable. However, there is some indication that this inelegant fix may be causing some difficulty.

Moving on to influence, a half-normal plot of the Cook statistics is shown in the first panel of Figure 6.5:

```
> halfnorm(cooks.distance(modpl))
```

Again we have some indication that Santa Cruz island is influential. We can examine the change in the fitted coefficients. For example, consider the change in the `Scruz` coefficient as shown in the second panel of Figure 6.5:

```
> plot(gali$coef[,5],ylab="Change in Scruz coef",xlab="Case no.")
```

We see a substantial change for case 25. If we compare the full fit to a model without this case, we find:

```
> modplr <- glm(Species ~ log(Area) + log(Elevation) + log(Nearest)
  + log(Scruz+0.1) + log(Adjacent), family=poisson, gala, subset=-25)
> cbind(coef(modpl),coef(modplr))
```

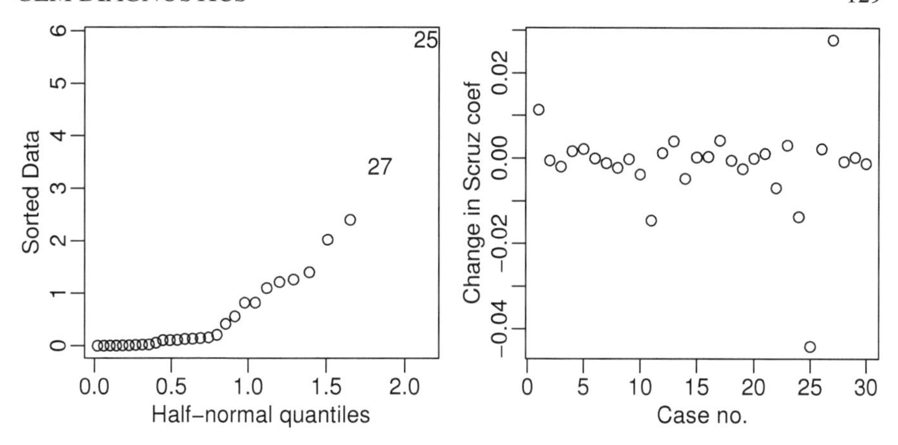

Figure 6.5 *Half-normal plot of the Cook statistics is shown on the left and an index plot of the change in the Scruz coefficient is shown on the right.*

```
                        [,1]      [,2]
(Intercept)         3.287941  3.050699
log(Area)           0.348445  0.334530
log(Elevation)      0.036421  0.059603
log(Nearest)       -0.040644 -0.052548
log(Scruz + 0.1)   -0.030045  0.015919
log(Adjacent)      -0.089014 -0.088516
```

We see a sign change for the `Scruz` coefficient. This is interesting since in the full model, the coefficient is more than twice the standard error way from zero indicating some significance. A simple solution is to add a larger amount, say 0.5, to `Scruz`.

Other than this user-introduced anomaly, we find no difficulty. Using our earlier discovery of the log transformation, some variable selection and allowing for remaining overdispersion, our final model is:

```
> modpla <- glm(Species ~ log(Area)+log(Adjacent), family=poisson, gala)
> dp <- sum(residuals(modpla,type="pearson")^2)/modpla$df.res
> summary(modpla,dispersion=dp)
Coefficients:
              Estimate Std. Error z value Pr(>|z|)
(Intercept)     3.2767     0.1794   18.26  < 2e-16
log(Area)       0.3750     0.0326   11.50  < 2e-16
log(Adjacent)  -0.0957     0.0249   -3.85  0.00012

(Dispersion parameter for poisson family taken to be 16.527)

    Null deviance: 3510.73  on 29  degrees of freedom
Residual deviance:  395.54  on 27  degrees of freedom
```

Notice that the deviance is much lower and the elevation variable is not used when compared with our model choice in Section 3.1.

This example concerned a Poisson GLM. Diagnostics for binomial GLMs are similar, but see Pregibon (1981) and Collett (2003) for more details.

Further Reading: The canonical book on GLMs is McCullagh and Nelder (1989). Other books include Dobson (1990), Lindsey (1997), Myers, Montgomery, and Vining (2002), Gill (2001) and Fahrmeir and Tutz (2001). For a Bayesian perspective, see Dey, Ghosh, and Mallick (2000).

Exercises

1. Consider the `orings` data from Chapter 2. Suppose that, in spite of all the drawbacks, we insist on fitting a model with an identity link, but with the binomial variance. Show how this may be done using a `quasi` family model using the `glm` function. (You will need need to consult the help pages for `quasi` and `glm` and in particular you will need to set good starting values for beta — if it doesn't work at the first attempt, try different values.) Describe how the fitted model differs from the standard logistic regression and give the predicted response at a temperature of 31°F.

2. Fit the `orings` data with a binomial response and a logit link as in Chapter 2.

   (a) Construct the appropriate test statistic for testing the effect of the temperature. State the appropriate null distribution and give the p-value.

   (b) Generate data under the null distribution for the previous test. Use the `rbinom` function with the average proportion of damaged O-rings. Recompute the test statistic and compute the p-value.

   (c) Repeat the process of the previous question 1000 times, saving the test statistic each time. Compare the empirical distribution of these simulated test statistics with the nominal null distribution stated in the first part of this question. Compare the critical values for a 5% level test computed using these two methods.

3. Fit the `orings` data with a binomial response and a logit link as in Chapter 2.

   (a) Construct the appropriate test statistic for testing the effect of the temperature. State the appropriate null distribution and give the p-value.

   (b) Generate a random permutation of the responses using `sample` and recompute the test statistic and compute the p-value.

   (c) Repeat the process of the previous question 1000 times, saving the test statistic each time. Compare the empirical distribution of these permuted data test statistics with the nominal null distribution stated in the first part of this question. Compare the critical values for a 5% level test computed using these two methods.

4. Data is generated from the exponential distribution with density $f(y) = \lambda \exp(-\lambda y)$ where $\lambda, y > 0$.

   (a) Identify the specific form of $\theta, \phi, a(), b()$ and $c()$ for the exponential distribution.

(b) What is the canonical link and variance function for a GLM with a response following the exponential distribution?

(c) Identify a practical difficulty that may arise when using the canonical link in this instance.

(d) When comparing nested models in this case, should an F or χ^2 test be used? Explain.

(e) Express the deviance in this case in terms of the responses y_i and the fitted values $\hat{\mu}_i$.

5. The Conway–Maxwell–Poisson distribution has probability function:

$$P(Y = y) = \frac{\lambda^y}{(y!)^\nu} \frac{1}{Z(\lambda, \nu)} \qquad y = 0, 1, 2, \ldots$$

where

$$Z(\lambda, \nu) = \sum_{i=0}^{\infty} \frac{\lambda^i}{(i!)^\nu}$$

Place this in exponential family form, identifying all the relevant components necessary for use in a GLM.

CHAPTER 7

Other GLMs

The binomial, Gaussian and Poisson GLMs are by far the most commonly used, but there are a number of less popular GLMs which are useful for particular types of data. The gamma and inverse Gaussian are intended for continuous, skewed responses. In some cases, we are interested in modeling both the mean and the dispersion of the response and so we present dual GLMs for this purpose. The quasi-GLM is a model that is useful for nonstandard responses where we are unwilling to specify the distribution but can state the link and variance functions.

7.1 Gamma GLM

The density of the gamma distribution is usually given by:

$$f(y) = \frac{1}{\Gamma(\nu)}\lambda^{\nu} y^{\nu-1} e^{-\lambda y} \qquad y > 0$$

where ν describes the shape and λ describes the scale of the distribution. However, for the purposes of a GLM, it is convenient to reparameterize by putting $\lambda = \nu/\mu$ to get:

$$f(y) = \frac{1}{\Gamma(\nu)}\left(\frac{\nu}{\mu}\right)^{\nu} y^{\nu-1} e^{-\left(\frac{y\nu}{\mu}\right)} \qquad y > 0$$

Now $EY = \mu$ and $\operatorname{var} Y = \mu^2/\nu = (EY)^2/\nu$. The dispersion parameter is $\phi = \nu^{-1}$. Here we plot a gamma density with three different values of the shape parameter ν (the scale parameter would just have a multiplicative effect) as seen in Figure 7.1:

```
> x <- seq(0,8,by=0.1)
> plot(x,dgamma(x,0.75),type="l",ylab="",xlab="",ylim=c(0,1.25),
  xaxs="i",yaxs="i")
> plot(x,dgamma(x,1.0),type="l",ylab="",xlab="",ylim=c(0,1.25),
  xaxs="i",yaxs="i")
> plot(x,dgamma(x,2.0),type="l",ylab="",xlab="",ylim=c(0,1.25),
  xaxs="i",yaxs="i")
```

The gamma distribution can arise in various ways. The sum of ν independent and identically distributed exponential random variables with rate λ has a gamma distribution. The χ^2 distribution is a special case of the gamma where $\lambda = 1/2$ and $\nu = df/2$.

The canonical parameter is $-1/\mu$, so the canonical link is $\eta = -1/\mu$. However, we typically remove the minus (which is fine provided we take account of this in any derivations) and just use the inverse link. We also have $b(\theta) = \log(1/\mu) = -\log(-\theta)$

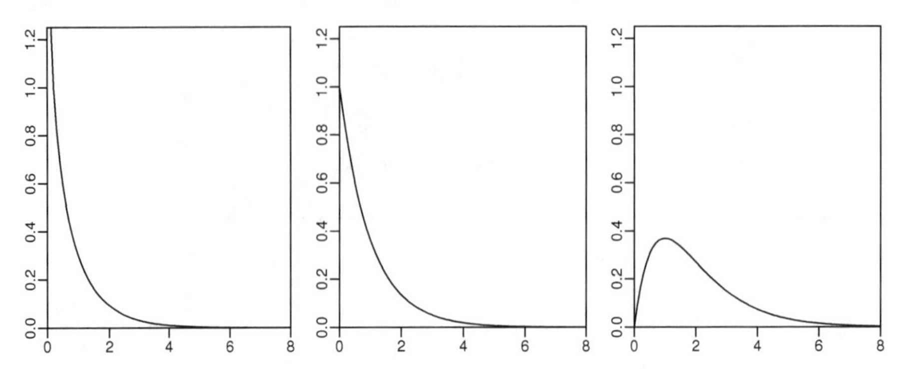

Figure 7.1 *The gamma density explored. In the first panel* $\nu = 0.75$ *and we see that the density is unbounded at zero. In the second panel,* $\nu = 1$ *which is the exponential density. In the third panel,* $\nu = 2$ *and we see a skewed distribution.*

and so $b''(\theta) = \mu^2$ is the variance function. The (unscaled) deviance is:

$$D(y,\hat{\mu}) = -2\sum\{\log y_i/\hat{\mu}_i - (y_i - \hat{\mu}_i)/\hat{\mu}_i\}$$

The utility of the gamma GLM arises in two different ways. Certainly, if we believe the response to have a gamma distribution, the model is clearly applicable. However, the model can also be useful in other situations where we may be willing to speculate on the relationship between the mean and the variance of the response but are not sure about the distribution. Indeed, it is possible to grasp the mean to variance relationship from graphical displays with relatively small datasets, while assertions about the response distribution would require a lot more data.

In the Gaussian linear model, var Y is constant as a function of the mean response. This is a fundamental assumption necessary for the optimality of least squares. However, sometimes contextual knowledge of the data or diagnostics show that var Y is nonconstant. When the form of nonconstancy is known exactly, then weighted least squares can be used. In practice, however, the form is often not known exactly. Alternatively, transformation of Y may lead to a constant variance model. The difficulty here is that while the original scale Y may be meaningful, $\log Y$ or $\sqrt{Y}$, for example, may not. An example is where a sum of the Ys may be of interest. In such a case, transformation would be a hindrance.

If var $Y \propto EY$, then $\sqrt{Y}$ is the variance stabilizing transform. If one wants to avoid a transformation, a GLM approach can be used. When Y has a Poisson distribution, then var $Y \propto EY$, suggesting the use of a Poisson GLM. Now one might object that the Poisson is a distribution for discrete data, which would seem to disallow its use for continuous responses. However, fitting a GLM only depends on the mean and variance of a distribution; the other moments are not used. This is important because it indicates that we need specify nothing more than the mean and variance. The distribution could be discrete or continuous and it would make no difference.

For some data, we might expect the standard deviation to increase linearly with the response. If so, the coefficient of variation, SD Y/EY, would be constant and var $Y \propto (EY)^2$. For example, measurements of larger objects do tend to have more error than smaller ones.

If we wanted to apply a Gaussian linear model, the log transform is indicated. This would imply a lognormal distribution for the original response. Alternatively, if $Y \sim$ gamma, then var $Y \propto (EY)^2$, so a gamma GLM is also appropriate in this situation. In a few cases, one may have some knowledge concerning the true distribution of the response which would drive the choice. However, in many cases, it would be difficult to distinguish between these two options on the basis of the data alone and the choice would be driven by the purpose and desired interpretation of the model.

There are three common choices of link function:

1. The canonical link is $\eta = \mu^{-1}$. Since $-\infty < \eta < \infty$, the link does not guarantee $\mu > 0$ which could cause problems and might require restrictions on β or on the range of possible predictor values. On the other hand the reciprocal link has some advantages. The Michaelis–Menten model has:

$$Ey = \mu = \frac{\alpha_0 x}{1 + \alpha_1 x}$$

   which can be represented after some reexpression as:

$$\eta = \alpha_1/\alpha_0 + 1/(\alpha_0 x) = \mu^{-1}$$

   As x increases, $\eta \to \alpha_1/\alpha_0$, which means that the mean μ will be bounded. The inverse link can be useful in such situations where we know the mean response to be bounded.

2. The log link, $\eta = \log \mu$, should be used when the effect of the predictors is suspected to be multiplicative on the mean. When the variance is small, this approach is similar to a Gaussian model with a logged response.

3. The linear link, $\eta = \mu$, is useful for modeling sums of squares or variance components which are χ^2. This is a special case of the gamma.

The general GLM procedures apply to the analysis and fitting. To estimate the dispersion, McCullagh and Nelder (1989) recommend the use of:

$$\hat{\phi} = \frac{1}{\hat{\nu}} = \frac{X^2}{n-p}$$

The maximum likelihood estimator and the usual estimator, $D/(n-p)$, are both sensitive to unusually small values of the response and are not consistent estimates of the coefficient of variation when the gamma distribution assumption does not hold.

Myers and Montgomery (1997) present data from a step in the manufacturing process for semiconductors. Four factors are believed to influence the resistivity of the wafer and so a full factorial experiment with two levels of each factor was run. Previous experience led to the expectation that resistivity would have a skewed distribution and so the need for transformation was anticipated. We start with a look at the data:

```
> data(wafer)
> summary(wafer)
 x1     x2     x3     x4           resist
 -:8    -:8    -:8    -:8    Min.    :166
```

```
 +:8    +:8    +:8    +:8    1st Qu.:201
                             Median :214
                             Mean   :229
                             3rd Qu.:259
                             Max.   :340
```

The application of the Box–Cox method or past experience suggests the use of a log transformation on the response. We fit the full model and then reduce it using AIC-based model selection:

```
> llmdl <- lm(log(resist) ~ .^2, wafer)
> rlmdl <- step(llmdl)
> summary(rlmdl)
Coefficients:
            Estimate Std. Error t value Pr(>|t|)
(Intercept)   5.3111     0.0476  111.53  4.7e-14
x1+           0.2009     0.0476    4.22  0.00292
x2+          -0.2107     0.0476   -4.43  0.00221
x3+           0.4372     0.0673    6.49  0.00019
x4+           0.0354     0.0476    0.74  0.47892
x1+:x3+      -0.1562     0.0673   -2.32  0.04896
x2+:x3+      -0.1782     0.0673   -2.65  0.02941
x3+:x4+      -0.1830     0.0673   -2.72  0.02635

Residual standard error: 0.0673 on 8 degrees of freedom
Multiple R-Squared: 0.947,Adjusted R-squared: 0.901
F-statistic: 20.5 on 7 and 8 DF,  p-value: 0.000165
```

We find a model with three two-way interactions, all with `x3`.

Now we fit the corresponding gamma GLM and again select the model using the AIC criterion. Note that the family must be specified as `Gamma` rather than `gamma` to avoid confusion with the Γ function. We use the log link to be consistent with the linear model. This must be specified as the default is the inverse link:

```
> gmdl <- glm(resist ~ .^2, family=Gamma(link=log), wafer)
> rgmdl <- step(gmdl)
> summary(rgmdl)
Coefficients:
            Estimate Std. Error t value Pr(>|t|)
(Intercept)   5.3120     0.0476  111.68  4.6e-14
x1+           0.2003     0.0476    4.21  0.00295
x2+          -0.2110     0.0476   -4.44  0.00218
x3+           0.4367     0.0673    6.49  0.00019
x4+           0.0354     0.0476    0.74  0.47836
x1+:x3+      -0.1555     0.0673   -2.31  0.04957
x2+:x3+      -0.1763     0.0673   -2.62  0.03064
x3+:x4+      -0.1819     0.0673   -2.70  0.02687

(Dispersion parameter for Gamma family taken to be 0.0045249)

    Null deviance: 0.697837  on 15  degrees of freedom
Residual deviance: 0.036266  on  8  degrees of freedom
```

```
AIC: 139.2
```

In this case, we see that the coefficients are remarkably similar to the linear model with the logged response. Even the standard errors are almost identical and the square root of the dispersion corresponds to the residual standard error of the linear model:

```
> sqrt(0.0045249)
[1] 0.067267
```

The maximum likelihood estimate of ϕ may be computed using the MASS package:

```
> library(MASS)
> gamma.dispersion(rgmdl)
[1] 0.0022657
```

We see that this gives a substantially smaller estimate, which would suggest smaller standard errors. However, it is not consistent with our experience with the Gaussian linear model in this example.

In this example, because the value of $\nu = 1/\phi$ is large (221), the gamma distribution is well approximated by a normal. Similarly, for the logged response linear model, a lognormal distribution with a small variance ($\sigma = 0.0673$) is also very well approximated by a normal. For this reason, there is not much to distinguish these two models. The gamma GLM has the advantage of modeling the response directly while the lognormal has the added convenience of working with a standard linear model.

Let us examine another example where there is more distinction between the two approaches. In Hallin and Ingenbleek (1983) data on payments for insurance claims for various areas of Sweden in 1977 are presented. The data is further subdivided by mileage driven, the bonus from not having made previous claims and the type of car. We have information on the number of insured, measured in policy-years, within each of these groups. Since we expect that the total amount of the claims for a group will be proportionate to the number of insured, it makes sense to treat the log of the number insured as an offset for similar reasons to those in Section 3.2. Attention has been restricted to data from Zone 1. After some model selection, a gamma GLM of the following form was found:

```
> data(motorins)
> motori <- motorins[motorins$Zone == 1,]
> gl <- glm(Payment ~ offset(log(Insured))+as.numeric(Kilometres)+
    Make+Bonus , family=Gamma(link=log), motori)
> summary(gl)
Coefficients:
                       Estimate Std. Error t value Pr(>|t|)
(Intercept)              6.5273     0.1777   36.72  < 2e-16
as.numeric(Kilometres)   0.1201     0.0311    3.85  0.00014
Make2                    0.4070     0.1782    2.28  0.02313
Make3                    0.1553     0.1796    0.87  0.38767
Make4                   -0.3439     0.1915   -1.80  0.07355
Make5                    0.1447     0.1810    0.80  0.42473
Make6                   -0.3456     0.1782   -1.94  0.05352
Make7                    0.0614     0.1824    0.34  0.73689
Make8                    0.7504     0.1873    4.01 0.000079
Make9                    0.0320     0.1782    0.18  0.85778
```

```
Bonus                     -0.2007     0.0215   -9.33  < 2e-16

(Dispersion parameter for Gamma family taken to be 0.55597)

    Null deviance: 238.97  on 294  degrees of freedom
Residual deviance: 155.06  on 284  degrees of freedom
AIC: 7168
```

In comparison, the lognormal model, where we have used the `glm` function for compatibility, looks like this:

```
> llg <- glm(log(Payment) ~ offset(log(Insured))+as.numeric(Kilometres)+
  Make+Bonus,family=gaussian ,  motori)
> summary(llg)
Coefficients:
                       Estimate Std. Error t value Pr(>|t|)
(Intercept)             6.51403    0.18634   34.96  < 2e-16
as.numeric(Kilometres)  0.05713    0.03265    1.75   0.0813
Make2                   0.36387    0.18686    1.95   0.0525
Make3                   0.00692    0.18824    0.04   0.9707
Make4                  -0.54786    0.20076   -2.73   0.0067
Make5                  -0.02179    0.18972   -0.11   0.9087
Make6                  -0.45881    0.18686   -2.46   0.0147
Make7                  -0.32118    0.19126   -1.68   0.0942
Make8                   0.20958    0.19631    1.07   0.2866
Make9                   0.12545    0.18686    0.67   0.5025
Bonus                  -0.17806    0.02254   -7.90  6.2e-14

(Dispersion parameter for gaussian family taken to be 0.61102)

    Null deviance: 238.56  on 294  degrees of freedom
Residual deviance: 173.53  on 284  degrees of freedom
AIC: 704.6
```

Notice that there are now important differences between the two models. We see that mileage class given by `Kilometers` is statistically significant in the gamma GLM, but not in the lognormal model. Some of the coefficients are quite different. For example, we see that for make 8, relative to the reference level of make 1, there are $\exp(0.7504) = 2.1178$ times as much payment when using the gamma GLM, while the comparable figure for the lognormal model is $\exp(0.20958) = 1.2332$.

These two models are not nested and have different distributions for the response, which makes direct comparison problematic. The AIC criterion, which is minus twice the maximized likelihood plus twice the number of parameters, has often been used as a way to choose between models. Smaller values are preferred. However, when computing a likelihood, it is common practice to discard parts that are not functions of the parameters. This has no consequence when models with same distribution for the response are compared since the parts discarded will be equal. For responses with different distributions, it is essential that all parts of the likelihood be retained. The large difference in AIC for these two models indicate that this precaution was not taken. Nevertheless, we note that the null deviance for both models is almost the same while the residual deviance is smaller for the gamma GLM. This improvement relative to the null indicates that the gamma GLM should be preferred here. Note that purely numerical comparisons such as this are risky and that some attention to residual diagnostics, scientific context and interpretation is necessary.

We compare the shapes of the distributions for the response using the dispersion estimates from the two models, as seen in Figure 7.2:

```
> x <- seq(0,5,by=0.05)
> plot(x,dgamma(x,1/0.55597,scale=0.55597),type="l",ylab="",
  xlab="",yaxs="i",ylim=c(0,1))
> plot(x,dlnorm(x,meanlog=-0.30551,sdlog=sqrt(0.61102)),type="l",
  ylab="",xlab="",yaxs="i",ylim=c(0,1))
```

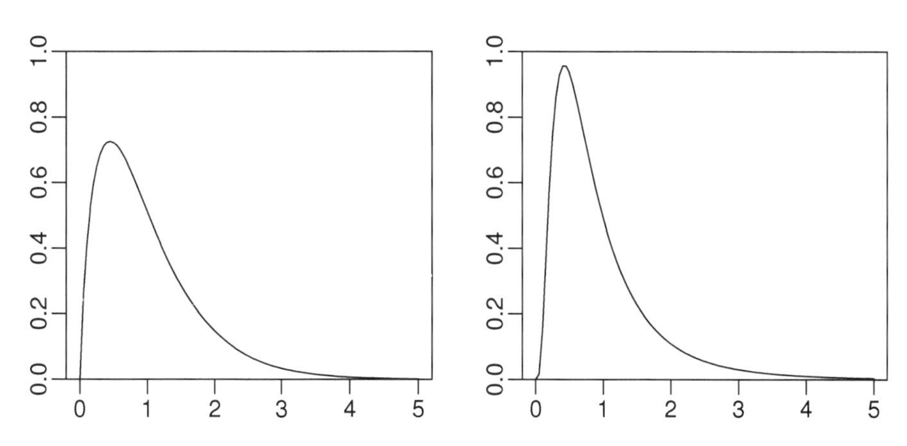

Figure 7.2 *Gamma density for observed shape of 1/0.55597 is shown on the left and lognormal density for an observed SD on the log scale of* $\sqrt{0.61102}$. *The means have been set to one in both cases.*

We see the greater peakedness of the lognormal indicating more small payments which are balanced by more large payments. The lognormal thus has higher kurtosis.

We may also make predictions from both models. Here is a plausible value of the predictors:

```
> x0 <- data.frame(Make="1",Kilometres=1,Bonus=1,Insured=100)
```

and here is predicted response for the gamma GLM:

```
> predict(gl,new=x0,se=T,type="response")
$fit
[1] 63061

$se.fit
[1] 9711.5
```

For the lognormal, we have:

```
> predict(llg,new=x0,se=T,type="response")
$fit
[1] 10.998
```

```
$se.fit
[1] 0.16145
```

so that the corresponding values on the original scale would be:

```
> c(exp(10.998),exp(10.998)*0.16145)
[1] 59754.5  9647.4
```

where we have used the delta method to estimate the standard error on the original scale.

7.2 Inverse Gaussian GLM

The density of an inverse Gaussian random variable, $Y \sim IG(\mu, \lambda)$ is:

$$f(y|\mu,\lambda) = (\lambda/2\pi y^3)^{1/2} \exp[-\lambda(y-\mu)^2/2\mu^2 y] \qquad y,\mu,\lambda > 0$$

The mean is μ and the variance is μ^3/λ. The canonical link is $\eta = 1/\mu^2$ and the variance function is $V(\mu) = \mu^3$. The deviance is given by:

$$D = \sum_i (y_i - \hat{\mu}_i)^2/(\mu_i^2 y_i)$$

Plots of the inverse Gaussian density for a range of values of the shape parameter, λ, are shown in Figure 7.3:

```
> library(SuppDists)
> x <- seq(0,8,by=0.1)
> plot(x,dinvGauss(x,1,0.5),type="l",ylab="",xlab="",ylim=c(0,1.5),
  xaxs="i",yaxs="i")
> plot(x,dinvGauss(x,1,1),type="l",ylab="",xlab="",ylim=c(0,1.5),
  xaxs="i",yaxs="i")
> plot(x,dinvGauss(x,1,5),type="l",ylab="",xlab="",ylim=c(0,1.5),
  xaxs="i",yaxs="i")
```

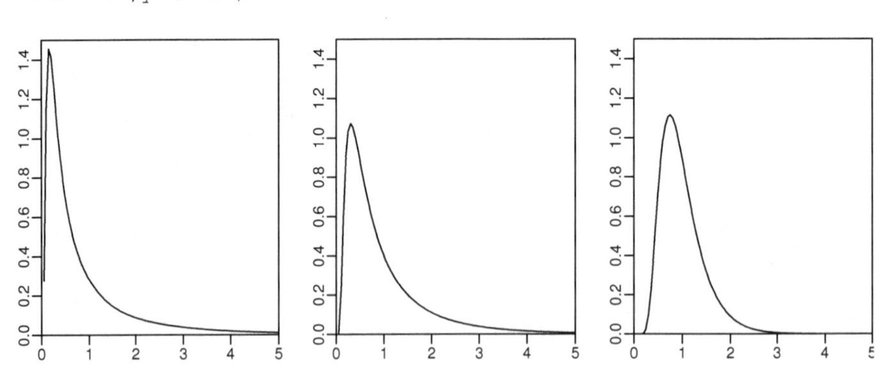

Figure 7.3 *Inverse Gaussian densities for $\lambda = 0.5$ on the left, $\lambda = 1$ in the middle and $\lambda = 5$ on the right. $\mu = 1$ in all three cases.*

The case of $\mu = 1$ is known as the `Wald` distribution. The inverse Gaussian has found application in the modeling of lifetime distributions with nonmonotone failure rates and in the first passage times of Brownian motions with drift. See Seshadri (1993) for a book-length treatment.

Notice that the variance function for the inverse Gaussian GLM increases more rapidly with the mean than the gamma GLM, making it suitable for data where this occurs.

In Whitmore (1986), some sales data on a range of products is presented for the projected, x_i, and actual, y_i, sales for $i = 1, \ldots, 20$. We consider a model, $y_i = \beta x_i$ where β would represent the relative bias in the projected sales. Since the sales vary over a wide range from small to large, a normal error would be unreasonable because Y is positive and violations of this constraint could easily occur. We start with a look at the normal model:

```
> data(cpd)
> lmod <- lm(actual ~ projected-1,cpd)
> summary(lmod)
Coefficients:
          Estimate Std. Error t value Pr(>|t|)
projected   0.9940     0.0172    57.9   <2e-16
> plot(actual ~ projected, cpd)
> abline(lmod)
```

Now consider the inverse Gaussian GLM where we must specify an identity link because we have $y_i = \beta x_i$:

```
> igmod <- glm(actual ~ projected-1,
  family=inverse.gaussian(link="identity"), cpd)
> summary(igmod)
Coefficients:
          Estimate Std. Error t value Pr(>|t|)
projected   1.1036     0.0614    18.0  2.2e-13

(Dispersion parameter for inverse.gaussian family taken to be 0.00017012)

    Null deviance:       Inf  on 20  degrees of freedom
Residual deviance: 0.0030616  on 19  degrees of freedom
> abline(igmod,lty=2)
```

We see that there is a clear difference in the estimates of the slope. The fits are shown in the first panel of Figure 7.4. We should check the diagnostics on the inverse Gaussian GLM:

```
> plot(residuals(igmod) ~ log(fitted(igmod)),ylab="Deviance residuals",
  xlab=expression(log(hat(mu))))
> abline(h=0)
```

We see in the second panel of Figure 7.4 that the variance of the residuals is decreasing with error indicating that the inverse Gaussian variance function is too strong for this data. We have used $\log(\hat{\mu})$ so that the points are more evenly spread horizontally making it easier, in this case, to see the variance relationship. A gamma GLM is a better choice here. In Whitmore (1986), a different variance function is used, but we do not pursue this here as this would not be a GLM.

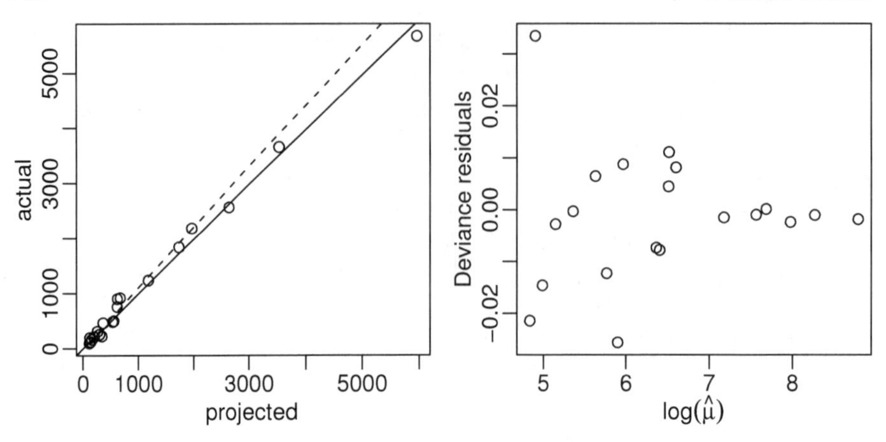

Figure 7.4 *Projected and actual sales are shown for 20 products on the left. The linear model fit is shown as a solid line and the inverse Gaussian GLM fit is shown with a dotted line. A residual-fitted plot for the inverse Gaussian GLM is shown on the right.*

7.3 Joint Modeling of the Mean and Dispersion

All models we have considered so far have modeled the mean response $\mu = EY$ where the variance takes a known form: var $Y_i = \phi V(\mu_i)$ where the dispersion parameter ϕ is the variance in the Gaussian model, the squared coefficient of variation in the gamma model and one in the binomial and Poisson models. We can generalize a little to allowing weights by letting $\phi \equiv \phi_i = \phi w_i$ when the weights are known.

In this section, we are interested in examples where ϕ_i varies with the covariates X. This is a particular issue that arises in industrial experiments. We wish to manufacture an item with a target mean or optimized response. We set the predictors to produce items as close as possible to the target mean or to optimize the mean. This requires a model for the mean. We would also prefer that the variance of the response be small at the chosen value of the predictors for production. So we need to model the variance as a function of the predictors.

We take, as an example, an experiment to determine which recipe will most reliably produce the best cake. The data comes from Box, Bisgaard, and Fung (1988) and is shown in Table 7.1. The objective is to bake a cake reliably no matter how incompetent the cook. For this data, we can see, by examination, that a response of 6.8 is possible for lower flour and shortening content and higher egg content, if the temperature is on the high side and the cooking time on the short side. However, we cannot be sure that the consumer will be able to set the temperature and cooking time correctly. Perhaps their oven is not correctly calibrated or they are just incompetent. If they happen to bake the cake for longer than the set time, they will produce a cake with a 3.5 rating. They will blame the product and not themselves and not buy that mix again. If on the other hand we produce the mix with high flour and eggs and low shortening, the worst the customer can do is a 5.2 and will do better than that for other combinations of time and temperature.

Design Vars				Environmental Vars				
			T	0	−	+	−	+
F	S	E	t	0	−	−	+	+
0	0	0		6.7	3.4	5.4	4.1	3.8
−	−	−		3.1	1.1	5.7	6.4	1.3
+	−	−		3.2	3.8	4.9	4.3	2.1
−	+	−		5.3	3.7	5.1	6.7	2.9
+	+	−		4.1	4.5	6.4	5.8	5.2
−	−	+		6.3	4.2	6.8	6.5	3.5
+	−	+		6.1	5.2	6.0	5.9	5.7
−	+	+		3.0	3.1	6.3	6.4	3.0
+	+	+		4.5	3.9	5.5	5.0	5.4

Table 7.1 *F=Flour, S=Shortening, E=Eggs, T=Oven temperature and t=Baking time. "+" indicates a higher-than-normal setting while "−" indicates a lower-than-normal setting. "0" indicates the standard setting.*

Here we need a combination of a high mean with respect to the design factors, flour, eggs and shortening, and a low variance with respect to the environmental factors, temperature and time. In this example, the answer is easily seen by inspection, but usually more formal model fitting methods will be needed.

Joint Model Specification: We use the standard GLM approach for the mean:

$$EY_i = \mu_i \qquad \eta_i = g(\mu_i) = \sum_j x_{ij}\beta_j \qquad \text{var } Y_i = \phi_i V(\mu_i) \qquad w_i = 1/\phi_i$$

Now the dispersion, ϕ_i, is no longer considered fixed. Suppose we find an estimate, d_i, of the dispersion and model it using a gamma GLM:

$$Ed_i = \phi_i \qquad \zeta_i = \log(\phi_i) = \sum_j z_{ij}\gamma_j \qquad \text{var } d_i = \tau\phi_i^2$$

Notice the connection between the two models. The model for the mean produce the response for the model for the dispersion, which in turn produces the weights for the mean model. In principle, something other than a gamma GLM could be used for the dispersion although since we wish to model a strictly positive, continuous and typically skewed dispersion, the gamma is the obvious choice. The dispersion predictors, Z are usually a subset of the mean model predictors X.

For unreplicated experiments, r_P^2 and r_D^2 are two possible choices for d_i. If replications are available, then a more direct estimate of dispersion would be possible. For more details on the formulation, estimation and inference for these kinds of model see McCullagh and Nelder (1989), Box and Meyer (1986), Bergman and Hynen (1997) and Nelder, Lee, Bergman, Hynen, Huele, and Engel (1998).

In the last three citations, data from a welding-strength experiment was analyzed. There were nine two-level factors and 16 unreplicated runs. Previous analyses have differed on which factors are significant for the mean. We found that two factors, `Drying` and `Material`, were apparently strongly significant, while the significance of others, including `Preheating`, was less clear. We fit a linear model for the mean using these three predictors:

```
> data(weldstrength)
> lmod <- lm(Strength ~ Drying + Material + Preheating, weldstrength)
> summary(lmod)
Coefficients:
            Estimate Std. Error t value Pr(>|t|)
(Intercept)   43.625      0.262  166.25  < 2e-16
Drying         2.150      0.262    8.19  2.9e-06
Material      -3.100      0.262  -11.81  5.8e-08
Preheating    -0.375      0.262   -1.43     0.18

Residual standard error: 0.525 on 12 degrees of freedom
Multiple R-Squared: 0.946,Adjusted R-squared: 0.932
F-statistic: 69.6 on 3 and 12 DF,  p-value: 7.39e-08
```

Following a suggestion of Smyth, Huele, and Verbyla (2001), we use the squared studentized residuals, $(y_i - \hat{y}_i)^2/(1-h_i)$, as the response in the dispersion with a gamma GLM using a log-link and weights of $1-h_i$. Again, we follow the suggestion of some previous authors as to which predictors are important for modeling the dispersion:

```
> h <- influence(lmod)$hat
> d <- residuals(lmod)^2/(1-h)
> gmod <- glm(d ~ Material+Preheating,family=Gamma(link=log),
  weldstrength,weights=1-h)
```

Now feedback the estimated weights to the linear model:

```
> w <- 1/fitted(gmod)
> lmod <- lm(Strength ~ Drying + Material + Preheating,
  weldstrength, weights=w)
```

We now iterate until convergence, where we find that:

```
> summary(lmod)
Coefficients:
            Estimate Std. Error t value Pr(>|t|)
(Intercept)   43.825      0.108  406.83  < 2e-16
Drying         1.869      0.045   41.53  2.5e-14
Material      -3.234      0.108  -30.03  1.2e-12
Preheating    -0.239      0.101   -2.35    0.036

Residual standard error: 1 on 12 degrees of freedom
Multiple R-Squared: 0.995,Adjusted R-squared: 0.994
F-statistic:  877 on 3 and 12 DF,  p-value: 2.56e-14
```

We note that `Preheating` is now significant in contrast to the initial mean model fit. The output for the dispersion model is:

```
> summary(gmod)
Coefficients:
            Estimate Std. Error t value  Pr(>|t|)
(Intercept)   -3.064      0.356   -8.60 0.0000010
Material      -3.037      0.413   -7.35 0.0000056
```

```
Preheating     2.904      0.413    7.03 0.0000089

(Dispersion parameter for Gamma family taken to be 0.50039)

    Null deviance: 57.919  on 15  degrees of freedom
Residual deviance: 20.943  on 13  degrees of freedom
```

The standard errors are not correct in this output and further calculation, described in Smyth, Huele, and Verbyla (2001), would be necessary. This would result in somewhat larger standard errors (about twice the size), but the two factors would still be significant.

7.4 Quasi-Likelihood

Suppose that we are able to specify the link and variance functions of the model for some new type of data, but that we do not have a strong idea about the appropriate distributional form for the response. For example, suppose that we specify an identity link and constant variance. This would be typical in the standard regression setting. We can use least squares to estimate the regression parameters. If we want to do some inference, then formally we need to assume a Gaussian distribution for the errors (or equivalently, the response). We know that the inference is fairly robust to nonnormality especially as the sample size gets larger. The important part of the model specification is the link and variance; the outcome is less sensitive to the distribution of the response.

The same effect holds for other GLMs. Provided we have a larger sample, the results are not sensitive to smaller deviations from the distributional assumptions. The link, variance and independence assumptions are far more important. Now suppose that we were to specify a link and variance function combination that does not correspond to any of the standard GLMs. An examination of the fitting procedure for GLMs reveals that only the link and variance functions are used and no distributional assumptions are necessary. This opens up new modeling possibilities because one might well be able to suggest reasonable link and variance functions but not know a suitable distribution.

Computation of $\hat{\beta}$ and standard errors is often not enough and some form of inference is required. To compute a deviance, we need a likelihood and to compute a likelihood we need a distribution. At this point, we need a suitable substitute for a likelihood that can be computed without assuming a distribution.

Let Y_i have mean μ_i and variance $\phi V(\mu_i)$. We assume that Y_i are independent. We define a score, U_i:

$$U_i = \frac{Y_i - \mu_i}{\phi V(\mu_i)}$$

Now:

$$EU_i = 0$$

$$\text{var } U_i = \frac{1}{\phi V(\mu_i)}$$

$$-E\frac{\partial U_i}{\partial \mu_i} = -E\frac{-\phi V(\mu_i) - (Y_i - \mu_i)\phi V'(\mu_i)}{[\phi V(\mu_i)]^2} = \frac{1}{\phi V(\mu_i)}$$

These properties are shared by the derivative of the log-likelihood, l'. This suggests that we can use U in place of l'. So we define:

$$Q_i = \int_{y_i}^{\mu_i} \frac{y_i - t}{\phi V(t)} dt$$

The intent is that Q should behave like the log-likelihood. We then define the log *quasi-likelihood* for all n observations as:

$$Q = \sum_{i=1}^{n} Q_i$$

The usual asymptotic properties expected of maximum likelihood estimators also hold for quasi-likelihood-based estimators as may be seen in McCullagh (1983).

Notice that the quasi-likelihood depends directly only on the variance function and that the choice of distribution also determines only the variance function. So the choice of variance function is associated with the random structure of the model while the link function determines the relationship with the systematic part of the model.

For the variance functions associated with the members of the exponential family distribution, the quasi-likelihood corresponds exactly to the log-likelihood. However, there is an advantage to using the quasi-likelihood approach for models with variance functions corresponding to the binomial and Poisson distribution. The regular GLMs assume $\phi = 1$ whereas the corresponding *quasi-binomial* and *quasi-Poisson* GLMs allow for the dispersion ϕ to be a free parameter which is useful in modeling overdispersion. One curious possibility is that some choices of $V(\mu)$ may not correspond to a known, or even any, distribution.

$\hat{\beta}$ is obtained by maximizing Q. Everything proceeds as in the standard GLMs except for the estimation of ϕ since the likelihood approach is not reliable here. We recommend:

$$\hat{\phi} = \frac{X^2}{n-p}$$

Although quasi-likelihood estimators are attractive because they require fewer assumptions, they are generally less efficient than the corresponding regular likelihood-based estimator. So if you have information about the distribution, you are advised to use it.

The inferential procedures are similar to those for standard GLMs. Recall that the regular deviance for a model is formed from the difference in log-likelihoods for the model and the saturated model:

$$D(y, \hat{\mu}) = -2\phi \sum_i (l(\hat{\mu}_i|y_i) - l(y_i|y_i))$$

so by analogy the quasi-deviance is $-2\phi Q$ because the contribution from the saturated model is zero. The ϕ cancels, so the quasi-deviance is just:

$$Q = -2\sum_i \int_{y_i}^{\mu_i} \frac{y_i - t}{V(t)} dt$$

In Allison and Cicchetti (1976), data on the sleep behavior of 62 mammals is presented. Suppose we are interested in modeling the proportion of sleep spent dreaming as a function of the other predictors: the weight of the body and the brain, the lifespan, the gestation period and the three constructed indices measuring vulnerability to predation, exposure while sleeping and overall danger:

```
> data(mammalsleep)
> mammalsleep$pdr <- with(mammalsleep, dream/sleep)
> summary(mammalsleep$pdr)
   Min. 1st Qu.  Median    Mean 3rd Qu.    Max.    NA's
  0.000   0.118   0.176   0.186   0.243   0.462  14.000
```

We notice that the proportion of time spent dreaming varies from zero up to almost half the time. A normal model seems inappropriate while transformations are problematic. We attempt to model the proportion response directly. A logit link seems sensible since the response is restricted between zero and one. Furthermore, we might expect the variance to be greater for moderate values of the proportion μ and less as μ approaches zero or one because of the nature of the measurements. This suggests a variance function of the approximate form $\mu(1-\mu)$. This corresponds to the binomial GLM with the canonical logit link and yet the response is not binomial. We propose a quasi-binomial:

```
> modl <- glm(pdr ~ log(body)+log(brain)+log(lifespan)+log(gestation)
          +predation+exposure+danger, family=quasibinomial, mammalsleep)
```

where we have logged many of the predictors because of skewness. Since we now have a free dispersion parameter, we must use F-tests to compare models:

```
> drop1(modl,test="F")
Single term deletions
               Df Deviance F value Pr(F)
<none>                1.57
log(body)       1     1.78    4.51 0.041
log(brain)      1     1.59    0.33 0.568
log(lifespan)   1     1.65    1.79 0.189
log(gestation)  1     1.62    1.15 0.292
predation       1     1.57    0.10 0.749
exposure        1     1.58    0.32 0.575
danger          1     1.58    0.31 0.579
```

We might eliminate `predation` as the least significant variable. Further sequential backward elimination results in:

```
> modl <- glm(pdr ~ log(body)+log(lifespan)+danger,
          family=quasibinomial, mammalsleep )
> summary(modl)
Coefficients:
              Estimate Std. Error t value Pr(>|t|)
(Intercept)    -0.4932     0.2913   -1.69  0.09796
log(body)       0.1463     0.0384    3.81  0.00046
log(lifespan)  -0.2866     0.1080   -2.65  0.01126
danger         -0.1732     0.0600   -2.89  0.00615
```

```
(Dispersion parameter for quasibinomial family taken to be 0.040654)

    Null deviance: 2.5088  on 44  degrees of freedom
Residual deviance: 1.7321  on 41  degrees of freedom
AIC: NA
```

Notice that the dispersion parameter is far less than the default value of one that we would see for a binomial. Furthermore, the AIC is not calculated since we do not have a true likelihood. We see that the proportion of time spent dreaming increases for heavier mammals that live less time and live in less danger. Notice that the relatively large residual deviance compared to the null deviance indicates that this is not a particularly well-fitting model.

The usual diagnostics should be performed. Here are two selected plots that have some interest:

```
> ll <- row.names(na.omit(mammalsleep[,c(1,6,10,11)]))
> halfnorm(cooks.distance(modl),labs=ll)
> plot(predict(modl),residuals(modl,type="pearson"),
  xlab="Linear Predictor", ylab="Pearson Residuals")
```

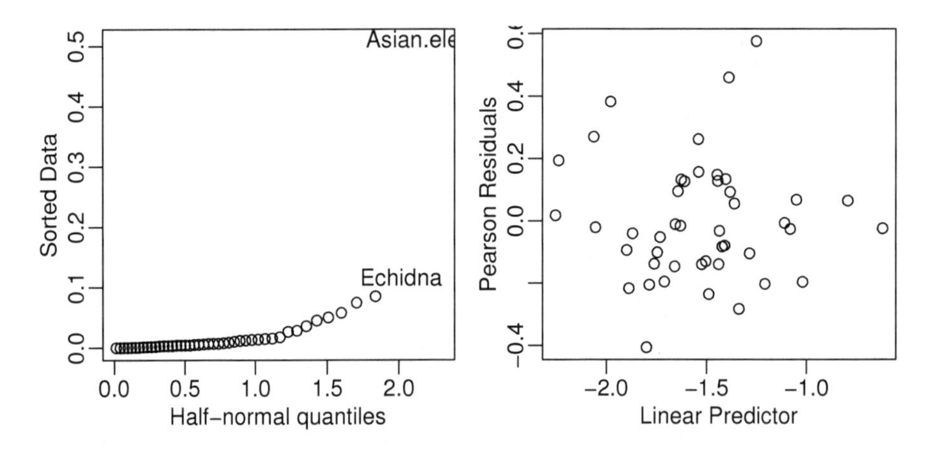

Figure 7.5 *A half-normal plot of the Cook statistics is shown on the left and a plot of the Pearson residuals against the fitted linear predictors is shown on the right.*

In the first panel of Figure 7.5, we see that the Asian elephant is quite influential and a fit without this case should be considered. In the second panel, we see that a pattern of constant variation indicating that our choice of variance function was reasonable. We used the Pearson residuals because these explicitly normalize the raw residuals using the variance function making the check more transparent. Even so, the deviance residuals would have served the same purpose.

Exercises

1. The relationship between corn yield (bushels per acre) and nitrogen (pounds per acre) fertilizer application were studied in Wisconsin. The data may be found in `cornnit`.

   (a) Using (Gaussian) linear model methods, represent the relationship between the yield response and the nitrogen predictor. You will need to find appropriate transformations for the data. Present a quantitative interpretation for the effect of nitrogen fertilizer on yield.

   (b) Now develop a GLM for the data that does not (explicitly) transform the response. Describe quantitatively the relationship between the response and the predictor and compare it to the linear model you found in the previous question.

2. An experiment was conducted as part of an investigation to combat the effects of certain toxic agents. The survival time of rats depended on the type of poison used and the treatment applied. The data is found in `rats`.

   (a) Construct a linear model for the data bearing in mind that some transformation of the response may be necessary and that the possibility of interactions needs to be considered. Interpret the effects of the poisons.

   (b) Build an inverse Gaussian GLM for this data. Select an appropriate link function and perform diagnostics to verify your choices. Interpret the effects of the poisons.

3. Components are attached to an electronic circuit card assembly by a wave-soldering process. The soldering process involves baking and preheating the circuit card and then passing it through a solder wave by conveyor. Defects arise during the process. Design is 2^{7-3} with 3 replicates. The data is found in `wavesolder`.

   Build a pair of models for the mean and dispersion in the number of defects. Investigate which factors are significant in the two models.

4. Data were collected from 39 students in a University of Chicago MBA class and presented in `happy`. Happiness was measured on a 10 point scale. The response could be viewed as ordinal, but a quasi-likelihood approach is also possible. Build a quasi-GLM selecting appropriate link and variance functions. Interpret the effect of the predictors.

5. The `leafblotch` data shows the percentage leaf area affected by leaf blotch on 10 varieties of barley at nine different sites. The data comes from Wedderburn (1974). The data is analyzed in McCullagh and Nelder (1989), which motivates the following questions:

   (a) Fit a quasi-GLM with a logit link and a $\mu(1-\mu)$ variance function. Construct a diagnostic plot that shows that this is not a good choice of variance function.

   (b) A better variance function is $\mu^2(1-\mu)^2$ and yet this is not one of the available choices in R. However, the effect may be obtained by the appropriate use of weighting. Define weights as a function of μ that, when used in conjunction with a variance function of $\mu(1-\mu)$, achieve the effect of a $\mu^2(1-\mu)^2$ variance function. Note that some iteration will be required in the fitting.

(c) Reprogram R to allow for a $\mu^2(1-\mu)^2$ function. (This is more challenging.)

6. One hundred twenty-five fruitflies were divided randomly into five groups of 25 each. The response was the longevity of the fruitfly in days. One group was kept solitary, while another was kept individually with a virgin female each day. Another group was given eight virgin females per day. As an additional control the fourth and fifth groups were kept with one or eight pregnant females per day. Pregnant fruitflies will not mate. The thorax length of each male was measured as this was known to affect longevity. The data is presented in `fruitfly`.

   Show how a gamma GLM may be used to model the lifetimes as a function of the predictors. Interpret your chosen model.

7. The `truck` data concerns an experiment to optimize the production of leaf springs for trucks. A heat treatment is designed so that the free height of the spring should come as close to eight inches as possible. We can vary five factors at two levels each. A 2^{5-1} fractional factorial experiment with three replicates was carried out. The data comes from Pignatiello and Ramberg (1985). McCullagh and Nelder (1989) recommended a Gaussian linear model for the mean response and gamma model for the variance. Fit these models using R and use them to select the best combination of factors given the purpose of the experiment.

CHAPTER 8

Random Effects

Grouped data arise in almost all areas of statistical application. Sometimes the grouping structure is simple, where each case belongs to single group and there is only one grouping factor. More complex datasets have a hierarchical or nested structure or include longitudinal or spatial elements. All such data share the common feature of correlation of observations within the same group and so analyses that assume independence of the observations will be inappropriate. The use of random effects is one common and convenient way to model such grouping structure.

A *fixed effect* is an unknown constant that we try to estimate from the data. Fixed effect parameters are commonly used in linear and generalized linear models as we have presented them earlier in this book. In contrast, a *random effect* is a random variable. It does not make sense to estimate a random effect; instead, we try to estimate the parameters that describe the distribution of this random effect.

Consider an experiment to investigate the effect of several drug treatments on a sample of patients. Typically, we are interested in specific drug treatments and so we would treat the drug effects as fixed. However, it makes most sense to treat the patient effects as random. It is often reasonable to treat the patients as being randomly selected from a larger collection of patients whose characteristics we would like estimate. Furthermore, we are not particularly interested in these specific patients, but in the whole population of patients. A random effects approach to modeling effects is more ambitious in the sense that it attempts to say something about the wider population beyond the particular sample. Blocking factors can often be viewed as random effects, because these often arise as a random sample of those blocks potentially available.

There is some judgment required in deciding when to use fixed and when to use random effects. Sometimes the choice is clear, but in other cases, reasonable statisticians may differ. In some analyses, random effects are used simply to induce a certain correlation structure in the data and there is sense in which the chosen levels represent a sample from a population. Gelman (2005) remarks on the variety of definitions for random effects and proposes a particular straightforward solution to the dilemma of whether to use fixed or random effects — he recommends always using random effects.

A *mixed effects* model has both fixed and random effects. A simple example of such a model would be a two-way analysis of variance (ANOVA):

$$y_{ijk} = \mu + \tau_i + v_j + \varepsilon_{ijk}$$

where the μ and τ_i are fixed effects and the error, ε_{ijk} and the random effects v_j are independent and identically distributed $N(0, \sigma^2)$ and $N(0, \sigma_v^2)$, respectively.

We would want to estimate the τ_i and test the hypothesis $H_0: \tau_i = 0\ \forall i$ while we would estimate σ_v^2 and test $H_0: \sigma_v^2 = 0$. Notice the difference: we need to estimate and test several fixed effect parameters while we need only estimate and test a single random effect parameter.

In the following sections, we consider estimation and inference for mixed-effect models and then illustrate the application to several common designs.

8.1 Estimation

This is not as simple as it was for fixed effects models, where least squares is an easily applied method with many good properties. Let's start with the simplest possible random effects model: a one-way ANOVA design with a factor at a levels:

$$y_{ij} = \mu + \alpha_i + \varepsilon_{ij} \qquad i = 1, \ldots, a \qquad j = 1, \ldots, n$$

where the αs and εs have mean zero, but variances σ_α^2 and σ_ε^2, respectively. These variances are known as the variance components. Notice that this induces a correlation between observations at the same level equal to:

$$\rho = \frac{\sigma_\alpha^2}{\sigma_\alpha^2 + \sigma_\varepsilon^2}$$

This is known as the *intraclass correlation coefficient* (ICC). In the limiting case when there is no variation between the levels, $\sigma_\alpha = 0$ so then $\rho = 0$. Alternatively, when the variation between the levels is much larger than that within the levels, the value of ρ will approach 1. This illustrates how random effects generate correlations between observations.

For simplicity, let there be an equal number of observations n per level. We can decompose the variation as follows:

$$\sum_{i=1}^{a}\sum_{j=1}^{n}(y_{ij} - \bar{y}_{\cdot\cdot})^2 = \sum_{i=1}^{a}\sum_{j=1}^{n}(y_{ij} - \bar{y}_{i\cdot})^2 + \sum_{i=1}^{a}\sum_{j=1}^{n}(\bar{y}_{i\cdot} - \bar{y}_{\cdot\cdot})^2$$

or $SST = SSE + SSA$. SSE is the residual sum of squares, SST is the total sum of squares (corrected for the mean) and SSA is the sum of squares due to α. These quantities are often displayed in an ANOVA table along with the degrees of freedom associated with each sum of squares. Dividing through by the respective degrees of freedom, we obtain the mean squares, MSE and MSA. Now we find that:

$$E(SSE) = a(n-1)\sigma_\varepsilon^2, \qquad E(SSA) = (a-1)(n\sigma_\alpha^2 + \sigma_\varepsilon^2)$$

which suggests using the estimators:

$$\hat{\sigma}_\varepsilon^2 = SSE/(a(n-1)) = MSE, \qquad \hat{\sigma}_\alpha^2 = \frac{SSA/(a-1) - \hat{\sigma}_\varepsilon^2}{n} = \frac{MSA - MSE}{n}$$

This method of estimating variance components can be used for more complex designs. The ANOVA table is constructed, the expected mean squares calculated and the variance components obtained by solving the resulting equations. These estimators are known as *ANOVA estimators*. These were the first estimators developed for variance components. They have the advantage of taking explicit forms suitable for hand calculation which was important in precomputing days, but they have a number of disadvantages:

1. The estimates can take negative values. For example, in our situation above, if $MSA < MSE$ then $\hat{\sigma}^2_\alpha < 0$. This is rather embarrassing since variances cannot be negative. Various fixes have been proposed, but these all take away from the original simplicity of the estimation method.
2. A balanced design has an equal number of observations per cell, where cell is defined as the finest subdivision of the data according to the factors. In such circumstances, the ANOVA decomposition into sums of squares is unique. For unbalanced data, this is not true and we must choose which ANOVA decomposition to use which will in turn affect the estimation of the variance components. Various rules have been suggested about how the decomposition should be done, but none of these have universal appeal.
3. The need for complicated algebraic calculations. Formulae for the simpler models are easy to find and coded in software. More complex models will require difficult and opaque constructions.

We would like a method that would avoid negative variances, work unambiguously for unbalanced data and that can be applied in a transparent and straightforward manner. Maximum likelihood (ML) estimation satisfies these requirements. This does require that we assume some distribution for the errors and the random effects. The usual assumption is normality; ML would work for other distributions, but these are almost never considered.

For a fixed effect model with normal errors, we can write:

$$y = X\beta + \varepsilon \qquad \text{or} \qquad y \sim N(X\beta, \sigma^2 I)$$

where X is an $n \times p$ model matrix and β is a vector of length p. We can generalize to a mixed effect model with a vector γ of q random effects with associated model matrix Z which has dimension $n \times q$. Then we can model the response y, given the value of the random effects as:

$$y = X\beta + Z\gamma + \varepsilon \qquad \text{or} \qquad y|\gamma \sim N(X\beta + Z\gamma, \sigma^2 I)$$

If we further assume that the random effects $\gamma \sim N(0, \sigma^2 D)$ then $\text{var}\, y = \text{var}\, Z\gamma + \text{var}\, \varepsilon = \sigma^2 ZDZ^T + \sigma^2 I$ and we can write the unconditional distribution of y as:

$$y \sim N(X\beta, \sigma^2(I + ZDZ^T))$$

If we knew D, then we could estimate β using generalized least squares; see, for example, Chapter 6 in Faraway (2004). However, the estimation of the variance components, D, is often one purpose of the analysis. Standard maximum likelihood is one method of estimation that can be used here. If we let $V = I + ZDZ^T$, then the joint density for the response is:

$$\frac{1}{2\pi^{n/2}|\sigma^2 V|^{1/2}} e^{-\frac{1}{2\sigma^2}(y-X\beta)^T V^{-1}(y-X\beta)}$$

so that the log-likelihood for the data is:

$$l(\beta,\sigma,D|y) = -\frac{n}{2}\log 2\pi - \frac{1}{2}\log|\sigma^2 V| - \frac{1}{2\sigma^2}(y - X\beta)^T V^{-1}(y - X\beta)$$

This can be optimized to find maximum likelihood estimates of β, σ^2 and D. This is straightforward in principle, but there may be difficulties in practice. More complex models involving larger numbers of random effects parameters can be difficult to estimate. One particular problem is that the variance cannot be negative so the MLE for the variance might be zero. This causes difficulties in the optimization when the unrestricted MLE has a maximum that is negative. This forces us to set that variance estimate to be zero where the derivative of the likelihood will not be zero. This complicates the numerical calculation.

Standard errors can be obtained using the usual large sample theory for maximum likelihood estimates. The variance can be estimated using the inverse of the negative second derivative of the log-likelihood evaluated at the MLE.

MLEs have some drawbacks. One particular problem is that they are biased. For example, consider an i.i.d. sample of normal data $x_1, \ldots, x_n$ then the MLE is:

$$\hat{\sigma}^2 = \frac{\sum_{i=1}^n (x_i - \bar{x})^2}{n}$$

A denominator of $n-1$ is needed for an unbiased estimator. Similar problems occur with the estimation of variance components. Given that the number of levels of a factor may not be large, the bias of the MLE of the variance component associated with that factor may be quite large. *Restricted maximum likelihood* (REML) estimators are an attempt to get round this problem. The idea is to find all independent linear combinations of the response, k, such that $k^T X = 0$. Form matrix K with columns k, so that:

$$K^T y \sim N(0, K^T V K)$$

We can then proceed to maximize the likelihood based on $K^T y$ which does not involve any of the fixed effect parameters. Once the random effect parameters have been estimated, it is simple enough to obtain the fixed effect parameter estimates. REML generally produces less biased estimates. For balanced data, the REML estimates are usually the same as the ANOVA estimates.

We illustrate the fitting methods using some data from an experiment to test the paper brightness depending on a shift operator described in Sheldon (1960). We start with a fixed effects one-way ANOVA:

```
> data(pulp)
> op <- options(contrasts=c("contr.sum", "contr.poly"))
> lmod <- aov(bright ~ operator, pulp)
> summary(lmod)
            Df Sum Sq Mean Sq F value Pr(>F)
operator     3  1.340   0.447     4.2  0.023
Residuals   16  1.700   0.106
> coef(lmod)
(Intercept)   operator1   operator2   operator3
      60.40       -0.16       -0.34        0.22
```

```
> options(op)
```

We have specified sum contrasts here instead of the default treatment contrasts to make the later connection to the corresponding random effects clearer. The `aov` function is just a wrapper for the standard `lm` function that produces results more appropriate for ANOVA models. We see that the operator effect is significant with a p-value of 0.023. The estimate of σ^2 is 0.106 and the estimated overall mean is 60.4. For sum contrasts, $\sum \alpha_i = 0$, so we can calculate the effect for the fourth operator as $0.16 + 0.34 - 0.22 = 0.28$.

Turning to the random effects model, we can compute the variance of the operator effects, σ^2_α, using the formula above as:

```
> (0.447-0.106)/5
[1] 0.0682
```

Now we demonstrate the maximum likelihood estimators. The original R package for fitting mixed effects models was `nlme` as described in Pinheiro and Bates (2000). More recently, Bates (2005) introduced an improved version with the package `lme4` which we shall use here:

```
> library(lme4)
> mmod <- lmer(bright ~ 1+(1|operator), pulp)
> summary(mmod)
Linear mixed-effects model fit by REML
Formula: bright ~ 1 + (1 | operator)
   Data: pulp
    AIC    BIC  logLik deviance REMLdeviance
 24.626 27.613 -9.3131   16.637       18.626
Random effects:
 Groups   Name        Variance Std.Dev.
 operator (Intercept) 0.0681   0.261
 Residual             0.1062   0.326
# of obs: 20, groups: operator, 4

Fixed effects:
            Estimate Std. Error DF t value Pr(>|t|)
(Intercept)   60.400      0.149 19     404   <2e-16
```

The model has fixed and random effects components. The fixed effect here is just the intercept represented by the first `1` in the model formula. The random effect is represented by `(1|operator)` indicating that the data is grouped by `operator` and the `1` indicating that the random effect is constant within each group. The parentheses are necessary to ensure that expression is parsed in the correct order.

The default fitting method is REML. We see that this gives identical estimates to the ANOVA method above — $\hat{\sigma}^2 = 0.106$, $\hat{\sigma}^2_\alpha = 0.068$ and $\hat{\mu} = 60.4$. For unbalanced designs, the REML and ANOVA estimators are not necessarily identical. The standard deviations are simply the square roots of the variances and not estimates of the uncertainty in the variances.

The maximum likelihood estimates may also be computed:

```
> smod <- lmer(bright ~ 1+(1|operator), pulp, method="ML")
> summary(smod)
```

```
Linear mixed-effects model fit by maximum likelihood
Formula: bright ~ 1 + (1 | operator)
   Data: pulp
    AIC    BIC  logLik deviance REMLdeviance
 22.512 25.499 -8.2558   16.512       18.738
Random effects:
 Groups   Name        Variance Std.Dev.
 operator (Intercept) 0.0482   0.219
 Residual             0.1118   0.334
# of obs: 20, groups: operator, 4

Fixed effects:
            Estimate Std. Error DF t value Pr(>|t|)
(Intercept)   60.400      0.129 19     467   <2e-16
```

As can be seen, the between-subjects variance, 0.0482, is smaller than with the REML method. Because the total variance is partitioned, this means the within-subjects variance, 0.1118, is larger than before.

8.2 Inference

Using standard likelihood theory, we may derive a test to compare two nested hypotheses, H_0 and H_1, by computing the likelihood ratio test statistic:

$$2(l(\hat{\beta}_1, \hat{\sigma}_1, \hat{D}_1|y) - l(\hat{\beta}_0, \hat{\sigma}_0, \hat{D}_0|y))$$

where $\hat{\beta}_0, \hat{\sigma}_0, \hat{D}_0$ are the MLEs of the parameters under the null hypothesis and $\hat{\beta}_1, \hat{\sigma}_1, \hat{D}_1$ are the MLEs of the parameters under the alternative hypothesis.

The null distribution of this test statistic is approximately chi-squared with degrees of freedom equal to difference in the dimensions of the two parameters spaces (the difference in the number of parameters when the models are identifiable).

Unfortunately, this test is only approximate and requires several assumptions — see a text such as Cox and Hinkley (1974) for more details. One crucial assumption is that the parameters under the null are not on the boundary of the parameter space. Since we are often interested in testing hypotheses about the random effects that take the form $H_0 : \hat{\sigma}^2 = 0$, this is a real concern. Furthermore, even if the conditions are met, the χ^2 approximation is sometimes poor.

Testing the fixed effects: If you plan to use the likelihood ratio test to compare two nested models that differ only in their fixed effects, you cannot use the REML estimation method. The reason is that REML estimates the random effects by considering linear combinations of the data that remove the fixed effects. If these fixed effects are changed, the likelihoods of the two models will not be directly comparable. Use ordinary maximum likelihood in this situation if you also wish to use the likelihood ratio test.

The *p*-values generated by the likelihood ratio test for fixed effects are approximate and unfortunately tend to be too small, thereby sometimes overstating the importance of some effects. We may use bootstrap methods to find more accurate *p*-values for the likelihood ratio test. The usual bootstrap approach is nonparametric in that no distribution is assumed. Since we are willing to assume normality for the errors and the random effects, we can use a technique called the *parametric bootstrap*. We generate data under the null model using the fitted parameter estimates. We compute the likelihood ratio statistic for this generated data. We repeat this many times and use this to judge the significance of the observed test statistic. This approach will be demonstrated below.

An alternative approach is to condition on the estimated values of the random effect parameters and then use standard F- or t-tests. This assumes that the covariance of the random part of the model, D, is equal to its estimated value and proceeds as one would for generalized least squares.

Testing the random effects: In most cases, a test of random effects will involve a hypothesis of the form $H_0 : \sigma^2 = 0$. The standard derivation of the asymptotic χ^2 distribution for the likelihood ratio statistic depends on the null hypothesis lying in the interior of the parameter space. This assumption is broken when we test if a variance is zero. The null distribution in these circumstances is unknown in general and we must resort to numerical methods if we wish for precise testing. If you do use the χ^2 distribution with the usual degrees of freedom, then the test will tend to be conservative — the p-values will tend to be larger than they should be. This means that if you observe a significant effect using the χ^2 approximation, you can be fairly confident that it is actually significant. Small, but not significant, p-values might spur one to use more accurate, but time-consuming, bootstrap methods.

Expected mean squares: Another method of hypothesis testing is based on the sums of squares found in the ANOVA decompositions. These tests are sometimes more powerful than their likelihood ratio test equivalents. However, the correct derivation of these tests usually requires extensive tedious algebra that must be recalculated for each type of model. Furthermore, the tests cannot be used (at least without complex and unsatisfactory adjustments) when the experiment is unbalanced.

Now let's demonstrate these methods on the `pulp` data. The fixed effect analysis shows that the operator effects are statistically significant with a p-value of 0.023. A random effects analysis using the expected mean squares approach yields exactly the same F-statistic for the one-way ANOVA.

We can also employ the likelihood ratio approach. Because we are testing the random effects, we can use either ML or REML. For fixed effects, we must use ML. In this example, the only fixed effect is the mean and there is no interest in testing that. We first fit the null model:

```
> nullmod <- lm(bright ~ 1, pulp)
```

As there are no random effects in this model, we must use `lm`. For models of the same class, we could use `anova` to compute the LRT and its p-value. Here, we need to compute this directly:

```
> as.numeric(2*(logLik(smod)-logLik(nullmod)))
[1] 2.5684
> pchisq(2.5684,1,lower=FALSE)
[1] 0.10902
```

The p-value is now well above the 5% significance level. We have used the MLE here — using REML produces a slightly different result, but still above 5%. We cannot say that this result is necessarily wrong, but the use of the χ^2 approximation does cause us to doubt the result.

We can use the parametric bootstrap approach to obtain a more accurate p-value. We need to estimate the probability, given that the null hypothesis is true, of observing an LRT of 2.5684 or greater. Under the null hypothesis, $y \sim N(\mu, \sigma^2)$. A simulation approach would generate data under this model, fit the null and alternative models and then compute the LRT. The process would be repeated a large number of times and the proportion of LRTs exceeding the observed value of 2.5684 would be used to estimate the p-value. In practice, we do not know the true values of μ and σ, but we can use the estimated values; this distinguishes the parametric bootstrap from the simulation approach. We can simulate responses under the null: under the null:

```
> y <- simulate(nullmod)
```

Now taking the data we generate, we fit both the null and alternative models and then compute the LRT. We repeat the process 1000 times:

```
> lrstat <- numeric(1000)
> for(i in 1:1000){
   y <- unlist(simulate(nullmod))
   bnull <- lm(y ~ 1)
   balt <- lmer(y ~ 1 + (1|operator),pulp,method="ML")
   lrstat[i] <- as.numeric(2*(logLik(balt)-logLik(bnull)))
  }
```

We may examine the distribution of the bootstrapped LRTs. We compute the proportion that are close to zero:

```
> mean(lrstat < 0.00001)
[1] 0.7
```

The LRT clearly does not have a χ^2 distribution. There is some discussion of this matter in Stram and Lee (1994), who propose a 50:50 mixture of a χ^2 and a mass at zero. Unfortunately, as we can see, the relative proportions of these two components vary from case to case. Crainiceanu and Ruppert (2004) give a more complete solution to the one-way ANOVA problem, but there is no general and exact result for this and more complex problems. The parametric bootstrap may be the simplest approach. The method we have used above is transparent and could be computed much more efficiently if speed is an issue.

Our estimate p-value is:

```
> mean(lrstat > 2.5684)
[1] 0.02
```

We should compute the standard error for this estimate by:

```
> sqrt(0.02*0.98/1000)
[1] 0.0044272
```

So we can be fairly sure it is under 5%. If in doubt, do some more replications to make sure; this only costs computer time. As it happens, this p-value is close to the fixed effects p-value.

In this example, the random and fixed effect tests gave similar outcomes. However, the hypotheses in random and fixed effects are intrinsically different. To generalize somewhat, it is easier to conclude there is an effect in a fixed effects model since the conclusion applies only to the levels of the factor used in the experiment, while for random effects, the conclusion extends to levels of the factor not considered. Since the range of the random effect conclusions is greater, the evidence necessarily has to be stronger.

8.3 Predicting Random Effects

In a fixed effects model, the effects are represented by parameters and it makes sense to estimate them. For example, in the one-way ANOVA model:

$$y_{ij} = \mu + \alpha_i + \varepsilon_{ij}$$

We can calculate $\hat{\alpha}_i$. We do need to resolve the identifiability problem with the αs and the μ, but once we decide on this, the meaning of the $\hat{\alpha}$s is clear enough. We can then proceed to make further inference such as multiple comparisons of these levels.

In a model with random effects, the αs are no longer parameters, but random variables. Using the standard normal assumption:

$$\alpha_i \sim N(0, \sigma_\alpha^2)$$

It does not makes sense to estimate the α's because they are random variables. So instead, we might think about the expected values. However:

$$E\alpha_i = 0 \qquad \forall i$$

which is clearly not very interesting. If one looks at this from a Bayesian point of view, as described in, for example, Gelman, Carlin, Stern, and Rubin (2003), we have a prior density on the αs and $E\alpha_i = 0$ is just the prior mean. Let f represent density, then the posterior density for α is given by:

$$f(\alpha_i|y) \propto f(y|\alpha_i)f(\alpha_i)$$

We can then find the posterior mean, denoted by $\hat{\alpha}$ as:

$$E(\alpha_i|y) = \int \alpha_i f(\alpha_i|y)d\alpha_i$$

For the general case, this works out to be:

$$\hat{\alpha} = DZ^T V^{-1}(y - X\beta)$$

Now a purely Bayesian approach would specify the parameters of the prior and we could simply compute this. We take an empirical Bayes point of view and substitute the MLEs into D, V and β to obtain the predicted random effects. These may be computed as:

```
> ranef(mmod)$operator
  (Intercept)
a    -0.12194
b    -0.25912
c     0.16767
d     0.21340
```

The predicted random effects are related to the fixed effects. We can show these for all operators as:

```
> (cc <- model.tables(lmod))
Tables of effects


 operator
operator
    a     b     c     d
-0.16 -0.34  0.22  0.28
```

and then compute the ratio to the random effects as:

```
> cc[[1]]$operator/ranef(mmod)$operator
  X.Intercept.
a       1.3121
b       1.3121
c       1.3121
d       1.3121
```

We see that the predicted random effects are exactly in proportion to the fixed effects. Typically, the predicted random effects are smaller and could be viewed as a type of *shrinkage* estimate.

Suppose we wish to predict a new value. If the prediction is to be made for a new operator or unknown operator, the best we can do is give $\hat{\mu} = 60.4$. If we know the operator, then we can combine this with our fixed effects to produce what are known as the *best linear unbiased predictors* (BLUPs) as follows:

```
> fixef(mmod)+ranef(mmod)$operator
  X.Intercept.
a       60.278
b       60.141
c       60.568
d       60.613
```

This means that we have more than one type of residual depending on what fitted values we use. The default predicted values and residuals are from the outermost level of grouping. The usual diagnostic plots are still worthwhile:

```
> qqnorm(resid(mmod),main="")
> plot(fitted(mmod),resid(mmod),xlab="Fitted",ylab="Residuals")
> abline(0,0)
```

The plots are shown in Figure 8.1 and indicate no particular problems. Random effects models are particular sensitive to outliers, depending as they do on variance components, which can be substantially inflated by unusual points. The QQ plot is one way to pick out outliers. We also need the normality for the testing. The residual-fitted plot is also important because we made the assumption that the error variance was constant.

If we had more than four groups, we could also look at the normality of the group level effects and check for constant variance also. With so few groups, it is not possible to do this. Also note that there is no point thinking about multiple comparisons. These are for comparing selected levels of a factor. For a random effect, the levels were randomly selected, so such comparisons have little value.

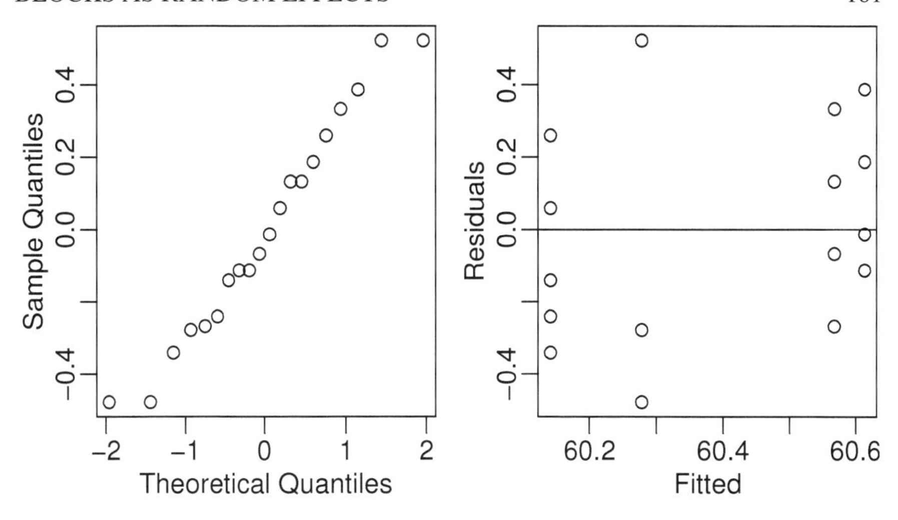

Figure 8.1 *Diagnostic plots for the one-way random effects model.*

8.4 Blocks as Random Effects

Blocks are properties of the experimental units. The blocks are either clearly defined by the conditions of the experiment or they are formed with the judgment of the experimenter. Sometimes, blocks represent groups of runs completed in the same period of time. Typically, we are not interested in the block effects specifically, but must account for their effect. It is therefore natural to treat blocks as random effects.

We illustrate with an experiment to compare four processes, A, B, C and D, for the production of penicillin. These are the treatments. The raw material, corn steep liquor, is quite variable and can only be made in blends sufficient for four runs. Thus a randomized complete block design is suggested by the nature of the experimental units. The data comes from Box, Hunter, and Hunter (1978). We start with the fixed effects analysis:

```
> data(penicillin)
> summary(penicillin)
 treat     blend          yield
 A:5   Blend1:4   Min.    :77
 B:5   Blend2:4   1st Qu.:81
 C:5   Blend3:4   Median :87
 D:5   Blend4:4   Mean    :86
       Blend5:4   3rd Qu.:89
                  Max.    :97
> op <- options(contrasts=c("contr.sum", "contr.poly"))
> lmod <- aov(yield ~ blend + treat, penicillin)
> summary(lmod)
            Df Sum Sq Mean Sq F value Pr(>F)
blend        4  264.0    66.0    3.50  0.041
treat        3   70.0    23.3    1.24  0.339
```

```
Residuals   12  226.0    18.8
> coef(lmod)
(Intercept)       blend1       blend2       blend3       blend4
         86            6           -3           -1            2
     treat1       treat2       treat3
         -2           -1            3
```

From this we see that there is no significant difference between the treatments, but there is between the blends. Now let's fit the data with a mixed model, where we have fixed treatment effects, but random blend effects. This seems natural since the blends we use can be viewed as having been selected from some notional population of blends.

```
> mmod <- lmer(yield ~ treat + (1|blend), penicillin)
> summary(mmod)
Linear mixed-effects model fit by REML
Formula: yield ~ treat + (1 | blend)
   Data: penicillin
    AIC    BIC  logLik deviance REMLdeviance
 118.60 124.58 -53.301   117.28       106.60
Random effects:
 Groups   Name        Variance Std.Dev.
 blend    (Intercept) 11.8     3.43
 Residual             18.8     4.34
# of obs: 20, groups: blend, 5

Fixed effects:
            Estimate Std. Error DF t value Pr(>|t|)
(Intercept)    86.00       1.82 16   47.34   <2e-16
treat1         -2.00       1.68 16   -1.19    0.251
treat2         -1.00       1.68 16   -0.59    0.560
treat3          3.00       1.68 16    1.78    0.093
> options(op)
```

We notice a few connections. The residual variance is the same in both cases: 18.8. This is because we have a balanced design and so REML is equivalent to the ANOVA estimator. The treatment effects are also the same as is the overall mean. The BLUPs for the random effects are:

```
> ranef(mmod)$blend
       (Intercept)
Blend1     4.28788
Blend2    -2.14394
Blend3    -0.71465
Blend4     1.42929
Blend5    -2.85859
```

which, as with the one-way ANOVA, are a shrunken version of the corresponding fixed effects. The usual diagnostics show nothing amiss.

We can test the significance of the fixed effects in two ways. We can use the ANOVA method, where we assume that the random effect parameters take their estimated values:

```
> anova(mmod)
Analysis of Variance Table
      Df Sum Sq Mean Sq Denom F value Pr(>F)
treat  3   70.0    23.3  16.0    1.24   0.33
```

The result is identical with the fixed effects analysis above.

We can also test for a treatment effect using the maximum-likelihood ratio method:

```
> amod <- lmer(yield ~ treat + (1|blend), penicillin, method="ML")
> nmod <- lmer(yield ~ 1 + (1|blend), penicillin, method="ML")
> anova(amod,nmod)
Data: penicillin
Models:
nmod: yield ~ 1 + (1 | blend)
amod: yield ~ treat + (1 | blend)
     Df   AIC   BIC logLik Chisq Chi Df Pr(>Chisq)
nmod  3 127.3 130.3  -60.7
amod  6 129.3 135.3  -58.6  4.05      3       0.26
```

Notice that we needed to use the ML method because comparison of models with different fixed effects is not valid when REML is used. This is because in the REML method, the likelihood of linear combination not involving the fixed effects parameters is maximized. When comparing models with different fixed effects, the linear combinations will be different and the obtained maximum likelihoods will not be comparable. The qualitative outcome of the test is the same as before, but the test itself is numerically different.

We can improve the accuracy with the parametric bootstrap approach. We can generate a response from the null model and use this to compute the LRT. We repeat this 1000 times, saving the LRT each time:

```
> lrstat <- numeric(1000)
> for(i in 1:1000){
   ryield <- unlist(simulate(nmod))
   nmodr <- lmer(ryield ~ 1 + (1|blend), penicillin, method="ML")
   amodr <- lmer(ryield ~ treat + (1|blend), penicillin, method="ML")
   lrstat[i] <- 2*(logLik(amodr)-logLik(nmodr))
  }
```

Under the standard likelihood theory, the LRT here should have a χ^2_3 distribution. We can do a QQ plot to check this:

```
> plot(qchisq((1:1000)/1001,3),sort(lrstat),xlab=expression(chi[3]^2),
  ylab="Simulated LRT")
> abline(0,1)
```

As can be seen in the first panel of Figure 8.2, the approximation is not particularly good. We can compute our estimated p-value as:

```
> mean(lrstat > 4.05)
[1] 0.336
```

which is much closer to the F-test result than the χ^2_3-based approximation.

We can also test the significance of the blends. As with a fixed effects analysis, we are typically not directly interested in size of the blocking effects. Once having

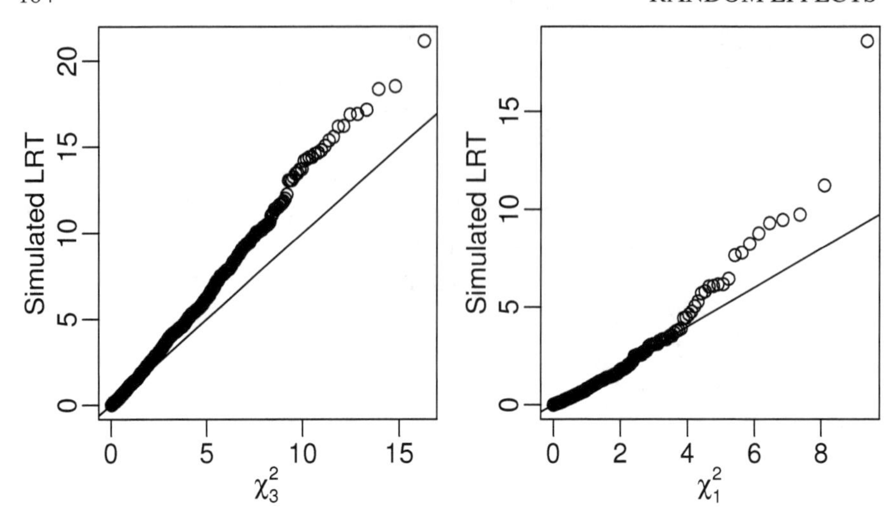

Figure 8.2 *Bootstrapped LRT approximations to the χ^2 distribution. QQ plots of the test statistics for the fixed effects are shown on the left and for the random effects on the right.*

decided to design the experiment with blocks, we must retain them in the model. However, we may wish to examine the blocking effects for information useful for the design of future experiments. We can fit the model with and without random effects and compute the LRT:

```
> rmod <- lmer(yield ~ treat + (1|blend), penicillin)
> nlmod <- lm(yield ~ treat, penicillin)
> 2*(logLik(rmod)-logLik(nlmod,REML=TRUE))
[1] 2.7629
```

We need to specify the nondefault REML option for null model to ensure that the LRT is computed correctly. Now we perform the parametric bootstrap much as before:

```
> lrstatf <- numeric(1000)
> for(i in 1:1000){
   ryield <-  unlist(simulate(nlmod))
   nlmodr <- lm(ryield ~ treat, penicillin)
   rmodr <- lmer(ryield ~ treat + (1|blend), penicillin)
   lrstatf[i] <- 2*(logLik(rmodr)-logLik(nlmodr,REML=TRUE))
  }
```

Again, the distribution is far from χ^2_1 which is clear when we examine the proportion of generated LRTs which are close to zero:

```
> mean(lrstatf < 0.00001)
[1] 0.551
```

Notice that this proportion is different from that observed for the one-way ANOVA illustrating the difficulty in finding a good approximation for the null distribution in such cases. We can also check the distribution of the nonzero LRTs:

```
> cs <- lrstatf[lrstatf > 0.00001]
> ncs <- length(cs)
> plot(qchisq((1:ncs)/(ncs+1),1),sort(cs),xlab=expression(chi[1]^2),
  ylab="Simulated LRT")
> abline(0,1)
```

We see in the right panel of Figure 8.2 that the distribution is close to χ_1^2 except for the tail. We can compute the estimated p-value as:

```
> mean(lrstatf > 2.7629)
[1] 0.043
```

So we find a significant blend effect. The p-value is close to that observed for the fixed effects analysis. Given that the p-value is close to 5%, we might wish to increase the number of bootstrap samples to increase our confidence in the result.

In this example, we saw no major advantage in modeling the blocks as random effects, so we might prefer to use the fixed effects analysis as it is simpler to execute. However, in subsequent analyses, we shall see that the use of random effects will be mandatory as equivalent results may not obtained from a purely fixed effects analysis.

8.5 Split Plots

Split plot designs originated in agriculture, but occur frequently in other settings. As the name implies, main plots are split into several subplots. The main plot is treated with a level of one factor while the levels of some other factor are allowed to vary with the subplots. The design arises as a result of restrictions on a full randomization. For example, a field may be divided into four subplots. It may be possible to plant different varieties in the subplots, but only one type of irrigation may be used for the whole field. Note the distinction between split plots and blocks. Blocks are features of the experimental units which we have the option to take advantage of in the experimental design. Split plots impose restrictions on what assignments of factors are possible. They impose requirements on the design that prevent a complete randomization. Split plots often arise in nonagricultural settings when one factor is easy to change while another factor takes much more time to change. If the experimenter must do all runs for each level of the hard-to-change factor consecutively, a split-plot design results with the hard-to-change factor representing the whole plot factor.

Consider the following example. In an agricultural field trial, the objective was to determine the effects of two crop varieties and four different irrigation methods. Eight fields were available, but only one type of irrigation may be applied to each field. The fields may be divided into two parts with a different variety planted in each half. The whole plot factor is the method of irrigation, which should be randomly assigned to the fields. Within each field, the variety is randomly assigned. Here is a summary of the data:

```
> data(irrigation)
> summary(irrigation)
     field   irrigation variety     yield
 f1     :2   i1:4       v1:8    Min.   :34.8
 f2     :2   i2:4       v2:8    1st Qu.:37.6
```

```
 f3      :2   i3:4                Median :40.1
 f4      :2   i4:4                Mean   :40.2
 f5      :2                       3rd Qu.:42.7
 f6      :2                       Max.   :47.6
 (Other):4
```

The irrigation and variety are fixed effects, but the field is clearly a random effect. We must also consider the interaction between field and variety, which is necessarily also a random effect because one of the two components is random. The fullest model that we might consider is:

$$y_{ijk} = \mu + i_i + v_j + (iv)_{ij} + f_k + (vf)_{jk} + \varepsilon_{ijk}$$

$\mu, i_i, v_j, (iv)_{ij}$ are fixed effects, the rest are random having variances σ_f^2, σ_{vf}^2 and σ_ε^2. Note that we have no $(if)_{ik}$ term in this model. It would not be possible to estimate such an effect since only one type of irrigation is used on a given field; the factors are not crossed.

We may fit this model as follows:

```
> lmod <- lmer(yield ~ irrigation * variety + (1|field) +(1|field:variety),
  data=irrigation)
> logLik(lmod)
'log Lik.' -22.697 (df=11)
```

A simpler model omits the variety by field interaction random effect:

$$y_{ijk} = \mu + i_i + v_j + (iv)_{ij} + f_k + \varepsilon_{ijk}$$

```
> lmodr <- lmer(yield ~ irrigation * variety + (1|field),data=irrigation)
> logLik(lmodr)
'log Lik.' -22.697 (df=10)
```

We see that although the model is simpler, the likelihood is the same. The reason for this is that it is not possible to distinguish the variety within field variation from the error variation. We would need more than one observation per variety within each field for us to separate these two variabilities. Now examine the output of the last model:

```
> summary(lmodr)
Linear mixed-effects model fit by REML
Formula: yield ~ irrigation * variety + (1 | field)
   Data: irrigation
    AIC    BIC  logLik deviance REMLdeviance
 65.395 73.121 -22.697   68.609       45.395
Random effects:
 Groups   Name        Variance Std.Dev.
 field    (Intercept) 16.20    4.02
 Residual              2.11    1.45
# of obs: 16, groups: field, 8

Fixed effects:
                       Estimate Std. Error DF t value  Pr(>|t|)
(Intercept)               38.50       3.03  8   12.73 0.0000014
```

```
irrigationi2              1.20     4.28  8    0.28    0.79
irrigationi3              0.70     4.28  8    0.16    0.87
irrigationi4              3.50     4.28  8    0.82    0.44
varietyv2                 0.60     1.45  8    0.41    0.69
irrigationi2:varietyv2   -0.40     2.05  8   -0.19    0.85
irrigationi3:varietyv2   -0.20     2.05  8   -0.10    0.92
irrigationi4:varietyv2    1.20     2.05  8    0.58    0.57
```

We can see that the largest variance component is that due to the field effect: $\hat{\sigma}_f = 4.02$ with $\hat{\sigma}_\varepsilon = 1.45$. We can check the fixed effects for significance:

```
> anova(lmodr)
Analysis of Variance Table
                   Df Sum Sq Mean Sq Denom F value Pr(>F)
irrigation          3   2.45    0.82  8.00    0.39   0.76
variety             1   2.25    2.25  8.00    1.07   0.33
irrigation:variety  3   1.55    0.52  8.00    0.25   0.86
```

So we see there is no evidence for a fixed effect for either irrigation or variety or their interaction. We should check the diagnostic plots to make sure there is nothing amiss:

```
> plot(fitted(lmodr),resid(lmodr),xlab="Fitted",ylab="Residuals")
> qqnorm(resid(lmodr),main="")
```

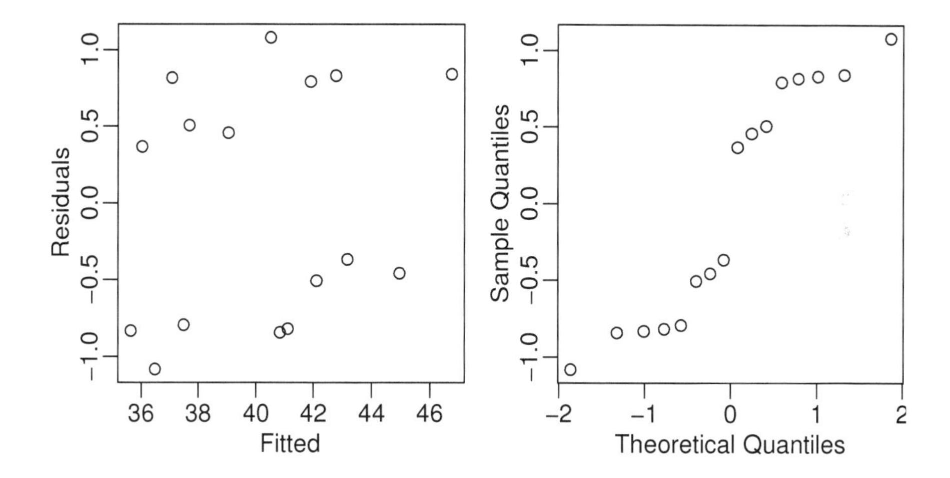

Figure 8.3 *Diagnostic plots for the split plot example.*

We can see in Figure 8.3 that there is no problem with the nonconstant variance, but that the residuals indicate a short-tailed distribution. This type of divergence from normality is unlikely to cause any major problems with the estimation and inference.

Sometimes analysts ignore the split-plot variable as in:

```
> mod <- lm(yield ~ irrigation * variety, data=irrigation)
> anova(mod)
Analysis of Variance Table
```

```
Response: yield
                  Df Sum Sq Mean Sq F value Pr(>F)
irrigation         3   40.2    13.4    0.73   0.56
variety            1    2.2     2.2    0.12   0.73
irrigation:variety 3    1.6     0.5    0.03   0.99
Residuals          8  146.5    18.3
```

The results will not be the same. This last model is incorrect because it fails to take into account the restrictions on the randomization introduced by the fields and the additional variability thereby induced.

8.6 Nested Effects

When the levels of one factor vary only within the levels of another factor, that factor is said to be *nested*. For example, when measuring the performance of workers at several different job locations, if the workers only work at one location, the workers are nested within the locations. If the workers work at more than one location, then the workers are *crossed* with locations.

Here is an example to illustrate nesting. Consistency between laboratory tests is important and yet the results may depend on who did the test and where the test was performed. In an experiment to test levels of consistency, a large jar of dried egg powder was divided up into a number of samples. Because the powder was homogenized, the fat content of the samples is the same, but this fact is withheld from the laboratories. Four samples were sent to each of six laboratories. Two of the samples were labeled as G and two as H, although in fact they were identical. The laboratories were instructed to give two samples to two different technicians. The technicians were then instructed to divide their samples into two parts and measure the fat content of each. So each laboratory reported eight measures, each technician four measures, that is, two replicated measures on each of two samples. The data comes from Bliss (1967):

```
> data(eggs)
> summary(eggs)
      Fat            Lab     Technician Sample
 Min.    :0.060   I  :8    one:24     G:24
 1st Qu.:0.307   II :8    two:24     H:24
 Median :0.370   III:8
 Mean    :0.388   IV :8
 3rd Qu.:0.430   V  :8
 Max.    :0.800   VI :8
```

Although the technicians have been labeled "one" and "two," they are two different people in each lab. Thus the technician factor is nested within laboratories. Furthermore, even though the samples are labeled "H" and "G," these are not the same samples across the technicians and the laboratories. Hence we have samples nested within technicians. Technicians and samples should be treated as random effects since we may consider these as randomly sampled. If the labs were specifically selected, then they should be taken a fixed effects. If, however, they were randomly selected from those available, then they should be treated as random effects. If the purpose of the study is come to some conclusion about consistency across laboratories, the latter approach is advisable.

For the purposes of this analysis, we will treat labs as random. So all our effects (except the grand mean) are random. The model is:

$$y_{ijkl} = \mu + L_i + T_{ij} + S_{ijk} + \varepsilon_{ijkl}$$

This can be fit using:

```
> cmod <- lmer(Fat ~ 1 + (1|Lab) + (1|Lab:Technician) +
  (1|Lab:Technician:Sample), data=eggs)
> summary(cmod)
Linear mixed-effects model fit by REML
Formula: Fat ~ 1 + (1 | Lab) + (1 | Lab:Technician) +
         (1 | Lab:Technician:Sample)
   Data: eggs
     AIC     BIC logLik deviance REMLdeviance
 -54.235 -44.879 32.118  -68.703      -64.235
Random effects:
 Groups                Name        Variance Std.Dev.
 Lab:Technician:Sample (Intercept) 0.00306  0.0554
 Lab:Technician        (Intercept) 0.00698  0.0835
 Lab                   (Intercept) 0.00592  0.0769
 Residual                          0.00720  0.0848
# of obs: 48, groups: Lab:Technician:Sample, 24;
          Lab:Technician, 12; Lab, 6

Fixed effects:
            Estimate Std. Error DF t value Pr(>|t|)
(Intercept)    0.388      0.043 47    9.02    8e-12
```

So we have $\hat{\sigma}_L = 0.077$, $\hat{\sigma}_T = 0.084$, $\hat{\sigma}_S = 0.055$ and $\hat{\sigma}_\varepsilon = 0.085$. So all four variance components are of a similar magnitude. The lack of consistency in measures of fat content can be ascribed to variance between labs, variance between technicians, variance in measurement due to different labeling and just plain measurement error. We can see if the model can be simplified by removing the lowest level of the variance components:

```
> cmodr <- lmer(Fat ~ 1 + (1|Lab) + (1|Lab:Technician), data=eggs)
> anova(cmod,cmodr)
Data: eggs
Models:
cmodr: Fat ~ 1 + (1 | Lab) + (1 | Lab:Technician)
cmod: Fat ~ 1 + (1 | Lab) + (1 | Lab:Technician) +
      (1 | Lab:Technician:Sample)
      Df   AIC   BIC logLik Chisq Chi Df Pr(>Chisq)
cmodr  4 -59.1 -51.6   33.5
cmod   5 -58.7 -49.3   34.4   1.6      1       0.21
```

We see that we cannot reject $H_0 : \sigma_S^2 = 0$. However, we know that this p-value is conservative and the true value will be somewhat lower. An examination of the reduced model is interesting:

```
> VarCorr(cmodr)
 Groups           Name        Variance Std.Dev.
 Lab:Technician (Intercept) 0.00800  0.0895
 Lab            (Intercept) 0.00592  0.0769
 Residual                   0.00924  0.0961
```

The variation due to samples has been absorbed into the other components.

As before, it is worth checking the accuracy of these p-values. We generate data under the null model and compute the LRT 1000 times:

```
> lrstat <- numeric(1000)
> for(i in 1:1000){
   rFat <- unlist(simulate(cmodr))
   nmod <- lmer(rFat ~ 1 + (1|Lab) + (1|Lab:Technician), data=eggs)
   amod <- lmer(rFat ~ 1 + (1|Lab) + (1|Lab:Technician) +
   (1|Lab:Technician:Sample), data=eggs)
   lrstat[i] <- 2*(logLik(amod)-logLik(nmod))
  }
```

As before, we can see that the LRT does not have a null χ^2 distribution because, in this case, 55% of the generated LRTs are close to zero:

```
> mean(lrstat < 0.00001)
[1] 0.55
```

We can estimate the p-value as:

```
> 2*(logLik(cmod)-logLik(cmodr))
[1] 1.6034
> mean(lrstat > 1.6034)
[1] 0.092
```

So we can reasonably say that the variation due to samples can be ignored. We may now test the significance of the variation between technicians. Using the same method above, this is found to be significant.

8.7 Crossed Effects

Effects are said to be crossed when they are not nested. In full factorial designs, effects are completely crossed because every level of one factor occurs with every level of another factor. However, in some other designs, crossing is less-than-complete. Even if just two levels of two factors occur in all four combinations, the factors are crossed. An example of less than complete crossing is a latin square design, where there is one treatment factor and two blocking factors. Although not all combinations of factors occur, the blocking factors are not nested. When at least some crossing occurs, methods for nested designs cannot be used. We consider a latin square example.

In an experiment reported by Davies (1954), four materials, A, B, C and D, were fed into a wear-testing machine. The response is the loss of weight in 0.1 mm over the testing period. The machine could process four samples at a time and past experience indicated that there were some differences due to the position of these four samples.

Also some differences were suspected from run to run. A fixed effects analysis of this dataset may be found in Faraway (2004). Four runs were made. The latin square structure of the design may be observed:

```
> data(abrasion)
> matrix(abrasion$material,4,4)
     [,1] [,2] [,3] [,4]
[1,] "C"  "A"  "D"  "B"
[2,] "D"  "B"  "C"  "A"
[3,] "B"  "D"  "A"  "C"
[4,] "A"  "C"  "B"  "D"
```

A fixed effects analysis of the data reveals:

```
> lmod <- aov(wear ~ material + run + position, abrasion)
> summary(lmod)
            Df Sum Sq Mean Sq F value  Pr(>F)
material     3   4622    1540   25.15 0.00085
run          3    986     329    5.37 0.03901
position     3   1468     489    7.99 0.01617
Residuals    6    367      61
```

All the effects are significant. However, we might regard the run and position as random effects. The appropriate model is then:

```
> mmod <- lmer(wear ~ material + (1|run) + (1|position), abrasion)
> anova(mmod)
Analysis of Variance Table
         Df Sum Sq Mean Sq Denom F value   Pr(>F)
material  3   4621    1540    12    25.1 0.000018
> summary(mmod)
Linear mixed-effects model fit by REML
Formula: wear ~ material + (1 | run) + (1 | position)
   Data: abrasion
    AIC    BIC  logLik MLdeviance REMLdeviance
 114.26 119.66 -50.128     120.43       100.26
Random effects:
 Groups   Name        Variance Std.Dev.
 run      (Intercept)  66.9     8.18
 position (Intercept) 107.1    10.35
 Residual              61.2     7.83
# of obs: 16, groups: run, 4; position, 4

Fixed effects:
            Estimate Std. Error DF t value Pr(>|t|)
(Intercept)   265.75       7.67 12   34.66  2.1e-13
materialB     -45.75       5.53 12   -8.27  2.7e-06
materialC     -24.00       5.53 12   -4.34  0.00097
materialD     -35.25       5.53 12   -6.37  3.6e-05
```

The `lmer` function is able to recognize that the run and position effects are crossed and fits the model appropriately. The F-test for the fixed effects is almost the same as the corresponding fixed effects analysis. The only difference is that the fixed effects analysis uses a denominator degrees of freedom of six while the random effects analysis is made conditional on the estimated random effects parameters which results in 12 degrees of freedom. The difference is not crucial here.

The significance of the random effects could be tested using the parametric bootstrap method. However, since the design of this experiment has already restricted the randomization to allow for these effects, there is no motivation to make these tests since we will not modify the analysis of this current experiment.

The fixed effects analysis was somewhat easier to execute, but the random effects analysis has the advantage of producing estimates of the variation in the blocking factors which will be more useful in future studies. Fixed effects estimates of the run effect for this experiment are only useful for the current study.

8.8 Multilevel Models

Multilevel models is a term used for models for data with hierarchical structure. The term is most commonly used in the social sciences. We can use the methodology we have already developed to fit some of these models.

We take as our example some data from the Junior School Project collected from primary (U.S. term is elementary) schools in inner London. The data is described in detail in Mortimore, Sammons, Stoll, Lewis, and Ecob (1988) and a subset is analyzed extensively in Goldstein (1995).

The variables in the data are the `school`, the `class` within the school (up to four), `gender`, `social` class of the father (I=1; II=2; III nonmanual=3; III manual=4; IV=5; V=6; Long-term unemployed=7; Not currently employed=8; Father absent=9), `raven`'s test in year 1, student `id` number, `english` test score, `math`ematics test score and school `year` (coded 0, 1, and 2 for years one, two and three). So there are up to three measures per student. The data was obtained from the *Multilevel Models project* at `http://www.ioe.ac.uk/multilevel/`.

We shall take as our response the math test score result from the final year and try to model this as a function of gender, social class and the Raven's test score from the first year which might be taken as a measure of ability when entering the school. We subset the data to ignore the math scores from the first two years:

```
> data(jsp)
> jspr <- jsp[jsp$year==2,]
```

We start with two plots of the data. Due to the discreteness of the score results, it is helpful to *jitter* (add small random perturbations) the scores to avoid overprinting:

```
> plot(jitter(math)~jitter(raven),data=jspr,xlab="Raven score",
  ylab="Math score")
> boxplot(math ~ social, data=jspr,xlab="Social class",ylab="Math score")
```

In Figure 8.4, we can see the positive correlation between the Raven's test score and the final math score. The maximum math score was 40 which reduces the variability at the upper end of the scale. We also see how the math scores tend to decline with social class.

One possible approach to analyzing these data is multiple regression. For example, we could fit:

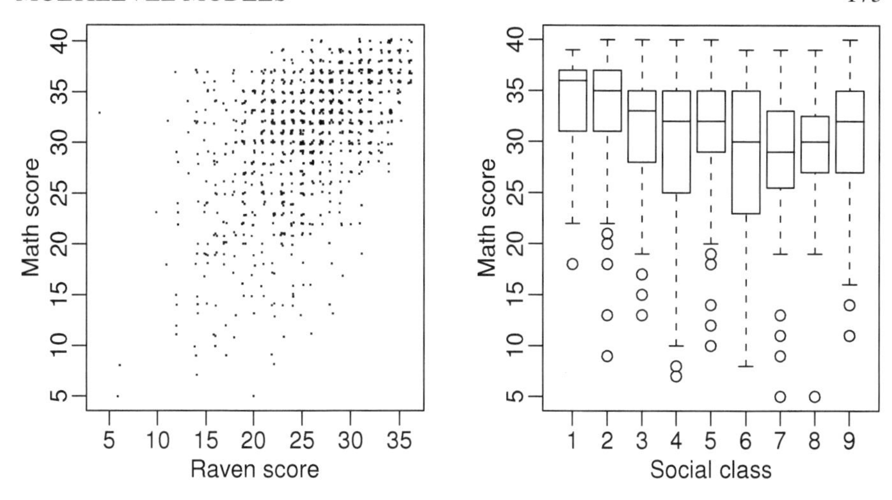

Figure 8.4 *Plots of the Junior School Project data.*

```
> glin <- lm(math ~ raven*gender*social,jspr)
> anova(glin)
Analysis of Variance Table

Response: math
                     Df  Sum Sq Mean Sq F value Pr(>F)
raven                 1   11481   11481  368.06 <2e-16
gender                1      44      44    1.41 0.2347
social                8     779      97    3.12 0.0017
raven:gender          1 0.01145 0.01145 0.00037 0.9847
raven:social          8     583      73    2.33 0.0175
gender:social         8     450      56    1.80 0.0727
raven:gender:social   8     235      29    0.94 0.4824
Residuals           917   28603      31
```

It would seem that gender effects can be removed entirely, giving us:

```
> glin <- lm(math ~ raven*social,jspr)
> anova(glin)
Analysis of Variance Table

Response: math
              Df Sum Sq Mean Sq F value Pr(>F)
raven          1  11481   11481  365.72 <2e-16
social         8    778      97    3.10 0.0019
raven:social   8    564      71    2.25 0.0222
Residuals    935  29351      31
```

This is a fairly large dataset, so even small effects can be significant. Even though the `raven:social` term is significant at the 5% level, we remove it to simplify interpretation:

```
> glin <- lm(math ~ raven+social,jspr)
> summary(glin)
Coefficients:
            Estimate Std. Error t value Pr(>|t|)
(Intercept)  17.0248     1.3745   12.39   <2e-16
raven         0.5804     0.0326   17.83   <2e-16
social2       0.0495     1.1294    0.04    0.965
social3      -0.4289     1.1957   -0.36    0.720
social4      -1.7745     1.0599   -1.67    0.094
social5      -0.7823     1.1892   -0.66    0.511
social6      -2.4937     1.2609   -1.98    0.048
social7      -3.0485     1.2907   -2.36    0.018
social8      -3.1175     1.7749   -1.76    0.079
social9      -0.6328     1.1273   -0.56    0.575

Residual standard error: 5.63 on 943 degrees of freedom
Multiple R-Squared: 0.291,Adjusted R-squared: 0.284
F-statistic: 42.9 on 9 and 943 DF,  p-value: <2e-16
```

We see that the final math score is strongly related to the entering Raven score and that the math scores of the lower social classes are lower, even after adjustment for the entering score. Of course, any regression analysis requires more investigation than this; there are diagnostics and transformations to be considered and more. However, even if we were to do this, there would still be a problem with this analysis: We are assuming that the 953 students in the dataset are independent observations. This is not a tenable assumption as the students come from 50 different schools. The number coming from each school varies:

```
> table(jspr$school)

 1  2  3  4  5  6  7  8  9 10 11 12 13 14 15 16 17 18 19 20 21
26 11 14 24 26 18 11 27 21  0 11 23 22 13  7 16  6 18 14 13 28
22 23 24 25 26 27 28 29 30 31 32 33 34 35 36 37 38 39 40 41 42
14 18 21 14 20 22 15 13 27 35 23 44 27 16 28 17 12 14 10 10 41
44 45 46 47 48 49 50
 5 11 15 33 63 22 14
```

It is highly likely that students in the same school (and perhaps) class will show some dependence. So we have somewhat less than 953 independent cases worth of information. Any analysis that pretends these are independent is likely to overstate the significance of the results. Furthermore, the analysis above tells us nothing about the variation between and within schools. People will certainly be interested in this. We could aggregate the results across schools but this would lose information and expose us to the dangers of an ecological regression.

We need an analysis that uses the individual-level information, but also reflects the grouping in the data. Our first model has fixed effects representing all interactions between `raven`, `social` and `gender` with random effects for the school and the class nested within the school:

```
> mmod <- lmer(math ~ raven*social*gender+(1|school)+(1|school:class),
  data=jspr)
```

```
> anova(mmod)
Analysis of Variance Table
                    Df Sum Sq Mean Sq Denom F value Pr(>F)
raven                1  10218   10218   917  374.40 <2e-16
social               8    616      77   917    2.82 0.0043
gender               1     22      22   917    0.79 0.3738
raven:social         8    577      72   917    2.64 0.0072
raven:gender         1      2       2   917    0.09 0.7639
social:gender        8    275      34   917    1.26 0.2605
raven:social:gender  8    187      23   917    0.86 0.5524
```

Again, it seems that gender is not important and so we simplify to:

```
> jspr$craven <- jspr$raven-mean(jspr$raven)
> mmod <- lmer(math ~ craven*social+(1|school)+(1|school:class),jspr)
> summary(mmod)
Linear mixed-effects model fit by REML
Formula: math ~ craven * social + (1 | school) + (1 | school:class)
   Data: jspr
    AIC    BIC  logLik deviance REMLdeviance
 5963.2 6065.2 -2960.6   5907.4       5921.2
Random effects:
 Groups       Name        Variance Std.Dev.
 school:class (Intercept)  1.18    1.08
 school       (Intercept)  3.15    1.77
 Residual                 27.14    5.21
# of obs: 953, groups: school:class, 90; school, 48

Fixed effects:
               Estimate Std. Error  DF t value Pr(>|t|)
(Intercept)     31.9112     1.1955 935   26.69   <2e-16
craven           0.6058     0.1885 935    3.21   0.0014
social2          0.0236     1.2722 935    0.02   0.9852
social3         -0.6307     1.3089 935   -0.48   0.6300
social4         -1.9670     1.1971 935   -1.64   0.1007
social5         -1.3585     1.3002 935   -1.04   0.2964
social6         -2.2687     1.3737 935   -1.65   0.0990
social7         -2.5518     1.4055 935   -1.82   0.0698
social8         -3.3950     1.8014 935   -1.88   0.0598
social9         -0.8313     1.2535 935   -0.66   0.5074
craven:social2  -0.1321     0.2058 935   -0.64   0.5212
craven:social3  -0.2243     0.2189 935   -1.02   0.3057
craven:social4   0.0358     0.1949 935    0.18   0.8542
craven:social5  -0.1503     0.2089 935   -0.72   0.4719
craven:social6  -0.0386     0.2326 935   -0.17   0.8682
craven:social7   0.3983     0.2318 935    1.72   0.0861
craven:social8   0.2560     0.2615 935    0.98   0.3279
craven:social9  -0.0810     0.2055 935   -0.39   0.6935
```

We centered the Raven score about its overall mean. This means that we can interpret the social effects as the predicted differences from social class one at the mean Raven score. If we did not do this, these parameter estimates would represent differences for raven=0 which is not very useful. We can see the math score is strongly related to the entering Raven score. We see that for the same entering score, the final math score tends to be lower as social class goes down. Note that class 9 here is when the father is absent and class 8 is not necessarily worse than 7, so this factor is not entirely ordinal. We also see the most substantial variation at the individual level with smaller amounts of variation at the school and class level.

We check the standard diagnostics first:

```
> qqnorm(resid(mmod),main="")
> plot(fitted(mmod),resid(mmod),xlab="Fitted",ylab="Residuals")
```

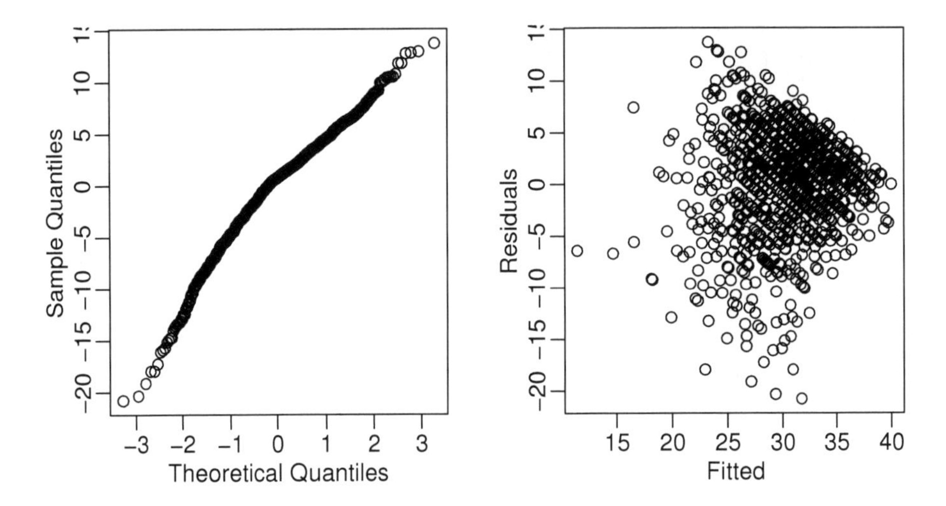

Figure 8.5 *Diagnostic plots for the Junior Schools Project model.*

In Figure 8.5, we see that the residuals are close to normal, but there is a clear decrease in the variance with an increase in the fitted values. This is due to the reduced variation in higher scores already observed. We might consider a transformation of the response to remove this effect.

We can also check the assumption of normally distributed random effects. We can do this at the school and class level:

```
> qqnorm(ranef(mmod)$school[[1]],main="School effects")
> qqnorm(ranef(mmod)$"school:class"[[1]],main="Class effects")
```

We see approximate normality in both cases with some evidence of short tails for the school effects. It is interesting to look at the sorted school effects:

```
> adjscores <- ranef(mmod)$school[[1]]
```

These represent a ranking of the schools adjusted for the quality of the intake and the social class of the students. The difference between the best and the worst is about 5 points on the math test. Of course, we must recognize that there is variability in these estimated effects before making any decisions about the relative strengths of

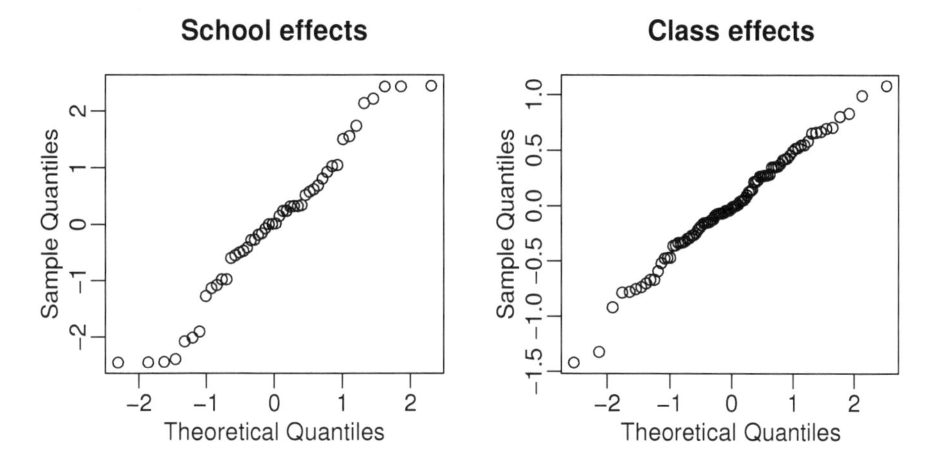

Figure 8.6 *QQ plots of the random effects at the school and class levels.*

these schools. Compare this with an unadjusted ranking that simply takes the average score achieved by the school, centered by the overall average:

```
> rawscores <- coef(lm(math ~ school-1,jspr))
> rawscores <- rawscores-mean(rawscores)
```

We compare these two measures of school quality in Figure 8.7:

```
> plot(rawscores,adjscores)
> sint <- c(9,14,29)
> text(rawscores[sint],adjscores[sint]+0.2,c("9","15","30"))
```

School 10 is listed but has no students hence the need to adjust the labeling. There are some interesting differences. School 15 looks best on the raw scores but after adjustment, it drops to 15th place. This is a school that apparently performs well, but when the quality of the incoming students is considered, its performance is not so impressive. School 30 illustrates the other side of the coin. This school looks average on the raw scores, but is doing quite well given the ability of the incoming students. School 9 is actually doing a poor job despite raw scores that look quite good.

It is also worth plotting the residuals and the random effects against the predictors. We would be interested in finding any inhomogeneity or signs of structure that might lead to an improved model.

Compositional effects: Fixed effect predictors in this example so far have been at the lowest level, the student, but it is not improbable that factors at the school or class level might be important predictors of success in the math test. We can construct some such predictors from the individual-level information; such factors are called *compositional effects*. For example, the average entering score for a school might be an important predictor. The ability of one's fellow students may have an impact on future achievement. We construct this variable:

```
> schraven <- lm(raven ~ school, jspr)$fit
```

and insert it into our model:

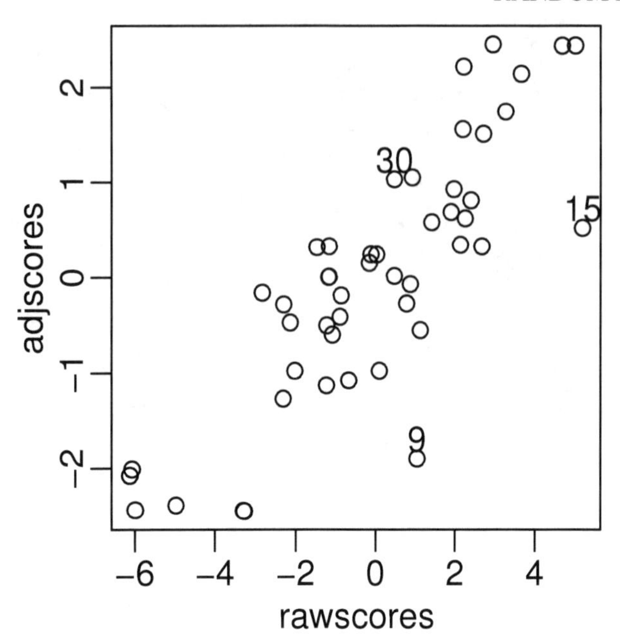

Figure 8.7 *Raw and adjusted school-quality measures. Three selected schools are marked.*

```
> mmodc <- lmer(math ~ craven*social+schraven*social+(1|school)+
  (1|school:class),jspr)
> anova(mmodc)
Analysis of Variance Table
                Df Sum Sq Mean Sq Denom F value Pr(>F)
craven           1  10166   10166   926  373.23 <2e-16
social           8    610      76   926    2.80 0.0045
schraven         1      5       5   926    0.18 0.6699
craven:social    8    561      70   926    2.58 0.0088
social:schraven  8    166      21   926    0.76 0.6353
```

We see that this new effect is not significant. We are not constrained to taking means. We might consider various quantiles or measures of spread as potential compositional variables.

Much remains to be investigated with this dataset. We have only used the simplest of error structures and we should investigate whether the random effects may also depend on some of the other covariates.

Further Reading: The classical approach to random effects can be found in many older books such as Snedecor and Cochran (1989) or Scheffé (1959). More recent books such as Searle, Casella, and McCulloch (1992) also focus on the ANOVA approach. A wide range of models are explicitly considered in Milliken and Johnson (1992). Multilevel models are covered in Goldstein (1995) and Raudenbush and Bryk

(2002). The predecessor to the `lme4` package was `nlme` which is described in Pinheiro and Bates (2000), but the book still contains much general material of interest.

Exercises

1. Use the `pulp` dataset for this question.

   (a) Analyze the data as a fixed effects model. Is the operator significant?

   (b) Analyze the data with operator as a random effect. What are the estimated variances?

   (c) Compute confidence intervals for these variances.

   (d) Compute the intraclass correlation coefficient.

   (e) Determine the significance of the operator effect using a likelihood ratio test taking care to compute the p-value accurately.

2. The `coagulation` dataset comes from a study of blood coagulation times. Twenty-four animals were randomly assigned to four different diets and the samples were taken in a random order.

   (a) A new animal is assigned to diet D. Predict the blood coagulation time for this animal along with an estimate of the variability in this prediction.

   (b) A new diet is given to a new animal. Predict the blood coagulation time for this animal along with an estimate of the variability in this prediction.

   (c) A new diet is given to the first animal in the dataset. Predict the blood coagulation time for this animal along with an estimate of the variability in this prediction. You may assume that the effects of the initial diet for this animal have washed out.

3. The `eggprod` dataset concerns an experiment where six pullets were placed into each of 12 pens. Four blocks were formed from groups of three pens based on location. Three treatments were applied. The number of eggs produced was recorded.

   (a) Fit a model for the number of eggs produced with the treatments as fixed effects and the blocks as random effects. Describe the estimated differences between the treatments.

   (b) Test for the significance of the treatment. Compute the p-value using both the χ^2 distribution and resampling methods.

4. Data on the cutoff times of lawnmowers may be found in the dataset `lawn`. 3 machines were randomly selected from those produced by manufacturers A and B. Each machine was tested twice at low speed and high speed.

   (a) Fit a mixed effects model with manufacturer and speed as main effects along with their interaction and machine nested in manufacturer as random effects. Write down the formula for the model. Explain why this model contains redundant terms. Discuss the choices for simplifying the model and the issues that would motivate the decision.

(b) Determine if the manufacturer term can be removed from the fixed part of the model.

(c) Determine if the manufacturer term can be removed from the random part of the model.

5. A number of growers supply broccoli to a food processing plant. The plant instructs the growers to pack the broccoli into standard-size boxes. There should be 18 clusters of broccoli per box and each cluster should weigh between 1.33 and 1.5 pounds. Because the growers use different varieties and methods of cultivation, there is some variation in the cluster weights. The plant manager selected three growers at random and then four boxes at random supplied by these growers. Three clusters were selected from each box. The data may be found in the `broccoli` dataset. The weight in grams of the cluster is given. Estimate and test the variance components.

6. An experiment was conducted to select the supplier of raw materials for production of a component. The breaking strength of the component was the objective of interest. Four suppliers were considered. The four operators can only produce one component each per day. A latin square design is used and the data is presented in `breaking`.

   (a) Explain why it would be natural to treat the operators and days as random effects but the suppliers as fixed effects.

   (b) Build a model to predict the breaking strength. Describe the variation from operator to operator and from day to day.

   (c) Test the significance of the supplier effect.

   (d) Is there a significant difference between the operators?

   (e) For the best choice of supplier, predict what proportion of components produced in the future will have a breaking strength exceeding 900.

7. An experiment was conducted to optimize the manufacture of semiconductors. The `semicond` data has the resistance recorded on the wafer as the response. The experiment was conducted during four different time periods denoted by ET and three different wafers during each period. The position on the wafer is a factor with levels 1 to 4. The `Grp` variable is a combination of ET and wafer. Analyze the data as a split plot experiment where ET and position are considered as fixed effects. Since the wafers are different in experimental time periods, the `Grp` variable should be regarded as the block or group variable. Determine the best model for the data and check all appropriate diagnostics.

8. Redo the Junior Schools Project data analysis in the text with the final year English score as the response. Highlight any differences from the analysis of the final year Math scores.

9. An experiment was conducted to determine the effect of recipe and baking temperature on chocolate cake quality. Fifteen batches of cake mix for each recipe were prepared. Each batch was sufficient for six cakes. Each of the six cakes was baked at a different temperature which was randomly assigned. Several measures

of cake quality were recorded of which breaking angle was just one. The dataset is presented as `choccake`.

Build an appropriate model for the data and write a report on the analysis.

CHAPTER 9

Repeated Measures and Longitudinal Data

In repeated measures designs, there are several individuals and measurements are taken repeatedly on each individual. When these repeated measurements are taken over time, it is called a *longitudinal* study or, in some applications, a *panel* study. Typically various covariates concerning the individual are recorded and the interest centers on how the response depends on the covariates over time. Often it is reasonable to believe that the response of each individual has several components: a fixed effect, which is a function of the covariates; a random effect, which expresses the variation between individuals; and an error, which is due to measurement or unrecorded variables.

Suppose each individual has response y_i, a vector of length n_i which is modeled conditionally on the random effects γ_i as:

$$y_i|\gamma_i \sim N(X_i\beta + Z_i\gamma_i, \sigma^2\Lambda_i)$$

Notice this is very similar to the model used in the previous chapter with the exception of allowing the errors to have a more general covariance Λ_i. As before, we assume that the random effects $\gamma_i \sim N(0, \sigma^2 D)$ so that:

$$y_i \sim N(X_i\beta, \Sigma_i)$$

where $\Sigma_i = \sigma^2(\Lambda_i + Z_i D Z_i^T)$. Now suppose we have M individuals and we can assume the errors and random effects between individuals are uncorrelated, then we can combine the data as:

$$y = \begin{bmatrix} y_1 \\ y_2 \\ \cdots \\ y_M \end{bmatrix} \qquad X = \begin{bmatrix} X_1 \\ X_2 \\ \cdots \\ X_M \end{bmatrix} \qquad \gamma = \begin{bmatrix} \gamma_1 \\ \gamma_2 \\ \cdots \\ \gamma_M \end{bmatrix}$$

and $\tilde{D} = diag(D, D, \ldots, D)$, $Z = diag(Z_1, Z_2, \ldots, Z_M)$, $\Sigma = diag(\Sigma_1, \Sigma_2, \ldots, \Sigma_M)$, and $\Lambda = diag(\Lambda_1, \Lambda_2, \ldots, \Lambda_M)$. Now we can write the model simply as

$$y \sim N(X\beta, \Sigma) \qquad \Sigma = \sigma^2(\Lambda + Z\tilde{D}Z^T)$$

The log-likelihood for the data is then computed as above and estimation, testing, standard errors and confidence intervals all follow using standard likelihood theory as before. In fact, there is no strong distinction between the methodology used in this and the previous chapter.

Of course, this general structure encompasses a wide range of possible models for different types of data. We explore some of these in the following three examples:

9.1 Longitudinal Data

The Panel Study of Income Dynamics (PSID), begun in 1968, is a longitudinal study of a representative sample of U.S. individuals described in Hill (1992). The study is conducted at the Survey Research Center, Institute for Social Research, University of Michigan, and is still continuing. There are currently 8700 households in the study and many variables are measured. We chose to analyze a random subset of this data, consisting of 85 heads of household who were aged 25–39 in 1968 and had complete data for at least 11 of the years between 1968 and 1990. The variables included were annual income, gender, years of education and age in 1968:

```
> data(psid)
> head(psid)
  age educ sex income year person
1  31   12   M   6000   68      1
2  31   12   M   5300   69      1
3  31   12   M   5200   70      1
4  31   12   M   6900   71      1
5  31   12   M   7500   72      1
6  31   12   M   8000   73      1
```

Now plot the data:

```
> library(lattice)
> xyplot(income ~ year | person, psid, type="l",
  subset=(person < 21),strip=FALSE)
```

The first 20 subjects are shown in Figure 9.1. We see that some individuals have a slowly increasing income, typical of someone in steady employment in the same job. Other individuals have more erratic incomes. We can also show how the incomes vary by sex. Income is more naturally considered on a log-scale:

```
> xyplot(log(income+100) ~ year | sex, psid, type="l", groups=person)
```

See Figure 9.2. We added \$100 to the income of each subject to remove the effect of some subjects having very low incomes for short periods of time. These cases distorted the plots without the adjustment. We see that men's incomes are generally higher and less variable while women's incomes are more variable, but are perhaps increasing more quickly. We could fit a line to each subject starting with the first:

```
> lmod <- lm(log(income) ~ I(year-78), subset=(person==1), psid)
> coef(lmod)
 (Intercept) I(year - 78)
    9.399957     0.084267
```

We have centered the predictor at the median value so that the intercept will represent the predicted log income in 1978 and not the year 1900 which would be nonsense. We now fit a line for all the subjects and plot the results:

```
> slopes <- numeric(85);intercepts <- numeric(85)
> for(i in 1:85){
     lmod <- lm(log(income) ~ I(year-78), subset=(person==i), psid)
     intercepts[i] <- coef(lmod)[1]
     slopes[i] <- coef(lmod)[2]
     }
```

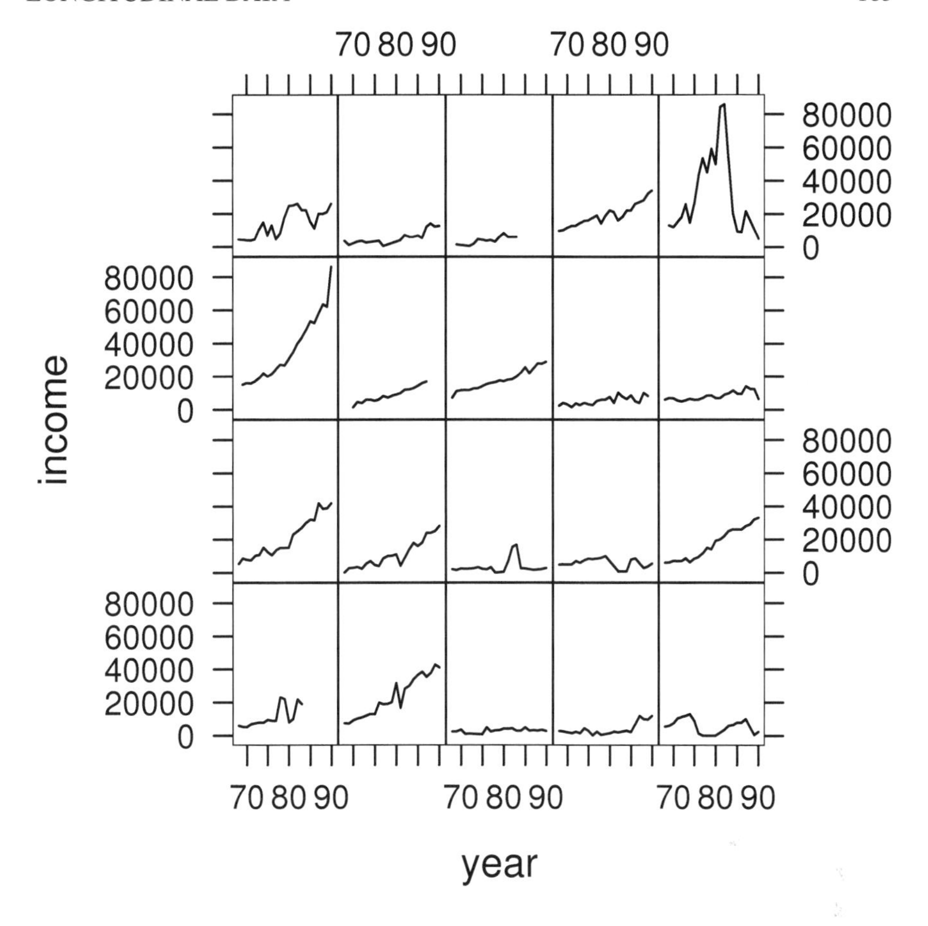

Figure 9.1 *The first 20 subjects in the PSID data. Income is shown over time.*

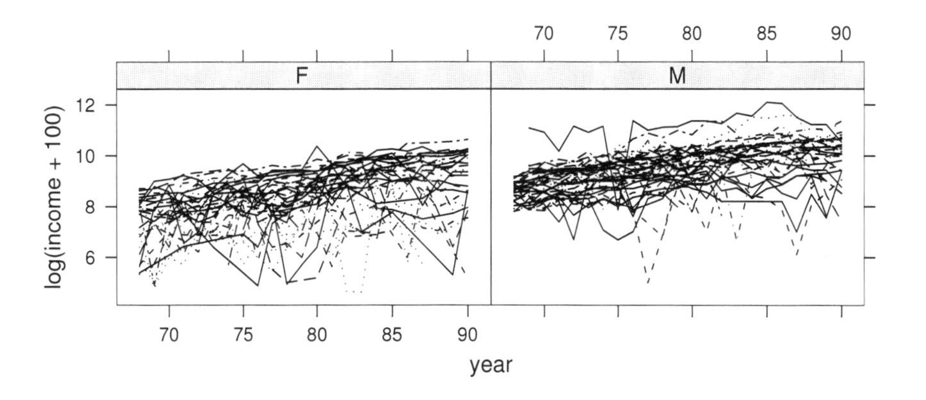

Figure 9.2 *Income change in the PSID data grouped by sex.*

```
> plot(intercepts,slopes,xlab="Intercept",ylab="Slope")
> psex <- psid$sex[match(1:85,psid$person)]
> boxplot(split(slopes,psex))
```

See Figure 9.3. We can simply compare the income growth rates for men and women:

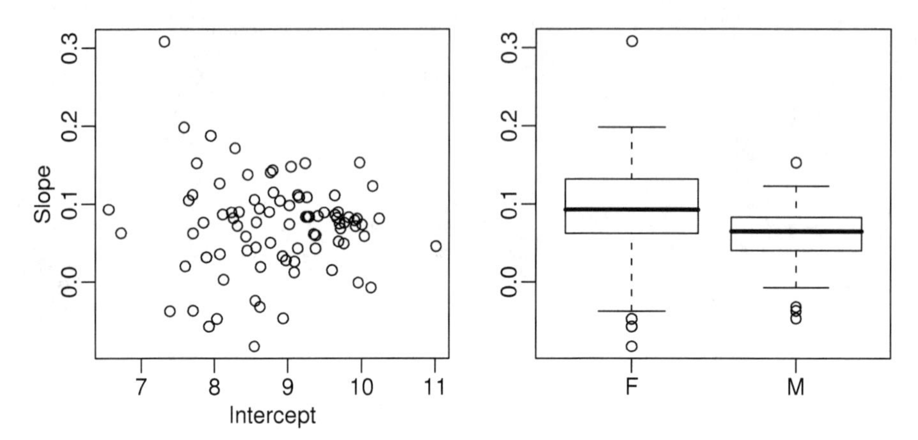

Figure 9.3 *Slopes and intercepts for the individual income growth relationships are shown on the left. A comparison of income growth rates by sex is shown on the right.*

```
> t.test(slopes[psex=="M"],slopes[psex=="F"])

Welch Two Sample t-test

data:  slopes[psex == "M"] and slopes[psex == "F"]
t = -2.3786, df = 56.736, p-value = 0.02077
alternative hypothesis: true difference in means is not equal to 0
95 percent confidence interval:
 -0.0591687 -0.0050773
sample estimates:
mean of x mean of y
 0.056910  0.089033
```

We see that women have a significantly higher growth rate than men. We can also compare the incomes at the intercept (which is 1978):

```
> t.test(intercepts[psex=="M"],intercepts[psex=="F"])

Welch Two Sample t-test

data:  intercepts[psex == "M"] and intercepts[psex == "F"]
t = 8.2199, df = 79.719, p-value = 3.065e-12
alternative hypothesis: true difference in means is not equal to 0
95 percent confidence interval:
 0.87388 1.43222
sample estimates:
```

```
mean of x mean of y
   9.3823    8.2293
```

We see that men have significantly higher incomes.

This is an example of a *response feature* analysis. It requires choosing an important characteristic. We have chosen two here: the slope and the intercept. For many datasets, this is not an easy choice and at least some information is lost by doing this.

Response feature analysis is attractive because of its simplicity. By extracting a univariate response for each individual, we are able to use a wide array of well-known statistical techniques. However, it is not the most efficient use of the data as all the additional information besides the chosen response feature is discarded. Notice that having additional data on each subject would be of limited value.

Suppose that the income change over time can be partly predicted by the subject's age, sex and educational level. We do not expect a perfect fit. The variation may be partitioned into two components. Clearly there are other factors that will affect a subject's income. These factors may cause the income to be generally higher or lower or they may cause the income to grow at a faster or slower rate. We can model this variation with a random intercept and slope, respectively, for each subject. We also expect that there will be some year-to-year variation within each subject. For simplicity, let us initially assume that this error is homogeneous and uncorrelated, that is, $\Lambda_i = I$. We also center the year to aid interpretation as before. We may express these notions in the model:

```
> library(lme4)
> psid$cyear <- psid$year-78
> mmod <- lmer(log(income) ~ cyear*sex +age+educ+(cyear|person),psid)
```

This model can be written as:

$$\begin{aligned}\log(\text{income})_{ij} &= \mu + \beta_y \text{year}_i + \beta_s \text{sex}_j + \beta_{ys} \text{sex}_j \times \text{year}_i + \beta_e \text{educ}_j + \beta_a \text{age}_j \\ &+ \gamma_j^0 + \gamma_j^1 \text{year}_i + \varepsilon_{ij}\end{aligned}$$

where i indexes the year and j indexes the individual. We have:

$$\begin{pmatrix} \gamma_k^0 \\ \gamma_k^1 \end{pmatrix} \sim N(0, \sigma^2 D)$$

The model summary is:

```
> summary(mmod)
Linear mixed-effects model fit by REML
Formula: log(income) ~ cyear * sex + age + educ + (cyear | person)
   Data: psid
    AIC    BIC  logLik MLdeviance REMLdeviance
 3839.8 3893.9 -1909.9     3785.6       3819.8
Random effects:
 Groups   Name        Variance Std.Dev. Corr
 person   (Intercept) 0.2817   0.531
          cyear       0.0024   0.049    0.187
 Residual             0.4673   0.684
# of obs: 1661, groups: person, 85
```

```
Fixed effects:
             Estimate Std. Error   DF t value  Pr(>|t|)
(Intercept)    6.6742     0.5433 1655   12.28   < 2e-16
cyear          0.0853     0.0090 1655    9.48   < 2e-16
sexM           1.1503     0.1213 1655    9.48   < 2e-16
age            0.0109     0.0135 1655    0.81     0.419
educ           0.1042     0.0214 1655    4.86 0.0000013
cyear:sexM    -0.0263     0.0122 1655   -2.15     0.032
```

Let's start with the fixed effects. We see that income increases about 10% for each additional year of education. We see that age does not appear to be significant. For females, the reference level in this example, income increases about 8.5% a year, while for men, it increases about 8.5-2.6=5.9% a year. We see that, for this data, the incomes of men are $\exp(1.15) = 3.16$ times higher.

We know the mean for males and females, but individuals will vary about this. The standard deviation for the intercept and slope are 0.531 and 0.049 ($\sigma\sqrt{D_{11}}$ and $\sigma\sqrt{D_{22}}$), respectively. These have a correlation of 0.19 ($cor(\gamma^0, \gamma^1)$). Finally, there is some additional variation in the measurement not so far accounted for having standard deviation of 0.684 ($sd(\varepsilon_{ijk})$). We see that the variation in increase in income is relatively small while the variation in overall income between individuals is quite large. Furthermore, given the large residual variation, there is a large year-to-year variation in incomes.

There is a wider range of possible diagnostic plots that can be made with longitudinal data than with a standard linear model. In addition to the usual residuals, there are random effects to be examined. We may wish to break the residuals down by sex as seen in the QQ plots in Figure 9.4:

```
> qqmath(~resid(mmod) | sex, psid)
```

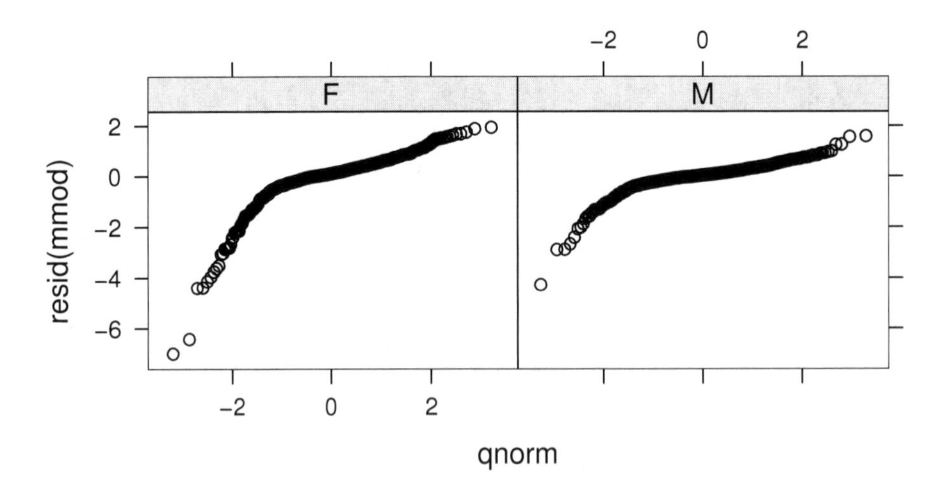

Figure 9.4 *QQ plots by sex.*

We see that the residuals are not normally distributed, but have a long tail for the lower incomes. We should consider changing the log transformation on the response. Furthermore, we see that there is greater variance in the female incomes. This suggests a modification to the model. We can make the same plot broken down by subject although there will be rather too many plots to be useful.

Plots of residuals and fitted values are also valuable. We have broken education into three levels: less than high school, high school or more than high school:

```
> xyplot(resid(mmod) ~ fitted(mmod) | cut(educ,c(0,8.5,12.5,20)),
  psid, layout=c(3,1),xlab="Fitted",ylab="Residuals")
```

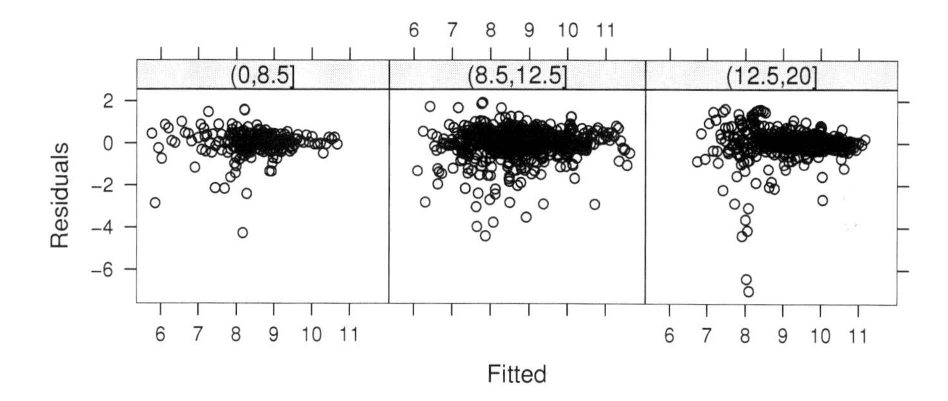

Figure 9.5 *Residuals vs. fitted plots for three levels of education: less than high school on the left, high school in the middle and more than high school on the right.*

See Figure 9.5. Again, we can see evidence that a different response transformation should be considered.

Plots of the random effects would also be useful here.

9.2 Repeated Measures

The acuity of vision for seven subjects was tested. The response is the lag in milliseconds between a light flash and a response in the cortex of the eye. Each eye is tested at four different powers of lens. An object at the distance of the second number appears to be at distance of the first number. The data is given in Table 9.1. The data comes from Crowder and Hand (1990) and was also analyzed by Lindsey (1999).

We start by making some plots of the data. We create a numerical variable representing the power to complement the existing factor so that we can see how the acuity changes with increasing power:

```
> data(vision)
> vision$npower <- rep(1:4,14)
> xyplot(acuity~npower|subject,vision,type="l",groups=eye,
  lty=1:2,layout=c(4,2))
```

See Figure 9.6. There is no apparent trend or difference between right and left eyes.

Power							
6/6	6/18	6/36	6/60	6/6	6/18	6/36	6/60
Left				Right			
116	119	116	124	120	117	114	122
110	110	114	115	106	112	110	110
117	118	120	120	120	120	120	124
112	116	115	113	115	116	116	119
113	114	114	118	114	117	116	112
119	115	94	116	100	99	94	97
110	110	105	118	105	105	115	115

Table 9.1 *Visual acuity of seven subjects measured in milliseconds of lag in responding to a light flash. The power of the lens causes an object six feet distant to appear at a distance of 6, 18, 36 or 60 feet.*

However, individual #6 appears anomalous with a large difference between the eyes. It also seems likely that the third measurement on the left eye is in error for this individual.

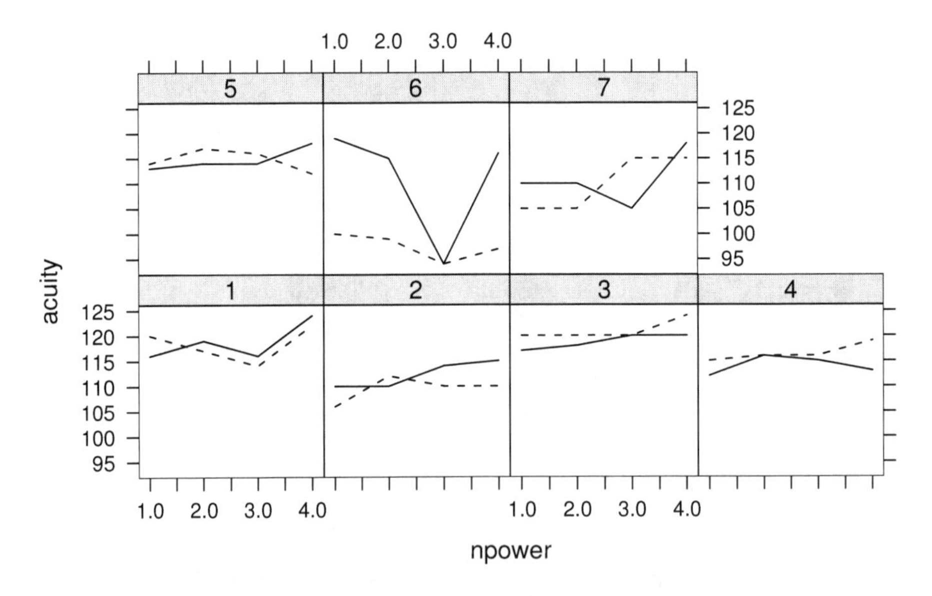

Figure 9.6 *Visual acuity profiles. The left eye is shown as a solid line and the right as a dashed line. Four powers of lens are displayed where 1=6/6, 2=6/18, 3=6/36 and 4=6/60.*

We must now decide how to model the data. The power is a fixed effect. In the model below, we have treated it as a nominal factor, but we could try fitting it in a quantitative manner. The subjects should be treated as random effects. Since we do not believe there is any consistent right-left eye difference between individuals, we should treat the eye factor as nested within subjects. We start with this model:

```
> mmod <- lmer(acuity~power+(1|subject)+(1|subject:eye),vision)
```

Note that if we did believe there was a consistent left vs. right eye effect, we would have used crossed random effects, putting `(1|eye)` in place of `(1|subject:eye)`.

We can write this (nested) model as:

$$y_{ijk} = \mu + p_j + s_i + e_{ik} + \varepsilon_{ijk}$$

where $i = 1, \ldots, 7$ runs over individuals, $j = 1, \ldots, 4$ runs over power and $k = 1, 2$ runs over eyes. The p_j term is a fixed effect, but the remaining terms are random. Let $s_i \sim N(0, \sigma_s^2)$, $e_{ik} \sim N(0, \sigma_e^2)$ and $\varepsilon_{ijk} \sim N(0, \sigma^2\Sigma)$ where we take $\Sigma = I$. The summary output is:

```
> summary(mmod)
Linear mixed-effects model fit by REML
Formula: acuity ~ power + (1 | subject) + (1 | subject:eye)
   Data: vision
    AIC    BIC  logLik MLdeviance REMLdeviance
 342.71 356.89 -164.35     339.22       328.71
Random effects:
 Groups      Name        Variance Std.Dev.
 subject:eye (Intercept) 10.3     3.21
 subject     (Intercept) 21.5     4.64
 Residual                16.6     4.07
# of obs: 56, groups: subject:eye, 14; subject, 7

Fixed effects:
            Estimate Std. Error DF t value Pr(>|t|)
(Intercept)  112.643      2.235 52   50.40   <2e-16
power6/18      0.786      1.540 52    0.51    0.612
power6/36     -1.000      1.540 52   -0.65    0.519
power6/60      3.286      1.540 52    2.13    0.038
```

We see that the estimated standard deviation for subjects is 4.64 and that for eyes for a given subject is 3.21. The residual standard deviation is 4.07. The random effects structure we have used here induces a correlation between measurements on the same subject and another between measurements on the same eye. We can compute these two correlations respectively as:

```
> 4.64^2/(4.64^2+3.21^2+4.07^2)
[1] 0.44484
> (4.64^2+3.21^2)/(4.64^2+3.21^2+4.07^2)
[1] 0.65774
```

As we might expect, there is a stronger correlation between observations on the same eye than between the left and right eyes of the same individual.

We can check for a power effect:

```
> anova(mmod)
Analysis of Variance Table
      Df Sum Sq Mean Sq Denom F value Pr(>F)
power  3  140.8    46.9  52.0    2.83  0.048
```

We see the result is just significant at the 5% level. We might expect some trend in acuity with power, but the estimated effects do not fit with this trend. While acuity is greatest at the highest power, 6/60, it is lowest for the second highest power, 6/36. A look at the data makes one suspect the measurement made on the left eye of the sixth subject at this power. If we omit this observation and refit the model, we find:

```
> mmodr <- lmer(acuity~power+(1|subject)+(1|subject:eye),vision,subset=-43)
> anova(mmodr)
Analysis of Variance Table
      Df Sum Sq Mean Sq Denom F value Pr(>F)
power  3   89.2    29.7  51.0     3.6  0.020
> summary(mmodr)
Fixed effects:
            Estimate Std. Error DF t value Pr(>|t|)
(Intercept)  112.643      1.880 51   59.91   <2e-16
power6/18      0.786      1.087 51    0.72   0.4731
power6/36      0.521      1.114 51    0.47   0.6418
power6/60      3.286      1.087 51    3.02   0.0039
```

Now the power effect is significant, but it appears this is due to an effect at the highest power only. We can test the hypothesis that the highest power has a higher acuity than the average of the first three levels by using Helmert contrasts:

```
> op <- options(contrasts=c("contr.helmert", "contr.poly"))
> mmodr <- lmer(acuity~power+(1|subject)+(1|subject:eye),vision,subset=-43)
> summary(mmodr)
Fixed effects:
            Estimate Std. Error DF t value Pr(>|t|)
(Intercept) 113.7911     1.7596 51   64.67   <2e-16
power1        0.3929     0.5436 51    0.72   0.4731
power2        0.0428     0.3242 51    0.13   0.8954
power3        0.7125     0.2228 51    3.20   0.0024
> options(op)
```

The Helmert contrast matrix is

```
> contr.helmert(4)
  [,1] [,2] [,3]
1   -1   -1   -1
2    1   -1   -1
3    0    2   -1
4    0    0    3
```

We can see that the third contrast (column) represents the difference between the average of the first three levels and the fourth level, scaled by a factor of three. In the output, we can see that this is statistically significant while the other two contrasts are not.

We finish with some diagnostic plots. The residuals and fitted values and the QQ plot of random effects for the eyes are shown in Figure 9.7:

```
> plot(resid(mmodr) ~ fitted(mmodr),xlab="Fitted",ylab="Residuals")
> abline(h=0)
> qqnorm(ranef(mmodr)$"subject:eye"[[1]],main="")
```

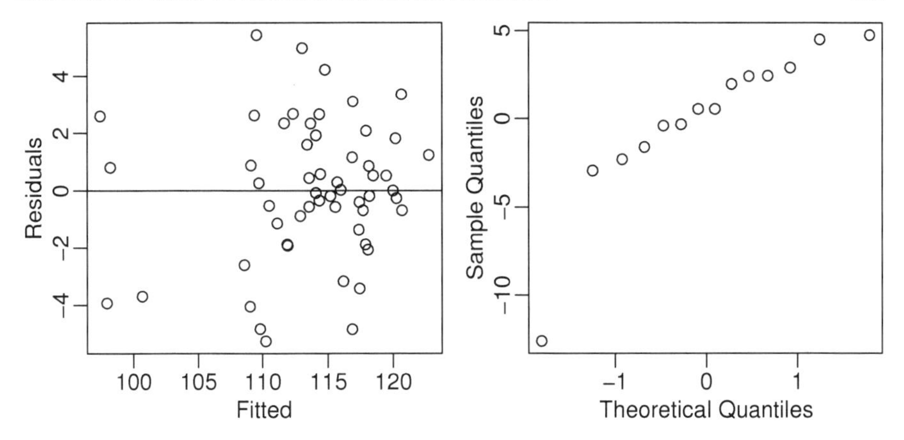

Figure 9.7 *Residuals vs. fitted plot is shown on the left and a QQ plot of the random effects for the eyes is shown on the right.*

The outlier corresponds to the right eye of subject #6. For further analysis, we should consider dropping subject #6. There are only seven subjects altogether, so we would certainly regret losing any data, but this may be unavoidable. Ultimately, we may need more data to make definite conclusions.

9.3 Multiple Response Multilevel Models

In Section 8.8, we analyzed some data from the Junior Schools Project. In addition to a math test, students also took a test in English. Although it would be possible to analyze the English test results in the same way that we analyzed the math scores, additional information may be obtained from analyzing them simultaneously. Hence we view the data as having a bivariate response with English and math scores for each student. The student is a nested factor within the class which is in turn nested within the school. We express the multivariate response for each individual by introducing an additional level of nesting at the individual level. So we might view this as just another nested model except that there is a fixed subject effect associated with this lowest level of nesting.

We set up the data in a format with one test score per line with an indicator `subject` identifying which type of test was taken. We scale the English and math test scores by their maximum possible values, 40 and 100, respectively, to aid comparison:

```
> data(jsp)
> jspr <- jsp[jsp$year==2,]
> mjspr <- data.frame(rbind(jspr[,1:6],jspr[,1:6]),
  subject=factor(rep(c("english","math"),c(953,953))),
  score=c(jspr$english/100,jspr$math/40))
```

We can examine the relationship between subject, gender and scores, as seen in Figure 9.8:

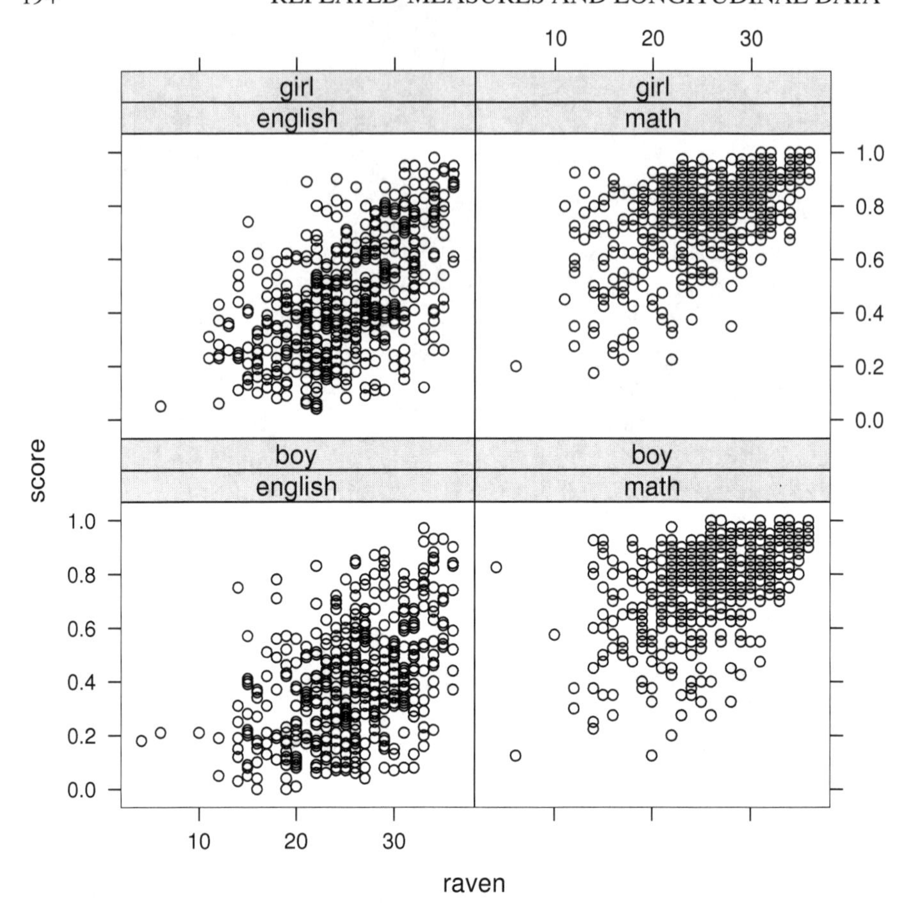

Figure 9.8 *Scores on test compared to raven score for subjects and genders.*

```
> xyplot(score ~ raven| subject*gender, mjspr)
```

We now fit a model for the data that includes all the variables of interest that incorporates some of the interactions that we suspect might be present:

```
> mjspr$craven <- mjspr$raven-mean(mjspr$raven)
> mmod <- lmer(score ~ subject*gender+craven*subject+social+
 (1|school)+(1|school:class)+(1|school:class:id),mjspr)
```

The model being fit for school i, class j, student k in subject l is:

$$\begin{aligned} score_{ijkl} = \quad & subject_l + gender_k + raven_k + social_k + (subject \times gender)_{lk} + \\ & (raven \times subject)_{lk} + school_i + class_j + student_k + \varepsilon_{ijkl} \end{aligned}$$

where the Raven score has been mean centered and school, class and student are random effects with the other terms, apart from ε, being fixed effects. The test on the fixed effects reveals:

```
> anova(mmod)
```

```
Analysis of Variance Table
               Df Sum Sq Mean Sq Denom F value  Pr(>F)
subject         1     54      54  1892 3953.67 < 2e-16
gender          1  0.101   0.101  1892    7.46  0.0064
craven          1      6       6  1892  444.63 < 2e-16
social          8      1   0.088  1892    6.47 2.5e-08
subject:gender  1  0.384   0.384  1892   28.23 1.2e-07
subject:craven  1  0.217   0.217  1892   15.99 6.6e-05
```

Both interactions are significant so no simplifications are indicated for this model. We might consider adding additional fixed effects, but we shall attempt to interpret only this model for now. The summary output:

```
> summary(mmod)
Linear mixed-effects model fit by REML
Formula: score ~ subject * gender + craven * subject + social + (1 | school) +
     (1 | school:class) + (1 | school:class:id)
   Data: mjspr
     AIC     BIC logLik MLdeviance REMLdeviance
 -1705.6 -1605.6 870.79    -1846.3      -1741.6
Random effects:
 Groups          Name        Variance Std.Dev.
 school:class:id (Intercept) 0.010252 0.1013
 school:class    (Intercept) 0.000582 0.0241
 school          (Intercept) 0.002231 0.0472
 Residual                    0.013592 0.1166
# of obs: 1906, groups: school:class:id, 953; school:class, 90; school, 48

Fixed effects:
                       Estimate  Std. Error   DF t value Pr(>|t|)
(Intercept)            0.441578    0.026459 1892   16.69  < 2e-16
subjectmath            0.366565    0.007710 1892   47.54  < 2e-16
gendergirl             0.063351    0.010254 1892    6.18  7.9e-10
craven                 0.017390    0.000925 1892   18.81  < 2e-16
social2                0.013754    0.027230 1892    0.51   0.6136
social3               -0.020768    0.028972 1892   -0.72   0.4736
social4               -0.070708    0.025868 1892   -2.73   0.0063
social5               -0.050474    0.028818 1892   -1.75   0.0800
social6               -0.087852    0.030672 1892   -2.86   0.0042
social7               -0.099408    0.031607 1892   -3.15   0.0017
social8               -0.081623    0.042352 1892   -1.93   0.0541
social9               -0.047337    0.027445 1892   -1.72   0.0847
subjectmath:gendergirl -0.059194    0.010706 1892   -5.53  3.7e-08
subjectmath:craven    -0.003720    0.000930 1892   -4.00  6.6e-05
```

Starting with the fixed effects, we see that the math subject scores were about 37% higher than the English scores. This may just reflect the grading scale and difficulty of the test and so perhaps nothing in particular should be concluded from this except, of course, that it is necessary to have this term in the model to control for this difference. Since gender has a significant interaction with subject, we must interpret these terms together. We see that on the English test, which is the reference level, girls score 6.3% higher than boys. On the math test, the difference is $6.3 - 5.9 = 0.4\%$ which is negligible. We see that the scores are strongly related to the entering Raven score although the relation is slightly less strong for math than English (slope is 0.0174 for English but $0.0174 - 0.0037 = 0.0137$ for math). We also see the declining performance as we move down the social class scale as we found in the previous analysis.

Moving to the random effects, we can see from Figure 9.9 that the standard deviation of the residual error in the math scores is smaller than that seen in the English scores. Perhaps this can be ascribed to the greater ease of consistent grading of math assignments or perhaps just greater variation is to be expected in English performance. The correlation between the English and math scores after adjusting for the other effects is also of interest. The last two terms in the model, $student_k + \varepsilon_{ijkl}$, represent a 2×2 covariance matrix for the residual scores for the two tests. We can compute the correlation as:

```
> 0.1013^2/(0.1013^2+0.1166^2)
[1] 0.43013
```

giving a moderate positive correlation between the scores. Various diagnostic plots can be made. An interesting one is:

```
> xyplot(residuals(mmod) ~ fitted(mmod)|subject,mjspr,pch=".",
  xlab="Fitted",ylab="Residuals")
```

as seen in Figure 9.9. There is somewhat greater variance in the verbal scores. The truncation effect of the maximum score is particularly visible for the math scores.

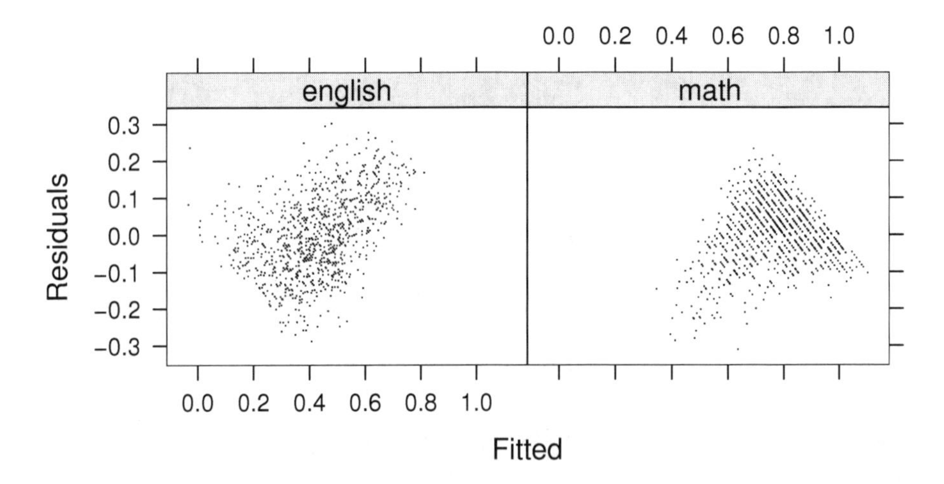

Figure 9.9 *Residuals vs. fitted plot broken down by subject.*

Further Reading: Longitudinal data analysis is explicitly covered in books by Verbeke and Molenberghs (2000), Fitzmaurice, Laird, and Ware (2004), Diggle, Heagerty, Liang, and Zeger (2002) and Frees (2004). Books stating repeated measures in the title, such as Lindsey (1999), cover much the same material.

Exercises

1. The `ratdrink` data consist of five weekly measurements of body weight for 27 rats. The first 10 rats are on a control treatment while seven rats have thyroxine added to their drinking water. Ten rats have thiouracil added to their water. Build a model for the rat weights that shows the effect of the treatment.
2. Data on housing prices in 36 metropolitan statistical areas (MSAs) over nine years from 1986–94 were collected and can be found in the dataset `hprice`. Find a good model for the data. Explain the effect of the predictors on housing prices. It is not necessary to present every part of your analysis. Present a compact description of how you found your model in five pages or less.
3. The `nepali` data is a subset from public health study on Nepalese children. Develop a model for the weight of the child as he or she ages. You may use `mage, lit, died, gender` and `alive` (but not `ht`) as predictors. Show how you developed your model and interpret your final model.
4. The `attenu` data gives peak accelerations measured at various observation stations for 23 earthquakes in California. The data has been used by various workers to estimate the attenuating affect of distance on ground acceleration.
   (a) Model the log of the acceleration as a function of the log of the distance while taking account of the magnitude of the quake.
   (b) Predict how the acceleration varied for an earthquake of magnitude 7.5. Express quantitatively the uncertainty in this prediction.
   (c) Predict how the acceleration varied for the first event where only one observation was available.
5. The `sleepstudy` data found in the `Matrix` package, which is loaded with `lme4`, describes the reaction times of subjects who are progressively sleep deprived. Form a model for the reaction times and describe the variation between individuals.

CHAPTER 10

Mixed Effect Models for Nonnormal Responses

10.1 Generalized Linear Mixed Models

Generalized linear mixed models (GLMM) combine the ideas of generalized linear models with the random effects modeling ideas of the previous two chapters. The response is a random variable, Y_i, taking observed values, y_i, for $i = 1, \ldots, n$, and follows an exponential family distribution as defined in Chapter 6:

$$f(y_i|\theta_i, \phi) = \exp\left[\frac{y_i\theta_i - b(\theta_i)}{a(\phi)} + c(y, \phi)\right]$$

Let $EY_i = \mu_i$ and let this be connected to the linear predictor η_i using the link function g by $\eta_i = g(\mu_i)$. Suppose for simplicity that we use the canonical link for g so that we may make the direct connection that $\theta_i = \mu_i$.

Now let the random effects, γ, have distribution $h(\gamma|V)$ for parameters V. The fixed effects are β. Conditional on the random effects, γ:

$$\theta_i = x_i^T \beta + z_i^T \gamma$$

where x_i and z_i are the corresponding rows from the design matrices, X and Z, for the respective fixed and random effects. Now the likelihood may be written as:

$$L(\beta, \phi, V|y) = \prod_{i=1}^{n} \int f(y_i|\beta, \phi, \gamma) h(\gamma|V) d\gamma$$

Typically the random effects are assumed normal: $\gamma \sim N(0, D)$. However, unless f is also normal, the integral remains in the likelihood, which becomes difficult to compute, particularly if the random effects structure is complicated.

A variety of approaches are available to approximating the likelihood using theoretical or numerical methods. A Bayesian approach is also possible. See Sinha (2004) for a recent approach that also contains a survey of past approaches. We investigate the issues through an example.

An experiment was conducted to study the effects of surface and vision on balance. The balance of subjects were observed for two different surfaces and for restricted and unrestricted vision. Balance was assessed qualitatively on an ordinal four-point scale based on observation by the experimenter. Forty subjects were studied, twenty males and twenty females ranging in age from 18 to 38, with heights given in cm and weights in kg. The subjects were tested while standing on foam or a normal surface and with their eyes closed or open or with a dome placed over their head. Each subject was tested twice in each of the surface and eye combinations for a total of 12 measures per subject. The data comes from Steele (1998) via the Australasian Data and Story Library (OzDASL).

For the purposes of this analysis, we will reduce the response to a two-point scale: whether the subject was judged completely stable or not. We start by defining this response:

```
> data(ctsib)
> ctsib$stable <- ifelse(ctsib$CTSIB==1,1,0)
> summary(ctsib)
   Subject          Sex            Age           Height          Weight
 Min.   : 1.0   female:240   Min.   :18.0   Min.   :142   Min.   : 44.2
 1st Qu.:10.8   male  :240   1st Qu.:21.8   1st Qu.:167   1st Qu.: 60.7
 Median :20.5                Median :25.5   Median :173   Median : 68.0
 Mean   :20.5                Mean   :26.8   Mean   :172   Mean   : 71.1
 3rd Qu.:30.3                3rd Qu.:33.0   3rd Qu.:180   3rd Qu.: 83.5
 Max.   :40.0                Max.   :38.0   Max.   :190   Max.   :102.0
 Surface       Vision          CTSIB          stable
 foam:240   closed:160   Min.   :1.00   Min.   :0.000
 norm:240   dome  :160   1st Qu.:2.00   1st Qu.:0.000
            open  :160   Median :2.00   Median :0.000
                         Mean   :1.92   Mean   :0.238
                         3rd Qu.:2.00   3rd Qu.:0.000
                         Max.   :4.00   Max.   :1.000
```

We could fit a binomial GLM ignoring the subject information entirely:

```
> gf <- glm(stable ~ Sex+Age+Height+Weight+Surface+Vision,binomial,data=ctsib)
> summary(gf)
Coefficients:
            Estimate Std. Error z value Pr(>|z|)
(Intercept)  7.27745    3.80399    1.91  0.05573
Sexmale      1.40158    0.51623    2.72  0.00663
Age          0.00252    0.02431    0.10  0.91739
Height      -0.09641    0.02684   -3.59  0.00033
Weight       0.04350    0.01800    2.42  0.01567
Surfacenorm  3.96752    0.44718    8.87  < 2e-16
Visiondome   0.36375    0.38322    0.95  0.34252
Visionopen   3.18750    0.41600    7.66  1.8e-14
```

However, there may be a significant subject effect. We could try including a fixed subject factor:

```
> gfs <- glm(stable ~ Sex+Age+Height+Weight+Surface+Vision+factor(Subject),
  binomial,data=ctsib)
> anova(gf,gfs,test="Chi")
Analysis of Deviance Table
  Resid. Df Resid. Dev  Df Deviance P(>|Chi|)
1       472        295
2       437        121  35      174   2.5e-20
```

We see strong evidence for a significant subject effect. However, when we examine the summary for this model, we see problems with identifiability and separability which prompt the use of bias-reduced logistic regression, as described in Section 2.8:

```
> library(brlr)
> modbr <- brlr(stable ~ Sex+Age+Height+Weight+Surface+Vision+
  factor(Subject), data=ctsib)
> summary(modbr)
Coefficients:
                  Value   Std. Error t value
(Intercept)       -44.619  50.432     -0.885
Sexmale             0.538   3.060      0.176
Age                -0.223   0.137     -1.625
Height              0.335   0.351      0.953
Weight             -0.255   0.186     -1.370
Surfacenorm         6.150   0.734      8.378
Visiondome          0.635   0.489      1.300
Visionopen          5.043   0.687      7.344
```

We find that the subject-specific variables, sex, age, height and weight, are no longer significant. This is because these predictors are constant for a given subject. We cannot completely unconfound these effects from the subject effects.

There are variety of ways of fitting GLMMs in R. We demonstrate the Penalized Quasi-Likelihood method implemented in the `MASS` package:

```
> library(MASS)
> gg <- glmmPQL(stable ~ Sex+Age+Height+Weight+Surface+Vision,
  random=~1|Subject,  family=binomial,data=ctsib)
> summary(gg)
Random effects:
 Formula: ~1 | Subject
        (Intercept) Residual
StdDev:      3.0608  0.59062

Variance function:
 Structure: fixed weights
 Formula: ~invwt
Fixed effects: stable ~ Sex + Age + Height + Weight + Surface + Vision
              Value Std.Error  DF t-value p-value
(Intercept) 15.5716   13.4985 437  1.1536  0.2493
Sexmale      3.3554    1.7526  35  1.9145  0.0638
Age         -0.0066    0.0820  35 -0.0810  0.9359
Height      -0.1908    0.0920  35 -2.0736  0.0455
Weight       0.0695    0.0629  35  1.1051  0.2766
Surfacenorm  7.7241    0.5736 437 13.4666  0.0000
Visiondome   0.7265    0.3259 437  2.2289  0.0263
Visionopen   6.4853    0.5440 437 11.9220  0.0000
```

The fit falls somewhere between the two above from the point of view of effect significance. Notice how there are more degrees of freedom for the experimental factors which do vary within individuals. This is expected. Compared to the fixed effect subject modeling, rather less of the variation is attributed to the GLMM. Here the SD for subjects is 3.06 while the SD of the subject effects from the GLM is:

```
> sd(coef(modbr)[9:43])
[1] 4.4407
```

This model can also be fit using the `lmer` function from the `lme4` package. Estimation of GLMMs is an active area of research and further study of the best methods of estimation is necessary.

10.2 Generalized Estimating Equations

The advantage of the quasi-likelihood approach compared to GLMs was that we did not need to specify the distribution of the response. We only needed to give the link function and the variance. We can adapt this approach for repeated measures and/or longitudinal studies. Let Y_i be a vector of random variables representing the responses on a given individual and let $EY_i = \mu_i$ which is then linked to the linear predictor $\eta = X\beta$ in some appropriate way. Let:

$$\text{var } Y_i \equiv \text{var } (Y_i; \beta, \alpha)$$

where α represents parameters that model the correlation structure within individuals. The parameters, β, may then be estimated setting the (multivariate) score function to zero and solving:

$$\sum_i \left(\frac{\partial \mu_i}{\partial \beta}\right)^T \text{var } (Y_i)^{-1}(Y_i - \mu_i) = 0$$

These equations can be regarded as the multivariate analogue of those used for the quasi-likelihood models described in Section 7.4. Since var Y also depends on α, we substitute any consistent estimate of α in this equation and still obtain an estimate as asymptotically efficient as if α were known. A similar set of equations can be derived representing the score with respect to α, which may be similarly solved.

These are called *generalized estimating equations* (GEE). Note that no specification of the distribution has been necessary which makes the fitting and specification much simpler. The estimates of β are consistent even if the variance is misspecified.

We reanalyze the stability dataset:

```
> data(ctsib)
> ctsib$stable <- ifelse(ctsib$CTSIB==1,1,0)
> library(gee)
> gg <- gee(stable ~ Sex+Age+Height+Weight+Surface+Vision,id=Subject,
  family=binomial,data=ctsib,corstr="exchangeable",scale.fix=TRUE)
```

We have specified the same fixed effects as in the corresponding GLMM earlier. The grouping variable is specified by the `id` argument. Only simple groups are allowed while nested grouping variables cannot be accommodated easily in this function. We must choose the correlation structure within each group. If we choose no correlation, then the problem reduces to a standard GLM. Several choices are available. For this data, it seems reasonable to assume that any pair of observations from the same subject have the same correlation. This is known as an *exchangeable* correlation or equivalently, *compound symmetry*. We have chosen to fix the scale parameter at the default value of 1 to ensure maximum compatibility with the GLMM fit. Otherwise, there would not be a strong reason to fix this. Let us now examine the output:

```
> summary(gg)
Model:
```

```
 Link:                       Logit
 Variance to Mean Relation: Binomial
 Correlation Structure:      Exchangeable

Coefficients:
             Estimate Naive S.E.  Naive z Robust S.E. Robust z
(Intercept)  8.602874   5.199006  1.65472    5.911263  1.45534
Sexmale      1.641080   0.701444  2.33957    0.902840  1.81769
Age         -0.011842   0.033022 -0.35862    0.047986 -0.24679
Height      -0.102020   0.036315 -2.80933    0.042336 -2.40976
Weight       0.043655   0.024630  1.77242    0.033964  1.28534
Surfacenorm  3.917254   0.412501  9.49636    0.566333  6.91687
Visiondome   0.358961   0.335804  1.06896    0.404175  0.88813
Visionopen   3.180126   0.377109  8.43292    0.460365  6.90783

Estimated Scale Parameter:  1
Number of Iterations:  4

Working Correlation
         [,1]    [,2]    [,3]    [,4]    [,5]    [,6]    [,7]
 [1,] 1.00000 0.21389 0.21389 0.21389 0.21389 0.21389 0.21389
 [2,] 0.21389 1.00000 0.21389 0.21389 0.21389 0.21389 0.21389
 [3,] 0.21389 0.21389 1.00000 0.21389 0.21389 0.21389 0.21389
 [4,] 0.21389 0.21389 0.21389 1.00000 0.21389 0.21389 0.21389
....rest deleted....
```

We can see from the working correlation that the estimated correlation between observations on the same subject is 0.21. The naive standard errors are based on the assumption that the proposed correlation structure is correct. However, GEE has the property that even if this structure is incorrect, the fixed effect estimates are still consistent. Nevertheless, the naive standard errors may be improved by the use of a *sandwich estimator*. This gives us the robust standard errors given above. These are typically, but not always, larger than the naive standard errors. The robust SEs should be used in practice. We see from the robust z-statistics that the height, surface and vision factors are significant. This corresponds to the result from GLMM.

There is one clear difference with the GLMM output: the estimates for the GEE are about half the size of the GLMM βs. This is to be expected. GLMMs model the data at the subject or individual level. The correlation between the measurements on the individual is generated by the random effect. Thus the βs for the GLMM represent the effect on an individual. A GEE models the data at the population level. The βs for a GEE represent the effect of the predictors averaged across all individuals with the same predictor values. GEEs do not use random effects but model the correlation at the marginal or correlation level.

Let's consider another GEE example. We have data from a clinical trial of 59 epileptics. For a baseline, patients were observed for 8 weeks and the number of seizures recorded. The patients were then randomized to treatment by the drug Progabide (31 patients) or to the placebo group (28 patients). They were observed for four 2-week periods and the number of seizures recorded. The data have been analyzed by many authors including Thall and Vail (1990), Breslow and Clayton (1993) and Diggle, Heagerty, Liang, and Zeger (2002). Does Progabide reduce the rate of seizures? Take a look at the first two patients:

```
> data(epilepsy)
> epilepsy[1:10,]
   seizures id treat expind timeadj age
1        11  1     0      0       8  31
2         5  1     0      1       2  31
3         3  1     0      1       2  31
4         3  1     0      1       2  31
5         3  1     0      1       2  31
6        11  2     0      0       8  30
7         3  2     0      1       2  30
8         5  2     0      1       2  30
9         3  2     0      1       2  30
10        3  2     0      1       2  30
```

Both were not treated (`treat=0`). The `expind` indicates the baseline period by 0 and the treatment period by 1. The length of these time periods is recorded in `timeadj`. We now compute the mean number of seizures *per week* broken down by the treatment and baseline vs. experimental period:

```
> with(epilepsy,by(seizures/timeadj,list(treat,expind),mean))
: 0
: 0
[1] 3.8482
---------------------------------------------------------
: 1
: 0
[1] 3.9556
---------------------------------------------------------
: 0
: 1
[1] 4.3036
---------------------------------------------------------
: 1
: 1
[1] 3.9839
```

We can tabulate this in Table 10.1.

	Baseline	Experiment
Placebo	3.85	4.30
Treatment	3.96	3.98

Table 10.1 *Seizures per week in epilepsy patients.*

We see that the rate of seizures in the treatment group actually increases during the period in which the drug was taken. The rate of seizures also increases even more in the placebo group. Perhaps some other factor is causing the rate of seizures to increase during the treatment period and the drug is actually having a beneficial effect. Now we make some plots to show the difference between the treatment and the control. The first plot shows the difference between the two groups during the experimental period only:

```
> y <- matrix(epilepsy$seizures,nrow=5)
> matplot(1:4,sqrt(y[-1,]),type="l",lty=epilepsy$treat[5*(1:59)]+1,
  xlab="Period",ylab="Sqrt(Seizures)")
```

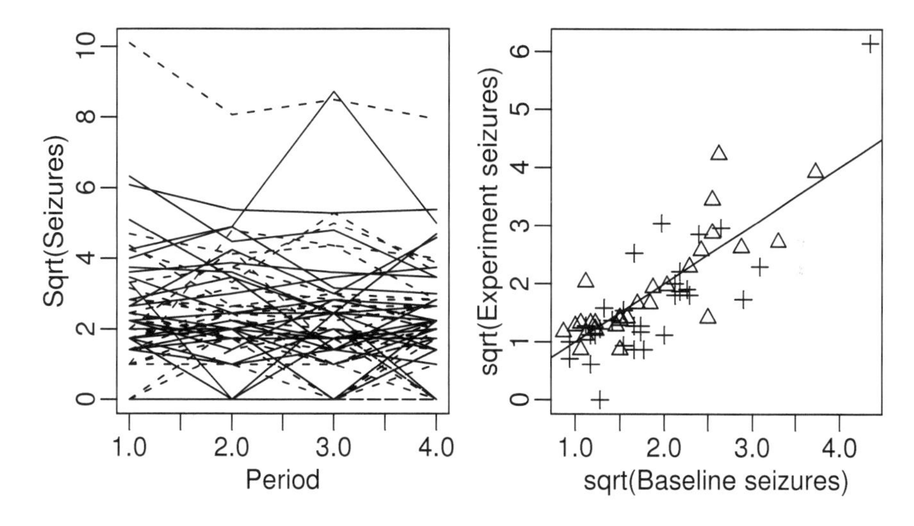

Figure 10.1 *Square root of seizures per 2-week period with treatment group shown as solid lines and the placebo group shown as dotted lines is shown in the plot on the left. The mean seizures per week is shown compared with the seizures per week during the baseline period is shown on the right where + indicates treated group.*

We compare the two groups in the left panel of Figure 10.1 and find little to choose between them. The square-root transform is used to stabilize the variance; this is often used with count data. Now we compare the average seizure rate to the baseline for the two groups:

```
> my <- apply(y[-1,],2,mean)/2
> plot(sqrt(epilepsy$seizures[epilepsy$expind == 0]/8),sqrt(my),
  pch=epilepsy$treat[5*(1:59)]+2,xlab="sqrt(Baseline seizures)",
  ylab="sqrt(Experiment seizures)")
> abline(0,1)
```

A treatment effect, if one exists, is not readily apparent. Now we fit the GEE model. An offset is necessary to account for the differing lengths of the baseline and treatment periods being 8 and 2 weeks, respectively. Patient #49 is unusual because of the high rate of seizures observed. We exclude this point. An AR(1) model for the correlation structure is most natural since consecutive measurements will be more correlated than measurements separated in time.

```
> g <- gee(seizures ~offset(log(timeadj))+expind+treat+I(expind*treat),
  id,family=poisson,corstr="AR-M",Mv=1,data=epilepsy,subset=(id!=49))
> summary(g)
Coefficients:
                    Estimate Naive S.E.  Naive z Robust S.E. Robust z
(Intercept)         1.320377    0.10354 12.75176     0.16065  8.21872
expind              0.142777    0.13932  1.02482     0.10769  1.32578
treat              -0.079402    0.14682 -0.54083     0.19716 -0.40273
I(expind * treat) -0.377546    0.21774 -1.73396     0.16839 -2.24210

Estimated Scale Parameter:  10.687
Number of Iterations:  3

Working Correlation
        [,1]    [,2]    [,3]    [,4]    [,5]
[1,] 1.00000 0.61854 0.38259 0.23665 0.14637
[2,] 0.61854 1.00000 0.61854 0.38259 0.23665
[3,] 0.38259 0.61854 1.00000 0.61854 0.38259
[4,] 0.23665 0.38259 0.61854 1.00000 0.61854
[5,] 0.14637 0.23665 0.38259 0.61854 1.00000
```

The drug does barely have a significant effect. The dispersion parameter is estimated as 10.687. This means that if we did not account for the overdispersion, the standard errors would be much larger. The AR(1) correlation structure can be seen in the working correlation where adjacent measurements have 0.62 correlation.

Further analysis would involve an investigation of alternative correlation structures, the age covariate and any trend during the experimental period. The analysis of this dataset is discussed in Diggle, Heagerty, Liang, and Zeger (2002).

Further Reading: McCulloch and Searle (2002) have some coverage of GLMMs as well as more material on GLMs. Hardin and Hilbe (2003) give a book-length treatment of GEEs. Diggle, Heagerty, Liang, and Zeger (2002) discuss both topics.

Exercises

1. The `ohio` data concern 536 children from Steubenville, Ohio and were taken as part of a study on the effects of air pollution. Children were in the study for four years from age seven to ten. The response was whether they wheezed or not. The variables are:

   resp an indicator of wheeze status (1 = yes, 0 = no)

   id an identifier for the child

   age 7 yrs = −2, 8 yrs = −1, 9 yrs = 0, 10 yrs = 1

   smoke an indicator of maternal smoking at the first year of the study (1 = smoker, 0 = nonsmoker)

   (a) Fit an appropriate GEE model and determine the effects of age and maternal smoking on wheezing.

(b) In your model, what indicates that a child who already wheezes is likely to continue to wheeze?

(c) What is the predicted probability that a 7 year-old with a smoking mother, wheezes?

(d) Repeat your analysis using a GLM where you assume that the observations are independent, that is, each single response value represents a different child. Indicate how the conclusions would differ and which results should be preferred.

(e) Sum the number of times wheezing is recorded for a child over the four measurements and model this as a function of the smoking status of the mother. This can be achieved as follows:

```
> nohio <- reshape(ohio,idvar="id",direction="wide",
  timevar="age",v.names="resp")
> nohio <- data.frame(smoke=nohio$smoke,
  wheeze=apply(nohio[,3:6],1,sum))
```

Now determine the effect of smoking. Compare this result to the previous analyses and discuss which is preferable.

2. The National Youth Survey collected a sample of 11–17 year-olds, 117 boys and 120 girls, asking questions about marijuana usage. The data is presented in `potuse`.

(a) Condense the levels of the response into whether the person did or did not use marijuana that year. Build a model for marijuana usage over the time period that takes account of sex differences.

(b) In your model, what describes correlation between marijuana usage one year and the next for a particular individual?

(c) What is the difference between boys and girls?

(d) Compute the predicted probability of usage by boys over time.

(e) Can you model the original three-level response in R?

3. Components are attached to an electronic circuit card assembly by a wave-soldering process. The soldering process involves baking and preheating the circuit card and then passing it through a solder wave by conveyor. Defects arise during the process. The design is 2^{7-3} with three replicates and the data is found in `wavesolder`. Assuming that the responses for each run are in time order, analyze the data. Is there any evidence of an effect due to the time order?

4. The `nitrofen` data in `boot` package come from an experiment to measure the reproductive toxicity of the pesticide nitrofen on a species of zooplankton called *Ceriodaphnia dubia*. Each animal produced three broods in which the number of live offspring was recorded. Fifty animals in total were used and divided into five batches. Each batch was treated in a solution with a different concentration of the pesticide.

   Build a model for the number of live offspring produced in the successive broods. Your model should describe how this number changes and is related within a given animal and how this relates to the concentration of pesticide.

5. The `toenail` data comes from a multicenter study comparing two oral treatments for toenail infection. Patients were evaluated for the degree of separation of the nail. Patients were randomized into two treatments and were followed over seven visits: four in the first year and yearly thereafter. The patients have not been treated prior to the first visit so this should be regarded as the baseline.

   Analyze the data to determine the difference between the two treatments and the progression of the infection over time.

CHAPTER 11

Nonparametric Regression

The generalized linear model was an extension of the linear model $y = X\beta + \varepsilon$ to allow the responses y from the exponential family. The mixed effect models allowed for a much more general treatment of ε. We now switch our attention to the linear predictor $\eta = X\beta$. We want to make this more flexible. There are a wide variety of available methods, but it is best to start with simple regression. The methods developed here form part of the solution to the multiple predictor problem.

We start with a simple regression problem. Given fixed $x_1, \ldots, x_n$, we observe $y_1, \ldots, y_n$ where:

$$y_i = f(x_i) + \varepsilon_i$$

where the ε_i are i.i.d. and have mean zero and unknown variance σ^2. The problem is to estimate the function f.

A parametric approach is to assume that $f(x)$ belong to a parametric family of functions: $f(x|\beta)$. So f is known up to a finite number of parameters. Some examples are:

$$\begin{aligned} f(x|\beta) &= \beta_0 + \beta_1 x \\ f(x|\beta) &= \beta_0 + \beta_1 x + \beta_2 x^2 \\ f(x|\beta) &= \beta_0 + \beta_1 x^{\beta_2} \end{aligned}$$

The parametric approach is quite flexible because we are not constrained to just linear predictors as in the first model of the three above. We can add many different types of terms such as polynomials and other functions of the variable to achieve flexible fits. Nonlinear models, such as the third case above, are also parametric in nature. Nevertheless, no matter what finite parametric family we specify, it will always exclude many plausible functions.

The nonparametric approach is to choose f from some smooth family of functions. Thus the range of potential fits to the data is much larger than the parametric approach. We do need to make some assumptions about f — that it has some degree of smoothness and continuity, for example, but these restrictions are far less limiting than the parametric way.

The parametric approach has the advantage that it is more efficient if the model is correct. If you have good information about the appropriate model family, you should prefer a parametric model. Parameters may also have intuitive interpretations. Nonparametric models do not have a formulaic way of describing the relationship between the predictors and the response; this often needs to be done graphically. This relates to another advantage of parametric models in that they reduce information necessary for prediction; you can write down the model formula, typically in a compact form. Nonparametric models are less easily communicated on paper. Parametric models also enable easy utilization of past experience.

The nonparametric approach is more flexible. In modeling new data, one often has very little idea of an appropriate form for the model. We do have a number of heuristic tools using diagnostic plots to help search for this form, but it would be easier to let the modeling approach take care of this search. Another disadvantage of the parametric approach is that one can easily choose the wrong form for the model and this results in bias. The nonparametric approach assumes far less and so is less liable to make bad mistakes. The nonparametric approach is particularly useful when little past experience is available

For our examples we will use three datasets, one real (data on Old Faithful) and two simulated, called exa and exb. The data comes from Härdle (1991). The reason we use simulated data is to see how well the estimates match the true function (which cannot usually be known for real data). We plot the data in the first three panels of Figure 11.1, using a line to mark the true function where known. For exa, the true function is $f(x) = \sin^3(2\pi x^3)$. For exb, it is constant zero, that is, $f(x) = 0$:

```
> data(exa)
> plot(y ~ x, exa,main="Example A",pch=".")
> lines(m ~ x, exa)
> data(exb)
> plot(y ~ x, exb,main="Example B",pch=".")
> lines(m ~ x, exb)
> data(faithful)
> plot(waiting ~ eruptions, faithful,main="Old Faithful",pch=".")
```

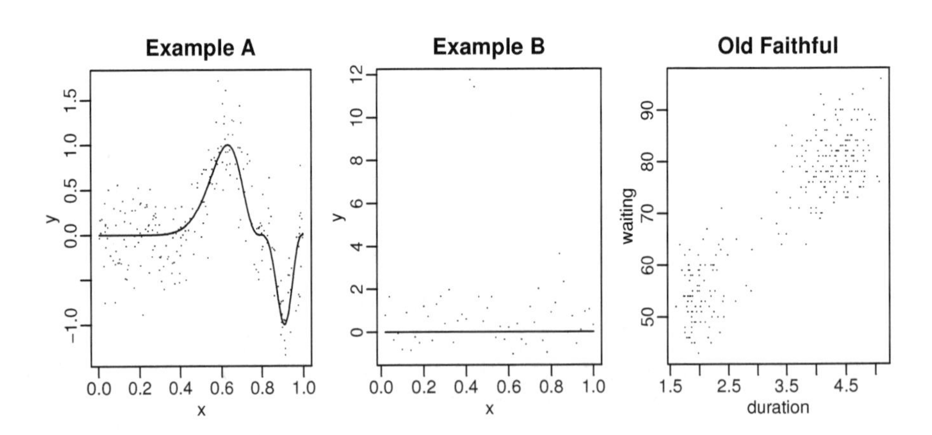

Figure 11.1 *Data examples. Example A has varying amounts of curvature, two optima and a point of inflexion. Example B has two outliers. The Old Faithful provides the challenges of real data.*

We now examine several widely used nonparametic regression estimators, also known as *smoothers*.

11.1 Kernel Estimators

In its simplest form, this is just a moving average estimator. More generally, our estimate of f, called $\hat{f}_\lambda(x)$, is:

$$\hat{f}_\lambda(x) = \frac{1}{n\lambda}\sum_{j=1}^{n} K\left(\frac{x-x_j}{\lambda}\right)Y_j = \frac{1}{n}\sum_{j=1}^{n} w_j Y_j \qquad \text{where} \qquad w_j = K\left(\frac{x-x_j}{\lambda}\right)/\lambda$$

K is a kernel where $\int K = 1$. The moving average kernel is rectangular, but smoother kernels can give better results. λ is called the bandwidth, window width or smoothing parameter. It controls the smoothness of the fitted curve.

If the xs are spaced very unevenly, then this estimator can give poor results. This problem is somewhat ameliorated by the *Nadaraya–Watson* estimator:

$$f_\lambda(x) = \frac{\sum_{j=1}^{n} w_j Y_j}{\sum_{j=1}^{n} w_j}$$

We see that this estimator simply modifies the moving average estimator so that it is a true weighted average where the weights for each y will sum to one.

It is worth understanding the basic asymptotics of kernel estimators. The optimal choice of λ gives:

$$\text{MSE}(x) = E(f(x) - \hat{f}_\lambda(x))^2 = O(n^{-4/5})$$

MSE stands for mean squared error and we see that this decreases at a rate proportional to $n^{-4/5}$ with the sample size. Compare this to the typical parametric estimator where $\text{MSE}(x) = O(n^{-1})$, but this only holds when the parametric model is correct. So the kernel estimator is less efficient. Indeed, the relative difference between the MSEs becomes substantial as the sample size increases. However, if the parametric model is incorrect, the MSE will be $O(1)$ and the fit will not improve past a certain point even with unlimited data. The advantage of the nonparametic approach is the protection against model specification error. Without assuming much stronger restrictions on f, nonparametric estimators cannot do better than $O(n^{-4/5})$.

The implementation of a kernel estimator requires two choices: the kernel and the smoothing parameter. For the choice of kernel, smoothness and compactness are desirable. We prefer smoothness to ensure that the resulting estimator is smooth, so for example, the uniform kernel will give stepped-looking fit that we may wish to avoid. We also prefer a compact kernel because this ensures that only data, local to the point at which f is estimated, is used in the fit. This means that the Gaussian kernel is less desirable, because although it is light in the tails, it is not zero, meaning in principle that the contribution of every point to the fit must be computed. The optimal choice under some standard assumptions is the Epanechnikov kernel:

$$K(x) = \begin{cases} \frac{3}{4}(1-x^2) & |x| < 1 \\ 0 & \text{otherwise} \end{cases}$$

This kernel has the advantage of some smoothness, compactness and rapid computation. This latter feature is important for larger datasets, particularly when resampling techniques like bootstrap are being used. Even so, any sensible choice of kernel will produce acceptable results, so the choice is not crucially important.

The choice of smoothing parameter λ is critical to the performance of the estimator and far more important than the choice of kernel. If the smoothing parameter is too small, the estimator will be too rough; but if it is too large, important features will be smoothed out.

We demonstrate the Nadaraya–Watson estimator next for a variety of choices of bandwidth on the Old Faithful data shown in Figure 11.2. We use the `ksmooth` function which is part of the R base package. This function lacks many useful features that can be found in some other packages, but it is adequate for simple use. The default uses a uniform kernel, which is somewhat rough. We have changed this to the normal kernel:

```
> plot(waiting ~ eruptions, faithful,main="bandwidth=0.1",pch=".")
> lines(ksmooth(faithful$eruptions,faithful$waiting,"normal",0.1))
> plot(waiting ~ eruptions, faithful,main="bandwidth=0.5",pch=".")
> lines(ksmooth(faithful$eruptions,faithful$waiting,"normal",0.5))
> plot(waiting ~ eruptions, faithful,main="bandwidth=2",pch=".")
> lines(ksmooth(faithful$eruptions,faithful$waiting,"normal",2))
```

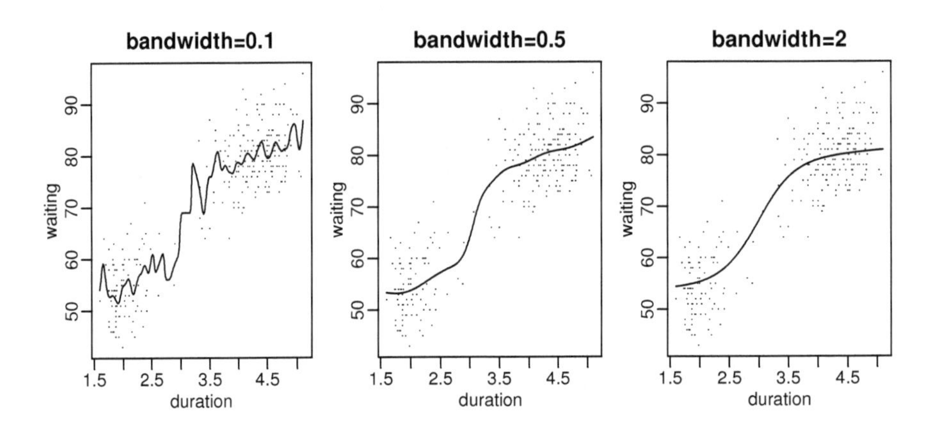

Figure 11.2 *Nadaraya–Watson kernel smoother with a normal kernel for three different bandwidths on the Old Faithful data.*

The central plot in Figure 11.2 is the best choice of the three. Since we do not know the true function relating waiting time and eruption duration, we can only speculate, but it does seem reasonable to expect that this function is quite smooth. The fit on the left does not seem plausible since we would not expect the mean waiting time to vary so much as a function of eruptions. On the other hand, the plot on the right is even smoother than the plot in the middle. It is not so easy to choose between these. Another consideration is that the eye can always visualize additional smoothing, but it is not so easy to imagine what a less smooth fit might look like. For this reason, we recommend picking the least smooth fit that does not show any implausible fluctuations. Of the three plots shown, the middle plot seems best. Smoothers are often used as a graphical aid in interpreting the relationship between variables. In such cases, visual selection of the amount of smoothing is effective because the user can employ background knowledge to make an appropriate choice and avoid serious mistakes.

You can choose λ interactively using this subjective method. Plot $\hat{f}_\lambda(x)$ for a range of different λ and pick the one that looks best as we have done above. You may need to iterate the choice of λ to focus your decision. Knowledge about what the true relationship might look like can be readily employed.

In cases where the fitted curve will be used to make numerical predictions of future values, the choice of the amount of smoothing has an immediate effect on the outcome. Even here subjective methods may be used. If this method of selecting the amount of smoothing seems disturbingly subjective, we should also understand that the selection of a family of parametric models for the same data would also involve a great deal of subjective choice although this is often not explicitly recognized. Statistical modeling requires us to use our knowledge of what general forms of relationship might be reasonable. It is not possible to determine these forms from the data in an entirely objective manner. Whichever methodology you use, some subjective decisions will be necessary. It is best to accept this and be honest about what these decisions are.

Even so, automatic methods for selecting the amount of smoothing are also useful. Selecting the amount of smoothing using subjective methods requires time and effort. When a large number of smooths are necessary, some automation is desirable. In other cases, the statistician will want to avoid the explicit appearance of subjectivity in the choice. Cross-validation (CV) is a popular general-purpose method. The criterion is:

$$CV(\lambda) = \frac{1}{n}\sum_{j=1}^{n}(y_j - \hat{f}_{\lambda(j)}(x_j))^2$$

where (j) indicates that point j is left out of the fit. We pick the λ that minimizes this criterion. True cross-validation is computationally expensive, so an approximation to it, known as *generalized cross-validation* or GCV, is sometimes used. There are also many other methods of automatically selecting the λ.

Our practical experience has been that automatic methods, such as CV, often work well, but sometimes produce estimates that are clearly at odds with the amount of smoothing that contextual knowledge would suggest. For this reason, we are unwilling to trust automatic methods completely. We recommend using them as a starting point for a possible interactive exploration of the appropriate amount of smoothing if time permits. They are also useful when a very large numbers of smooths are needed such as in the additive modeling approach described in Chapter 12.

When smoothing is used to determine whether f has certain features such as multiple maximums (called *bump hunting*) or monotonicity, special methods are necessary to choose the amount of smoothing since this choice will determine the outcome of the investigation.

The `sm` library, described in Bowman and Azzalini (1997), allows the computation of the cross-validated choice of bandwidth. For example, we find the CV choice of bandwidth for the Old Faithful and plot the result:

```
> library(sm)
> hm <- hcv(faithful$eruptions,faithful$waiting,display="lines")
```

```
> sm.regression(faithful$eruptions,faithful$waiting,h=hm,
  xlab="eruptions",ylab="waiting")
```

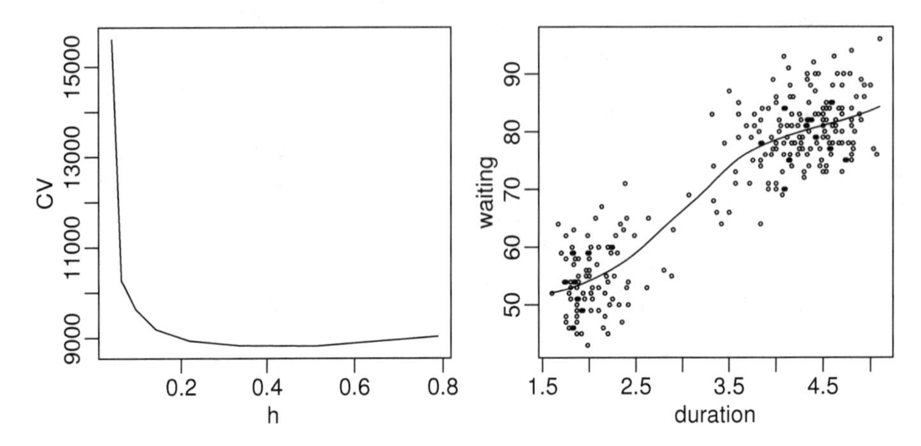

Figure 11.3 *The cross-validation criterion shown as a function of the smoothing parameter is shown in the first panel. The minimum occurs at a value of 0.424. The second panel shows the kernel estimate with this value of the smoothing parameter.*

We see the criterion plotted in the first panel of Figure 11.3. Notice that the function is quite flat in the region of the minimum indicating that a wide range of choices will produce acceptable results. The resulting choice of fit is shown in the second panel of the figure. The `sm` package uses a Gaussian kernel where the smoothing parameter is the standard deviation of the kernel.

We repeat the exercise for Example A; the plots are shown in Figure 11.4:

```
> hm <- hcv(exa$x,exa$y,display="lines")
> sm.regression(exa$x,exa$y,h=hm,xlab="x",ylab="y")
```

For Example B:

```
> hm <- hcv(exb$x,exb$y,display="lines")
hcv: boundary of search area reached.
Try readjusting hstart and hend.
hstart:  0.014122
hend  :  0.28244

             h     cv
[1,] 0.014122 171.47
[2,] 0.021665 190.99
[3,] 0.033237 219.87
[4,] 0.050990 242.99
[5,] 0.078226 258.13
[6,] 0.120008 272.24
[7,] 0.184108 284.39
[8,] 0.282445 288.48
```

we find that the CV choice is at the lower boundary of suggested bandwidths. We can look at a smaller range:

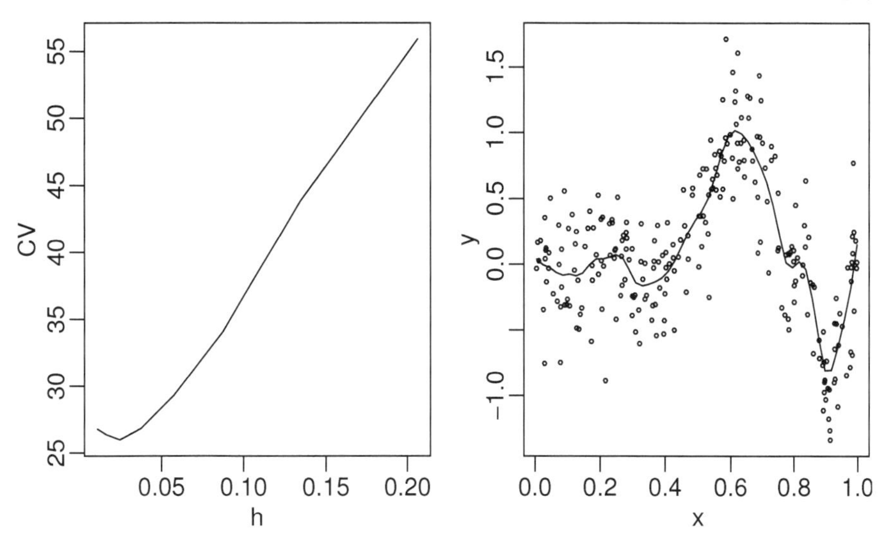

Figure 11.4 *Cross-validation selection of smoothing for Example A. The cross-validation criterion is shown on left — the minimum is at h=0.022. The fit from this choice of h is shown on the right.*

```
> hm <- hcv(exb$x,exb$y,display="lines",hstart=0.005)
```

However, bandwidths this small represent windows that include only a single point, making cross-validation impractical. Choosing such a small bandwidth as in:

```
> sm.regression(exb$x,exb$y,h=0.005)
```

gives us a dramatic undersmooth because the regression exactly fits the data.

11.2 Splines

Smoothing Splines: The model is $y_i = f(x_i) + \varepsilon_i$, so in the spirit of least squares, we might choose $\hat{f}$ to minimize the MSE: $\frac{1}{n}\sum(y_i - f(x_i))^2$. The solution is $\hat{f}(x_i) = y_i$. This is a "join the dots" regression that is almost certainly too rough. Instead, suppose we choose $\hat{f}$ to minimize a modified least squares criterion:

$$\frac{1}{n}\sum(Y_i - f(x_i))^2 + \lambda \int [f''(x)]^2 dx$$

where $\lambda > 0$ is the smoothing parameter and $\int [f''(x)]^2 dx$ is a *roughness penalty*. When f is rough, the penalty is large, but when f is smooth, the penalty is small. Thus the two parts of the criterion balance fit against smoothness. This is the *smoothing spline* fit.

For this choice of roughness penalty, the solution is of a particular form: $\hat{f}$ is a cubic spline. This means that $\hat{f}$ is a piecewise cubic polynomial in each interval (x_i, x_{i+1}) (assuming that the x_is are sorted). It has the property that $\hat{f}$, $\hat{f}'$ and $\hat{f}''$ are continuous. Given that we know the form of the solution, the estimation is reduced to the parametric problem of estimating the coefficients of the polynomials. This can be done in a numerically efficient way.

Several variations on the basic theme are possible. Other choices of roughness penalty can be considered, where penalties on higher-order derivatives lead to fits with more continuous derivatives. We can also use weights by inserting them in the sum of squares part of the criterion. This feature is useful when smoothing splines are means to an end for some larger procedure that requires weighting. A robust version can be developed by modifying the sum of squares criterion to:

$$\sum \rho(y_i - f(x_i)) + \lambda \int [f''(x)]^2 dx$$

where $\rho(x) = |x|$ is one possible choice.

In R, cross-validation is used to select the smoothing parameter by default. We show this default choice of smoothing for our three test cases:

```
> plot(waiting ~ eruptions, faithful,pch=".")
> lines(smooth.spline(faithful$eruptions,faithful$waiting))
> plot(y ~ x,exa,pch=".")
> lines(exa$x,exa$m)
> lines(smooth.spline(exa$x,exa$y),lty=2)
> plot(y ~ x,exb,pch=".")
> lines(exb$x,exb$m)
> lines(smooth.spline(exb$x,exb$y),lty=2)
```

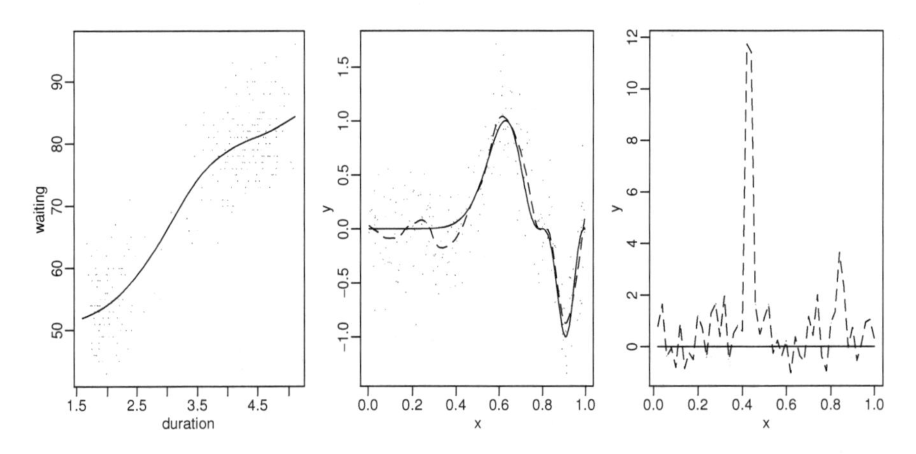

Figure 11.5 *Smoothing spline fits. For Examples A and B, the true function is shown as solid and the spline fit as dotted.*

The fits may be seen in Figure 11.5. The fit for the Old Faithful data looks reasonable. The fit for Example A does a good job of tracking the hills and valleys but overfits in the smoother region. The default choice of smoothing parameter given by CV is a disaster for Example B as the data is just interpolated. This illustrates the danger of blindly relying on automatic bandwidth selection methods.

Regression Splines: Regression splines differ from smoothing splines in the following way: For regression splines, the knots of the B-splines used for the basis are typically much smaller in number than the sample size. The number of knots chosen controls the amount of smoothing. For smoothing splines, the observed unique x values are the knots and λ is used to control the smoothing. It is arguable whether the regression spline method is parametric or nonparametric, because once the knots are chosen, a parametric family has been specified with a finite number of parameters. It is the freedom to choose the number of knots that makes the method nonparametric. One of the desirable characteristics of a nonparametric regression estimator is that it should be consistent for smooth functions. This can be achieved for regression splines if the number of knots is allowed to increase at an appropriate rate with the sample size.

We demonstrate some regression splines here. We use piecewise linear splines in this example, which are constructed and plotted as follows:

```
> rhs <- function(x,c) ifelse(x>c,x-c,0)
> curve(rhs(x,0.5),0,1)
```

where the spline is shown in the first panel of Figure 11.6. Now we define some knots for Example A:

```
> knots <- 0:9/10
> knots
 [1] 0.0 0.1 0.2 0.3 0.4 0.5 0.6 0.7 0.8 0.9
```

and compute a design matrix of splines with knots at these points for each x:

```
> dm  <- outer(exa$x,knots,rhs)
> matplot(exa$x,dm,type="l",col=1)
```

where the basis functions are shown in the second panel of Figure 11.6. Now we

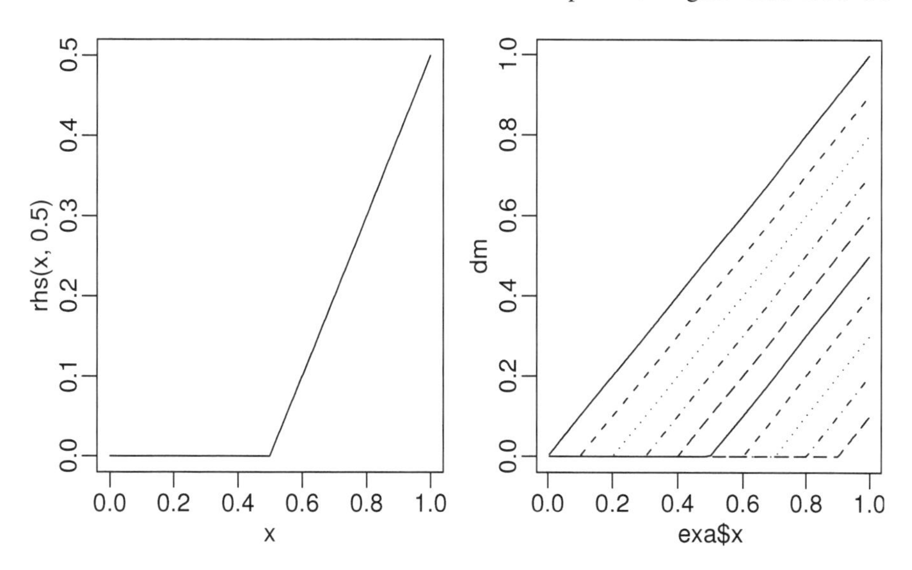

Figure 11.6 *One basis function for linear regression splines shown on the left and the complete set shown on the right.*

compute and display the regression fit:

```
> g <- lm(exa$y ~ dm)
> plot(y ~ x, exa,pch=".",xlab="x",ylab="y")
> lines(exa$x,predict(g))
```

where the plot is shown in the first panel of Figure 11.7. Because the basis functions are piecewise linear, the fit is also piecewise linear. A better fit may be obtained by adjusting the knots so that they are denser in regions of greater curvature:

```
> newknots <- c(0,0.5,0.6,0.65,0.7,0.75,0.8,0.85,0.9,0.95)
> dmn  <- outer(exa$x,newknots,rhs)
> gn <- lm(exa$y ~ dmn)
> plot(y ~x, exa,pch=".",xlab="x",ylab="y")
> lines(exa$x,predict(gn))
```

where the plot is shown in the second panel of Figure 11.7. We obtain a better fit

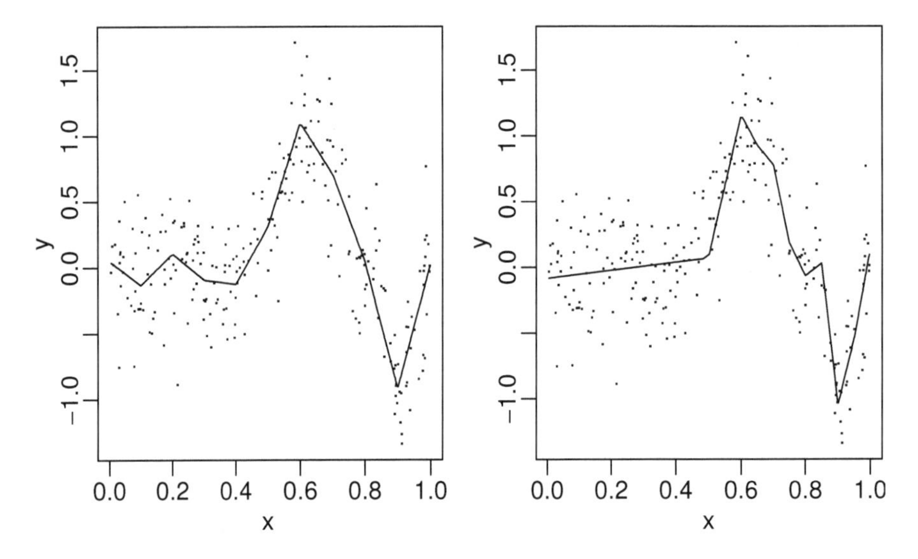

Figure 11.7 *Evenly spaced knots fit shown on the left and knots spread relative to the curvature on the right.*

but only by using our knowledge of the true curvature. This knowledge would not be available for real data, so more practical methods place the knots adaptively according to the *estimated* curvature.

One can achieve a smoother fit by using higher-order splines. The `bs()` function can be used to generate the appropriate spline basis. The default is cubic B-splines. We display 12 cubic B-splines evenly spaced on the [0,1] interval. The splines close to the boundary take a different form as seen in the first panel of Figure 11.8:

```
> library(splines)
> matplot(bs(seq(0,1,length=1000),df=12),type="l",ylab="",col=1)
```

We can now use least squares to determine the coefficients. We then display the fit as seen in the second panel of Figure 11.8:

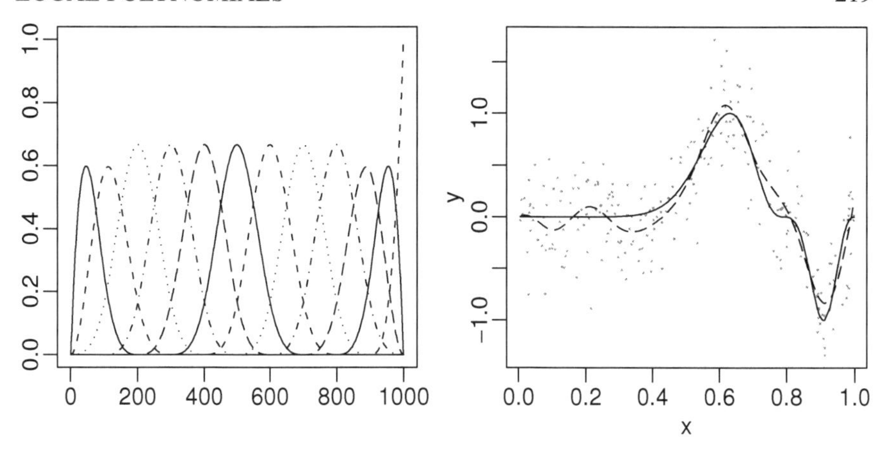

Figure 11.8 *A cubic B-spline basis is shown in the left panel and the resulting fit to the Example A data is shown in the right panel.*

```
> sml <- lm(y ~ bs(x,12),exa)
> plot(y ~ x, exa, pch=".")
> lines(m ~ x, exa)
> lines(predict(sml) ~ x, exa, lty=2)
```

We see a smooth fit, but again we could do better by placing more knots at the points of high curvature and fewer in the flatter regions.

11.3 Local Polynomials

Both the kernel and spline methods have been relatively vulnerable to outliers as seen by their performance on Example B. The fits can be improved with some manual intervention, either to remove the outliers or to increase the smoothing parameters. However, smoothing is frequently just a small part of an analysis and so we might wish to avoid giving each smooth individual attention. Furthermore, habitual removal of outliers is an ad hoc strategy that is better replaced with a method that deals with long-tailed errors gracefully. The local polynomial method combines robustness ideas from linear regression and local fitting ideas from kernel methods.

First we select a window. We then fit a polynomial to the data in that window using robust methods. The predicted response at the middle of the window is the fitted value. We then simply slide the window over the range of the data, repeating the fitting process as the window moves. The most well-known implementation of this type of smoothing is called *lowess* or *loess* and is due to Cleveland (1979).

As with any smoothing method, there are choices to be made. We need to choose the order of the polynomial fit. A quadratic allows us to capture peaks and valleys in the function. However, a linear term also performs well and is the default choice in the `loess` function. As with most smoothers, it is important to pick the window width well. The default choice takes three quarters of the data and may not be a good choice as we shall see below.

For the Old Faithful data, the default choice is satisfactory, as seen in the first panel of Figure 11.9:

```
> plot(waiting ~ eruptions, faithful,pch=".")
> f <- loess(waiting ~ eruptions, faithful)
> i <- order(faithful$eruptions)
> lines(f$x[i],f$fitted[i])
```

For Example A, the default choice is too large. The choice that minimizes the integrated squared error between the estimated and true function requires a span (proportion of the range) of 0.22. Both fits are seen in the middle panel of Figure 11.9:

```
> plot(y ~ x, exa, pch=".")
> lines(exa$x,exa$m,lty=1)
> f <- loess(y ~ x,exa)
> lines(f$x,f$fitted,lty=2)
> f <- loess(y ~ x,exa,span=0.22)
> lines(f$x,f$fitted,lty=5)
```

In practice, the true function is, of course, unknown and we would need to select the span ourselves, but this optimal choice does at least show how well loess can do in the best of circumstances. The fit is similar to that for smoothing splines.

For Example B, the optimal choice of span is one (that is all the data). This is not surprising since the true function is a constant and so maximal smoothing is desired. We can see that the robust qualities of loess prevent the fit from becoming too distorted by the two outliers even with the default choice of smoothing span:

```
> plot(y ~ x,exb,pch=".")
> f <- loess(y ~ x, exb)
> lines(f$x,f$fitted,lty=2)
> f <- loess(y ~ x, exb,span=1)
> lines(f$x,f$fitted,lty=5)
> lines(exb$x,exb$m)
```

11.4 Wavelets

Regression splines are an example of a basis function approach to fitting. We approximate the curve by a family of basis functions, $\phi_i(x)$, so that $\hat{f}(x) = \sum_i c_i \phi_i(x)$. Thus the fit requires estimating the coefficients, c_i. The choice of basis functions will determine the properties of the fitted curve. The estimation of c_i is particularly easy if the basis functions are orthogonal.

Examples of orthogonal bases are orthogonal polynomials and the Fourier basis. The disadvantage of both these families is that the basis functions are not compactly supported so that the fit of each basis function depends on the whole data. This means that these fits lack the desirable local fit properties that we have seen in previously discussed smoothing methods. Although Fourier methods are popular for some applications, they are not typically used for general-purpose smoothing.

Cubic B-splines are compactly supported, but they are not orthogonal. *Wavelets* have the advantage that they are compactly supported and can be defined so as to possess the orthogonality property. They also possess the *multiresolution* property

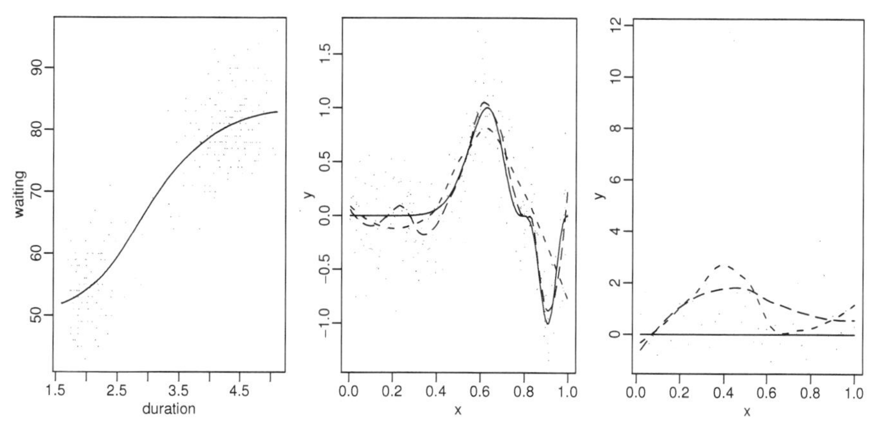

Figure 11.9 *Loess smoothing: Old Faithful data is shown in the left panel with the default amount of smoothing. Example A data is shown in the middle and B in the right panel. The true function is shown as a solid line along with the default choice (dotted) and respective optimal amounts of smoothing (dashed) are also shown.*

which allows them to fit the grosser features of the curve while focusing on the finer detail where necessary.

We begin with the simplest type of wavelet: the *Haar* basis. The *mother wavelet* for the Haar family is defined on the interval $[0,1)$ as:

$$w(x) = \begin{cases} 1 & x \leq 1/2 \\ -1 & x > 1/2 \end{cases}$$

We generate the members of family by dilating and translating this function. The next two members of the family are defined on $[0,1/2)$ and $[1/2,1)$ by rescaling the mother wavelet to these two intervals. The next four members are defined on the quarter intervals in the same way. We can index the family members by level j and within the level by k so that each function will be defined on the interval $[k/2^j,(k+1)/2^j)$ and takes the form:

$$h_n(x) = 2^{j/2} w(2^j x - k)$$

where $n = 2^j + k$ and $0 \leq k \leq 2^j$. We can see by simply plotting these functions that they are orthogonal. They are also orthonormal, because they integrate to 1. Furthermore, they have a local basis where the support becomes narrower as the level is increased. Computing the coefficients is particularly quick because of these properties.

Wavelet fitting can be implemented using the `wavethresh` package. The first step is to make the wavelet decomposition. We will illustrate this with Example A:

```
> library(wavethresh)
> wds <- wd(exa$y,filter.number=1,bc="interval")
```

The filter number specifies the complexity of the family. The Haar basis is the simplest available but is not the default choice. We can now show the mother wavelet and wavelet coefficients:

```
> draw(filter.number=1,family="DaubExPhase")
> plot(wds)
```

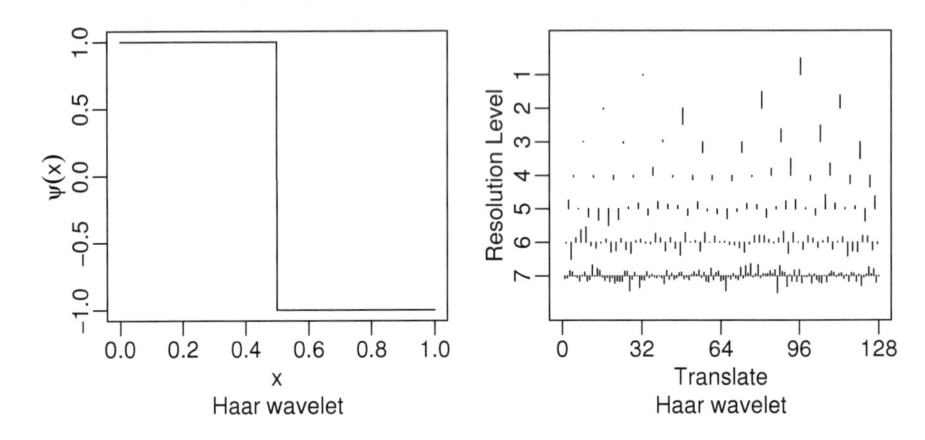

Figure 11.10 *Haar mother wavelet and wavelet coefficients from decomposition for Example A.*

We can see the Haar mother wavelet in the left panel of Figure 11.10. We see the wavelet decomposition in the right panel.

Suppose we wanted to compress the data into a more compact format. Smoothing can be viewed as a form of compression because it retains the important features of the data while discarding some of the finer detail. The smooth is described in terms of the fitted coefficients which are fewer than the number of data points. The method would be called *lossy* since some information about the original data is lost.

For example, suppose we want to smooth the data extensively. We could throw away all the coefficients of level four or higher and then reconstruct the function as follows:

```
> wtd <- threshold(wds,policy="manual",value=9999)
> fd <- wr(wtd)
```

Only level-three and higher coefficients are retained. There are only $2^3 = 8$ of these. The thresholding here applies to level four and higher only by default. Any coefficient less than 9999 in absolute value is set to zero — that is, all of them in this case. The `wr()` inverts the wavelet transform. We now plot the result as seen in the first panel of Figure 11.11:

```
> plot(y ~ x, exa,pch=".")
> lines(m ~ x, exa)
> lines(fd ~ x, exa, lty=5, lwd=2)
```

We see that the fit consists of eight constant fits; we expect this since Haar basis is piecewise constant and we have thrown away the higher-order parts leaving just eight coefficients.

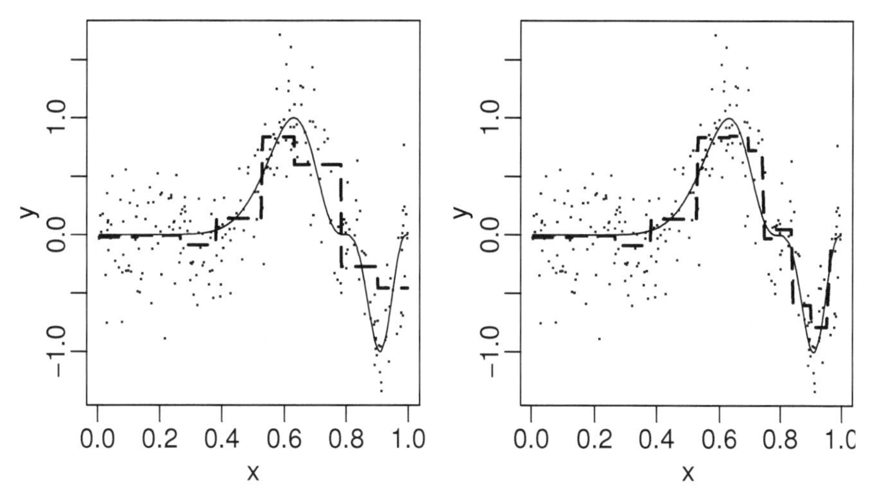

Figure 11.11 *Thresholding and inverting the transform. In the left panel all level-four and above coefficients are zeroed. In the right, the coefficients are thresholded using the default method. The true function is shown as a solid line and the estimate as a dashed line.*

Instead of simply throwing away higher-order coefficients, we could zero out only the small coefficients. We choose the threshold using the default method:

```
> wtd2 <- threshold(wds)
> fd2 <- wr(wtd2)
```

Now we plot the result as seen in the second panel of Figure 11.11.

```
> plot(y ~ x, exa,pch=".")
> lines(m ~ x,exa)
> lines(fd2 ~ x, exa, lty=5, lwd=2)
```

Again, we see a piecewise constant fit, but now the segments are of varying lengths. Where the function is relatively flat, we do not need the detail from the higher-order terms. Where the function is more variable, the finer detail is helpful.

We could view the thresholded coefficients as a compressed version of the original data (or signal). Some information has been lost in the compression, but the thresholding algorithm ensures that we tend to keep the detail we need, while throwing away noisier elements.

Even so, the fit is not particularly good because the fit is piecewise constant. We would like to use continuous basis functions while retaining the orthogonality and multiresolution properties. Families of such functions were discovered recently and described in Daubechies (1991). We illustrate such a form on our data:

```
> wds <- wd(exa$y,filter.number=2,bc="interval")
> draw(filter.number=2,family="DaubExPhase")
> plot(wds)
```

The mother wavelet takes an unusual form. The function is not explicitly defined, but is implicitly computed from the method for making the wavelet decomposition. Now we try the default thresholding and reconstruct the fit:

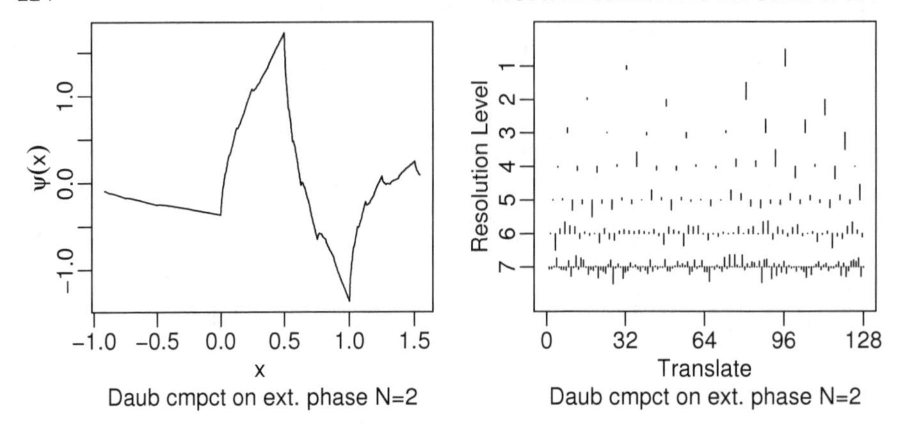

Figure 11.12 *Mother wavelet is shown in the left panel — the Daubechies orthonormal compactly supported wavelet N=2 from the extremal phase family. The right panel shows the wavelet coefficients.*

```
> wtd <- threshold(wds)
> fd <- wr(wtd)
> plot(y ~ x, exa,pch=".")
> lines(m ~ x,exa)
> lines(fd ~ x, exa, lty=2)
```

We can see the fit in Figure 11.13. Although the fit follows the true function quite well, there is still some roughness.

Example A does not illustrate the true strengths of the wavelet approach which is not well suited to smoothly varying functions with the default choice of wavelet. It comes into its own on functions with discontinuities and higher dimensions such as two-dimensional image data.

11.5 Other Methods

Nearest Neighbor: In this method, we set $\hat{f}_\lambda(x) =$ average of λ nearest neighbors of x. We let the window width vary to accommodate the same number of points. We need to pick the number of neighbors to be used; cross-validation can be used for this purpose.

Variable Bandwidth: When the true f varies a lot, a small bandwidth is good, but where it is smooth, a large bandwidth is preferable. This motivates this estimator, suitable for evenly spaced data:

$$\hat{f}_{\lambda(x)}(x) = \frac{1}{\lambda(x)} \sum_{i=1}^{n} K\left(\frac{x-x_j}{\lambda(x)}\right) y_i$$

This is an appealing idea in principle, but it is not easy to execute in practice, because it requires prior knowledge of the relative smoothness of the function across the range of the data. A pilot estimate may be used, but it has been difficult to make this work in practice.

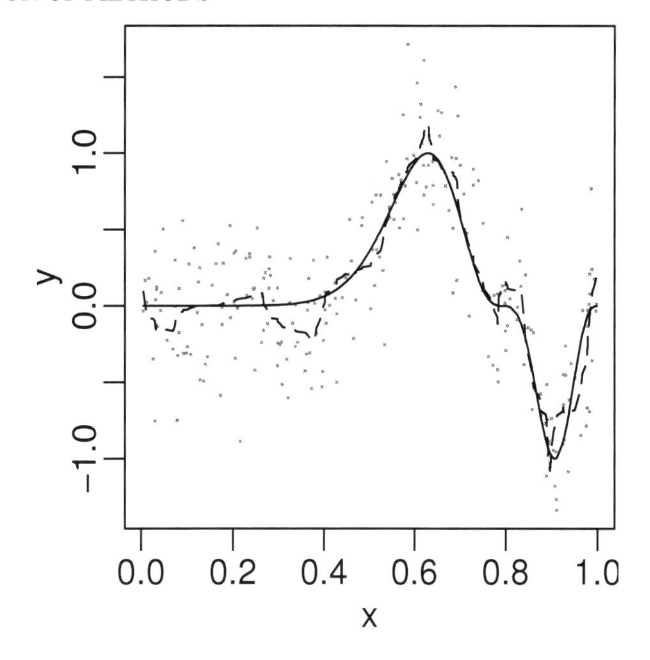

Figure 11.13 *Daubechies wavelet N=2 thresholded fit to the Example A data.*

Running Medians: Nonparametric regression is more robust than parametric regression in a model sense, but that does not mean that it is robust to outliers. Local averaging-based methods are sensitive to outliers, so medians can be useful. We let $N(x,\lambda) = \{i : x_i \text{ is one of the } \lambda\text{-nearest neighbors of } x\}$ then:

$$\hat{f}_\lambda(x) = \text{median}\{Y_i \text{ such that } i \in N(x,\lambda)\}$$

This method is robust to outliers, but produces a rough-looking fit. One might want to smooth again using another method. This is called *twicing*.

Others: The construction of alternate smoothing methods has long been a popular topic of interest for statisticians and researchers in other fields. Because no definitive solution is possible, this has encouraged the development of a wide range of methods. Some are intended for general use while others are customized for a particular application.

11.6 Comparison of Methods

In the univariate case, we can describe three situations. When there is very little noise, interpolation (or at most, very mild smoothing) is the best way to recover the relation between x and y. When there is a moderate amount of noise, nonparametric methods are most effective. There is enough noise to make smoothing worthwhile but also enough signal to justify a flexible fit. When the amount of noise becomes larger, parametric methods become relatively more attractive. There is insufficient signal to justify anything more than a simple model.

It is not reasonable to claim that any one smoother is better than the rest. The best choice of smoother will depend on the characteristics of the data and knowledge about the true underlying relationship. The choice will also depend on whether the fit is to be made automatically or with human intervention. When only a single dataset is being considered, it's simple enough to craft the fit and intervene if a particular method produces unreasonable results. If a large number of datasets are to be fit automatically, then human intervention in each case may not be feasible. In such cases, a reliable and robust smoother may be needed.

We think the loess smoother makes a good all-purpose smoother. It is robust to outliers and yet can produce smooth fits. When you are confident that no outliers are present, smoothing splines is more efficient than local polynomials.

11.7 Multivariate Predictors

Given $\mathbf{x}_1, \ldots, \mathbf{x}_n$ where $\mathbf{x} \in \mathbb{R}^p$, we observe:

$$y_i = f(\mathbf{x}) + \varepsilon_i \quad i = 1, \ldots n$$

Many of the methods discussed previously extend naturally to higher dimensions, for example, the Nadaraya–Watson estimator becomes:

$$f_\lambda(\mathbf{x}) = \frac{\sum_{j=1}^n K(\frac{\mathbf{x}-\mathbf{x}_j}{\lambda}) Y_j}{\sum_{j=1}^n K(\frac{\mathbf{x}-\mathbf{x}_j}{\lambda})}$$

where the kernel K is typically spherically symmetric. The spline idea can be used with the introduction of thin plate splines and local polynomials can be naturally extended.

We can illustrate kernel smoothing in two dimensions:

```
> data(savings)
> y <- savings$sr
> x <- cbind(savings$pop15,savings$ddpi)
> sm.regression(x,y,h=c(1,1),xlab="pop15",ylab="growth",zlab="savings rate")
> sm.regression(x,y,h=c(5,5),xlab="pop15",ylab="growth",zlab="savings rate")
```

Developing such estimators is not so difficult but there are problems: Because nonparametric fits are quite complex, we need to visualize them to make sense of them and yet this cannot be done easily for more than two predictors. Most nonparametric regression methods rely on local smoothing; local averaging is the crudest example of this. However, to maintain a stable average we need sufficient points in the window. For data in high dimensions, the window will need to be wide to capture sufficient points to average. You need an extremely large number of points to cover a high-dimensional space to high density. This is known as the "curse of dimensionality," a term coined by Bellman (1961). In truth, it should not be called a curse, but rather a blessing, since information on additional variables should have some value, even if it is inconvenient. Our challenge is to make use of this information. Nonpara-

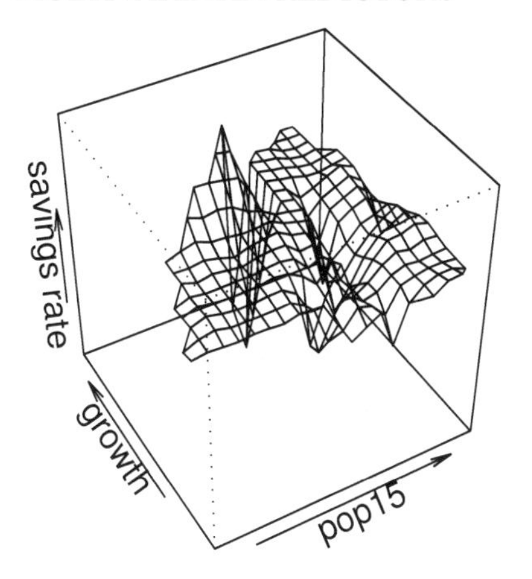

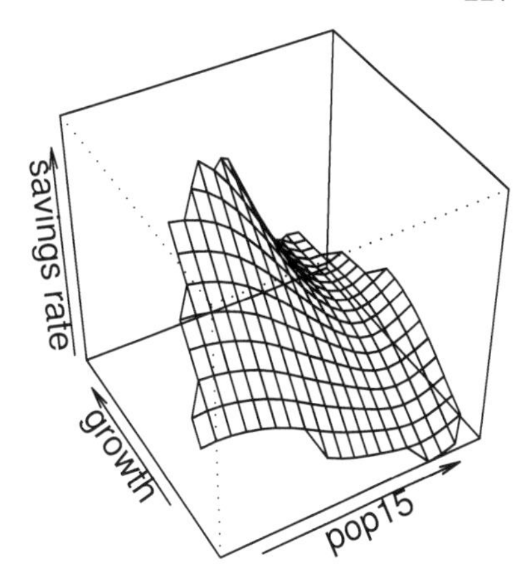

Figure 11.14 *Smoothing savings rate as a function growth and population under 15. Plot on the left is too rough while that on the right seems about right.*

metric regression fits are hard to interpret in higher dimensions where visualization is difficult. Simply extending the one-dimensional method is not effective.

The methods we describe in the following chapters impose additional restrictions on the fitted function to make the problem feasible and the results easier to interpret.

Further Reading: For a general review of smoothing methods, see Simonoff (1996). For books on specific methods of smoothing, see Loader (1999), Wahba (1990), Bowman and Azzalini (1997), Wand and Jones (1995) and Eubank (1988). The application of nonparametric regression to the goodness of fit problem may be found in Hart (1997).

Exercises

We have introduced kernel smoothers, splines, local polynomials, wavelets and other smoothing methods in this chapter. Apply these methods to the following datasets. You must choose the amount of smoothing you think is appropriate. Compare the fits from the methods. Comment on the features of the fitted curves. Comment on the advantage of the nonparametric approach compared to a parametric one for the data and, in particular, whether the nonparametric fit reveals structure that a parametric approach would miss.

1. The dataset `teengamb` concerns a study of teenage gambling in Britain. Take the variables `gamble` as the response and `income` as the predictor. Does a transformation of the data before smoothing help?
2. The dataset `uswages` is drawn as a sample from the Current Population Survey in 1988. Predict the `wage` from the years of `educ`ation. Compute the mean wage for each number of years of education and compare the result to the smoothed fit. Take the square root of the absolute value of the residuals from your chosen fit and smooth these as a function of `educ`.

3. The dataset `prostate` is from a study of 97 men with prostate cancer who were due to receive a radical prostatectomy. Predict the `lweight` using the `age`. How do the methods deal with the outlier?
4. The dataset `divusa` contains data on divorces in the United States from 1920 to 1996. Predict `divorce` from `year`. Predict `military` from `year`. There really were more military personnel during the Second World War, so these points are not outliers. How well do the different smoothers respond to this?
5. The `aatemp` data comes from the U.S. Historical Climatology network. They are the annual mean temperatures (in degrees Fahrenheit) in Ann Arbor, Michigan, going back about 150 years. Fit a smooth to the temperature as a function of year. Does the smooth help determine whether the temperature is changing over time?

CHAPTER 12

Additive Models

Suppose we have a response y and predictors $x_1, \ldots, x_p$. A linear model takes the form:

$$y = \beta_0 + \sum_{j=1}^{p} \beta_j X_j + \varepsilon$$

We can include transformations and combinations of the predictors among the xs, so this model can be very flexible. However, it can often be difficult to find a good model, given the wide choice of transformations available. We can try a systematic approach of fitting a family of transformations. For example, we can try polynomials of the predictors, but particularly if we include interactions, the number of terms becomes very large, perhaps greater than the sample size. Alternatively, we can use more interactive and graphical approaches that reward intuition over brute force. However, this requires some skill and effort on the part of the analyst. It is easy to miss important structure; a particular difficulty is that the methods only consider one variable at a time, when the secret to finding good transformations may require that variables be considered simultaneously.

We might try a nonparametric approach by fitting:

$$y = f(x_1, \ldots, x_p) + \varepsilon$$

This avoids the necessity of parametric assumptions about the form of the function f, but for p bigger than two or three, it is simply impractical to fit such models due to large sample size requirements, as discussed at the end of the previous chapter.

A good compromise between these extremes is the *additive model*:

$$y = \beta_0 + \sum_{j=1}^{p} f_j(X_j) + \varepsilon$$

where the f_j are smooth arbitrary functions. Additive models were introduced by Stone (1985).

Additive models are more flexible than the linear model, but still interpretable since the functions f_j can be plotted to give a sense of the marginal relationship between the predictor and the response. Of course, many linear models discovered during a data analysis take an additive form where the transformations are determined in an ad hoc manner by the analyst. The advantage of the additive model approach is that the best transformations are determined simultaneously and without parametric assumptions regarding their form.

In its basic form, the additive model will do poorly when strong interactions exist. In this case we might consider adding terms like $f_{ij}(x_i x_j)$ or even $f_{ij}(x_i, x_j)$ if there is sufficient data. Categorical variables can be easily accommodated within the model using the usual regression approach. For example:

$$y = \beta_0 + \sum_{j=1}^{p} f_j(X_j) + Z\gamma + \varepsilon$$

where Z is the design matrix for the variables that will not be modeled additively, where some may be quantitative and others qualitative. The γ are the associated regression parameters. We can also have an interaction between a factor and a continuous predictor by fitting a different function for each level of that factor. For example, we might have f_{male} and f_{female}.

There are at least three ways of fitting additive models in R. The `gam` package originates from the work of Hastie and Tibshirani (1990). The `mgcv` package is part of the recommended suite that comes with the default installation of R and is based on methods described in Wood (2000). The `gam` package allows more choice in the smoothers used while the `mgcv` package has an automatic choice in the amount of smoothing as well as wider functionality. The `gss` package of Gu (2002) takes a spline-based approach.

The fitting algorithm depends on the package used. The *backfitting* algorithm is used in the `gam` package. It works as follows:

1. We initialize by setting $\beta_0 = \bar{y}$ and $f_j(x) = \hat{\beta}_j x$ where $\hat{\beta}$ is some initial estimate, such as the least squares, for $j = 1, \ldots p$.

2. We cycle $j = 1, \ldots, p, 1, \ldots, p, 1, \ldots$

$$f_j = S(x_j, y - \beta_0 - \sum_{i \neq j} f_i(X_i))$$

   where $S(x, y)$ means the smooth on the data (x, y). The choice of S is left open to the user. It could be a nonparametric smoother like splines or loess, or it could be a parametric fit, say linear or polynomial. We can even use different smoothers on different predictors with differing amounts of smoothing.

The algorithm is iterated until convergence. Hastie and Tibshirani (1990) show that convergence is assured under some rather loose conditions. The term $y - \beta_0 - \sum_{i \neq j} f_i(x_i)$ is a partial residual — the result of fitting everything except x_j, making the connection to linear model diagnostics.

The `mgcv` package employs a penalized smoothing spline approach. Suppose we represent $f_j(x) = \sum_i \beta_i \phi_i(x)$ for a family of spline basis functions, ϕ_i. We impose a penalty $\int [f_j''(x)]^2 dx$ which can be written in the form $\beta_j^T S_j \beta_j$, for a suitable matrix S_j that depends on the choice of basis. We then maximize:

$$\log L(\beta) - \sum_j \lambda_j \beta_j^T S_j \beta_j$$

where $L(\beta)$ is likelihood with respect to β and the λ_js control the amount of smoothing for each variable. GCV is used select the λ_js.

12.1 Additive Models Using the gam Package

We use data from a study of the relationship between atmospheric ozone concentration, O_3 and other meteorological variables in the Los Angeles Basin in 1976. To simplify matters, we will reduce the predictors to just three: temperature measured at El Monte, temp, inversion base height at LAX, ibh, and inversion top temperature at LAX, ibt. A number of cases with missing variables have been removed for simplicity. The data were first presented by Breiman and Friedman (1985). First we fit a simple linear model for reference purposes:

```
> data(ozone)
> olm <- lm(O3 ~ temp + ibh + ibt, ozone)
> summary(olm)
Coefficients:
             Estimate Std. Error t value  Pr(>|t|)
(Intercept) -7.727982   1.621662   -4.77 0.0000028
temp         0.380441   0.040158    9.47   < 2e-16
ibh         -0.001186   0.000257   -4.62 0.0000055
ibt         -0.005821   0.010179   -0.57      0.57

Residual standard error: 4.75 on 326 degrees of freedom
Multiple R-Squared: 0.652,      Adjusted R-squared: 0.649
F-statistic:  204 on 3 and 326 DF,  p-value: <2e-16
```

Note that ibt is not significant in this model. One task among others in a regression analysis is to find the right transforms on the predictors. Additive models can help here. We fit an additive model using a Gaussian response as the default.

```
> library(gam)
> amgam <- gam(O3 ~ lo(temp) + lo(ibh) + lo(ibt), data=ozone)
> summary(amgam)
(Dispersion Parameter for gaussian family taken to be 18.664)

    Null Deviance: 21115 on 329 degrees of freedom
Residual Deviance: 5935.1 on 318 degrees of freedom
AIC: 1916.0

Number of Local Scoring Iterations: 2

DF for Terms and F-values for Nonparametric Effects

             Df Npar Df Npar F    Pr(F)
(Intercept) 1.0
lo(temp)    1.0     2.5   7.45  0.00025
lo(ibh)     1.0     2.9   7.62 0.000082
lo(ibt)     1.0     2.7   7.84 0.000099
```

We have used the loess smoother here by specifying lo in the model formula for all three predictors. Compared to the linear model, the R^2 is:

```
> 1-5935.1/21115
[1] 0.71892
```

So the fit is a little better. However, the loess fit does use more degrees of freedom. We can compute the equivalent degrees of freedom by an analogy to linear models. For linear smoothers, the relationship between the observed and fitted values may be written as $\hat{y} = Py$. The trace of P then estimates the effective number of parameters. For example, in linear regression, the projection matrix is $X(X^TX)^{-1}X^T$ whose trace is equal to the rank of X or the number of identifiable parameters. This notion can be used to obtain the degrees of freedom for additive models.

The gam package uses a score test for the predictors. However, the p-values are only approximate at best and should be viewed with some skepticism. It is generally better to fit the model without the predictor of interest and then construct the F-test:

```
> amgamr <- gam(O3 ~ lo(temp) + lo(ibh) , data=ozone)
> anova(amgamr,amgam,test="F")
Analysis of Deviance Table

Model 1: O3 ~ lo(temp) + lo(ibh)
Model 2: O3 ~ lo(temp) + lo(ibh) + lo(ibt)
  Resid. Df Resid. Dev      Df Deviance    F Pr(>F)
1    321.67       6045
2    318.00       5935    3.66        109 1.6   0.18
```

Again the p-value is an approximation, but we can see there is some evidence that ibt is not significant. We now examine the fit:

```
> plot(amgam,residuals=TRUE,se=TRUE,pch=".")
```

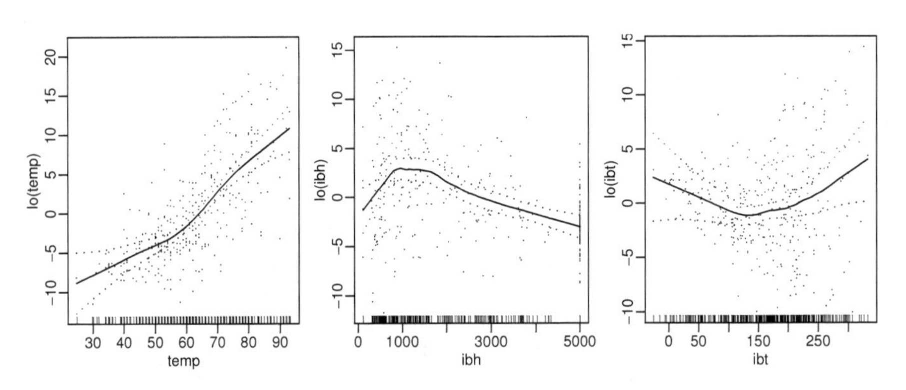

Figure 12.1 *Transformations on the predictors chosen by the* gam *fit on the ozone data. Partial residuals and approximate 95% pointwise confidence bands are shown.*

We see the transformations chosen in Figure 12.1. For ibt, a constant function would fit between the confidence bands. This reinforces the conclusion that this predictor is not significant. For temperature, we can see a change in the slope around 60°, while for ibh, there is a clear maximum. The partial residuals allow us to check for outliers or influential points. We see no such problems here. However, the use of the loess smoother is recommended where such problems arise and is perhaps the best feature of the gam package relative to the other options.

12.2 Additive Models Using mgcv

Another method of fitting additive models is provided by the mgcv package of Wood (2000). We demonstrate its use on the same data. Splines are the only choice of smoother, but the appropriate amount of smoothing is internally chosen by default. In contrast, the gam package relies on the user to select this. There are advantages and disadvantages to automatic smoothing selection. On the positive side, it avoids the work and subjectivity of making the selection by hand, but on the negative side, automatic selection can fail and human intervention may sometimes be necessary. We follow a similar analysis:

```
> library(mgcv)
> ammgcv <- gam(O3 ~ s(temp)+s(ibh)+s(ibt),data=ozone)
> summary(ammgcv)
Parametric coefficients:
               Estimate  std. err.    t ratio    Pr(>|t|)
(Intercept)      11.776     0.2382      49.44    <2e-16
Approximate significance of smooth terms:
                edf         chi.sq      p-value
s(temp)        3.386        88.047      <2e-16
 s(ibh)        4.174        37.559      4.28e-07
 s(ibt)        2.112        4.2263      0.134

R-sq.(adj) =   0.708   Deviance explained = 71.7%
GCV score = 19.346   Scale est. = 18.72      n = 330
```

We see that the R^2 is similar to the gam fit. We also have additional information concerning the significance of the predictors. We can also examine the transformations used:

```
> plot(ammgcv)
```

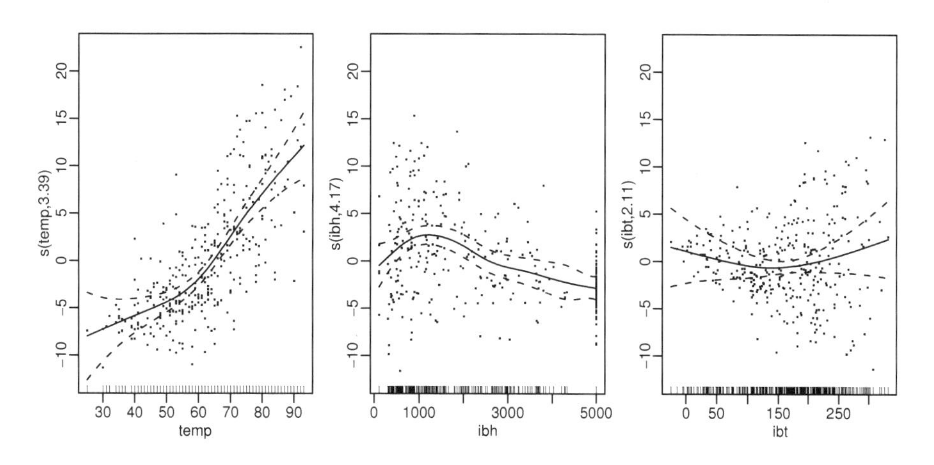

Figure 12.2 *Transformation functions for the model fit by mgcv. Note how the same scale has been deliberately used on all three plots. This allows us to easily compare the relative contribution of each variable.*

The chosen transformations are again similar. We see that ibt does not appear to be significant. We might also be interested in whether there really is a change in the trend for temperature. We test this by fitting a model with a linear term in temperature and then make the F-test:

```
> am1 <- gam(O3 ~ s(temp)+s(ibh),data=ozone)
> am2 <- gam(O3 ~ temp+s(ibh),data=ozone)
> anova(am2,am1,test="F")
Analysis of Deviance Table

Model 1: O3 ~ temp + s(ibh)
Model 2: O3 ~ s(temp) + s(ibh)
  Resid. Df Resid. Dev      Df Deviance     F Pr(>F)
1    323.67       6950
2    320.97       6054    2.70       896 17.6   8e-10
```

The p-value is only approximate, but it certainly seems there really is a change in the trend.

You can also do bivariate transformations with mgcv. For example, suppose we suspect that there is an interaction between temperature and IBH. We can fit a model with this term:

```
> amint <- gam(O3 ~ s(temp,ibh)+s(ibt),data=ozone)
> summary(amint)
Parametric coefficients:
              Estimate  std. err.    t ratio    Pr(>|t|)
(Intercept)     11.776     0.2409      48.88    <2e-16

Approximate significance of smooth terms:
                   edf        chi.sq      p-value
s(temp,ibh)      6.346        120.46      <2e-16
     s(ibt)      2.917        36.081      1.57e-07

R-sq.(adj) =  0.702   Deviance explained =   71%
GCV score = 19.767   Scale est. = 19.152    n = 330
```

We compare this to the previous additive model:

```
> anova(ammgcv,amint,test="F")
Analysis of Deviance Table

Model 1: O3 ~ s(temp) + s(ibh) + s(ibt)
Model 2: O3 ~ s(temp, ibh) + s(ibt)
  Resid. Df Resid. Dev      Df Deviance     F Pr(>F)
1   319.327       5978
2   319.737       6124  -0.409      -146 19.0 0.0014
```

We see that the supposedly more complex model with the bivariate fit actually fits worse than the model with univariate functions. This is because fewer degrees of freedom have been used to fit the bivariate function than the two corresponding univariate functions. In spite of the output p-value, we suspect that there is no interaction effect, because the fitting algorithm is able to fit the bivariate function so simply. We now graphically examine the fit as seen in Figure 12.3:

```
> plot(amint)
> vis.gam(amint,theta=-45,color="gray")
```

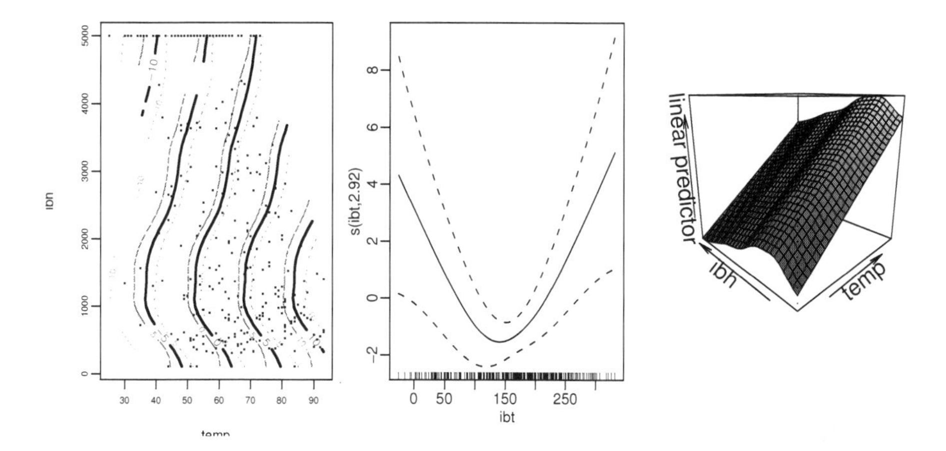

Figure 12.3 *The bivariate contour plot for temperature and* `ibh` *is shown in the left panel. The middle panel shows the univariate transformation on* `ibt` *while the right panel shows a perspective view of the information on the left panel.*

Given that the contours appear almost parallel and the perspective view looks like it could be constructed with piece of paper rippled in one direction, we conclude that there is no significant interaction. One interesting side effect is that `ibt` is now significant.

One use for additive models is as an exploratory tool for standard parametric regression modeling. We can use the fitted functions to help us find suitable simple transformations of the predictors. One idea here is to model the `temp` and `ibh` effects using piecewise linear regression (also known as "broken stick" or segmented regression). We define the right and left "hockey-stick" functions:

```
> rhs <- function(x,c) ifelse(x > c, x-c, 0)
> lhs <- function(x,c) ifelse(x < c, c-x, 0)
```

and now fit a parametric model using cutpoints of 60 and 1000 for `temp` and `ibh`, respectively. We pick the cutpoints using the plots:

```
> olm2 <- lm(O3 ~ rhs(temp,60)+lhs(temp,60)+rhs(ibh,1000)+lhs(ibh,1000),
  ozone)
> summary(olm2)
Coefficients:
                Estimate Std. Error t value Pr(>|t|)
(Intercept)    11.603832   0.622651   18.64  < 2e-16
rhs(temp, 60)   0.536441   0.033185   16.17  < 2e-16
lhs(temp, 60)  -0.116173   0.037866   -3.07   0.0023
rhs(ibh, 1000) -0.001486   0.000198   -7.49  6.7e-13
lhs(ibh, 1000) -0.003554   0.001314   -2.71   0.0072

Residual standard error: 4.34 on 325 degrees of freedom
```

```
Multiple R-Squared: 0.71,      Adjusted R-squared: 0.706
F-statistic:  199 on 4 and 325 degrees of freedom,     p-value:   0
```

Compare this model to the first linear model we fit to this data. The fit is better and about as good as the additive model fit. It is unlikely we could have discovered these transformation without the help of the intermediate additive models. Furthermore, the linear model has the advantage that we can write the prediction formula in a compact form.

We can use additive models for building a linear model as above, but they can be used for inference in their own right. For example, we can predict new values with standard error:

```
> predict(ammgcv,data.frame(temp=60,ibh=2000,ibt=100),se=T)
$fit
[1] 11.013

$se.fit
[1] 0.97278
```

If we try to make predictions for predictor values outside the original range of the data, we will need to linearly extrapolate the spline fits. This is dangerous for all the usual reasons:

```
> predict(ammgcv,data.frame(temp=120,ibh=2000,ibt=100),se=T)
$fit
[1] 35.511

$se.fit
[1] 5.7261
```

We see that the standard error is much larger although this likely does not fully reflect the uncertainty.

We should also check the usual diagnostics:

```
> plot(predict(ammgcv),residuals(ammgcv),xlab="Predicted",ylab="Residuals")
> qqnorm(residuals(ammgcv),main="")
```

We can see in Figure 12.4 that although the residuals look normal, there is some nonconstant variance.

Now let's see the model for the full dataset. We found that the `ibh` and `ibt` terms were insignificant and so we removed them:

```
> amred <- gam(O3 ~ s(vh)+s(wind)+s(humidity)+s(temp)+s(dpg)+
  s(vis)+s(doy),data=ozone)
> summary(amred)
Approximate significance of smooth terms:
                 edf        chi.sq      p-value
       s(vh)       1        20.497      0.00000852
     s(wind)       1        6.5571      0.0109
 s(humidity)       1        14.608      0.00016
     s(temp)   5.769        87.825      7.36e-15
      s(dpg)   3.312        59.782      1.26e-11
      s(vis)   2.219        20.731      0.00006
      s(doy)   4.074        106.69      <2e-16
```

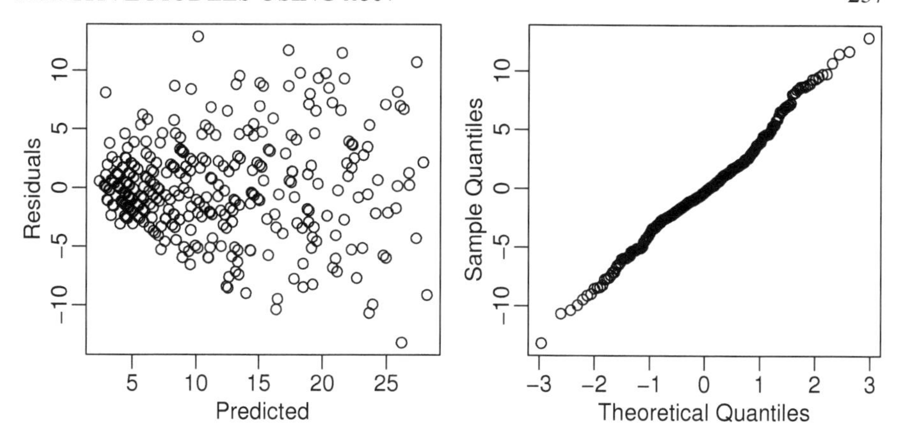

Figure 12.4 *Residuals plots for the additive model.*

```
R-sq.(adj) =  0.793   Deviance explained = 80.5%
GCV score = 14.113   Scale est. = 13.285    n = 330
```

We will compare this to the results of different modeling approaches that we will present later. We can see that we achieve a good fit with an R^2 of 80.5%, but at the cost of using effectively 19.4 parameters including the intercept.

Also for future reference, here is the linear model with all insignificant terms removed:

```
> alm <- lm(O3 ~ vis+doy+ibt+humidity+temp,data=ozone)
Coefficients:
             Estimate Std. Error t value Pr(>|t|)
(Intercept) -10.01786    1.65306   -6.06  3.8e-09
vis          -0.00820    0.00369   -2.22    0.027
doy          -0.01020    0.00245   -4.17  3.9e-05
ibt           0.03491    0.00671    5.21  3.4e-07
humidity      0.08510    0.01435    5.93  7.7e-09
temp          0.23281    0.03607    6.45  4.0e-10

Residual standard error: 4.43 on 324 degrees of freedom
Multiple R-Squared: 0.699,      Adjusted R-squared: 0.694
F-statistic:  150 on 5 and 324 DF,  p-value: <2e-16
```

We can see that the fit is substantially worse, but uses only six parameters. Of course, we may be able to improve this fit with some manual data analysis. We could look for good transformations and check for outliers and influential points. However, since we want to compare different modeling techniques, we want to avoid making subjective interventions for the sake of a fair comparison.

12.3 Generalized Additive Models

In generalized linear models:

$$\eta = X\beta \qquad EY = \mu \qquad g(\mu) = \eta \qquad Var(Y) \propto V(\mu)$$

The approach is readily extended to additive models to form generalized additive models (GAM). The fitting process is different in the mgcv and gam packages. The mgcv package takes a likelihood approach, so the implementation of the fitting algorithm is conceptually straightforward. The gam package uses a backfitting approach as described below.

Recalling the GLM fitting method described in Section 6.2, the iterative reweighted least squares (IRWLS) fitting algorithm starts from some reasonable μ_0, forms the "adjusted dependent variate" $z_0 = \hat{\eta}_0 + (y - \hat{\mu}_0)\frac{d\eta}{d\mu}|_{\hat{\eta}_0}$ and weights $w_0^{-1} = \left(\frac{d\eta}{d\mu}\right)^2 |_{\hat{\eta}_0} V(\hat{\mu}_0)$. It then regresses z on X using weights w using weighted least squares to get $\hat{\eta}_1$. The process is repeated until convergence.

In generalized additive models the linear predictor becomes:

$$\beta_0 + \sum_{j=1}^{p} f_j(X_j)$$

and we just add an iteration step to estimate the f_js. There are two levels of iteration: the GLM part where z and w are computed and the additive model part. We need to use a smoother that understands weights like loess or splines.

The ozone data has a response with relatively small integer values. Furthermore, the diagnostic plot in Figure 12.4 shows nonconstant variance. This suggests that a Poisson response might be suitable. We fit this using the mgcv package:

```
> gammgcv <- gam(O3 ~ s(temp)+s(ibh)+s(ibt),family=poisson,
  scale=-1,data=ozone)
> summary(gammgcv)
Parametric coefficients:
             Estimate  std. err.    t ratio    Pr(>|t|)
(Intercept)    2.2927    0.02304      99.51    <2e-16

Approximate significance of smooth terms:
              edf       chi.sq      p-value
s(temp)      3.803      79.802      8.19e-15
 s(ibh)      3.779      48.471      2.59e-09
 s(ibt)      1.422     0.94684      0.465

R-sq.(adj) =   0.712   Deviance explained = 72.9%
GCV score = 1.5025    Scale est. = 1.4569     n = 330
```

We have set scale=-1 because negative values for this parameter indicate that the dispersion should be estimated rather than fixed at one. Since we do not truly believe the response is Poisson, it seems wise to allow for overdispersion. The default of not specifying scale would fix the dispersion at one. We see that the estimated dispersion is indeed somewhat bigger than one. We see that IBT is not significant. We can check the transformations on the predictors as seen in Figure 12.5:

```
> plot(gammgcv,residuals=TRUE)
```

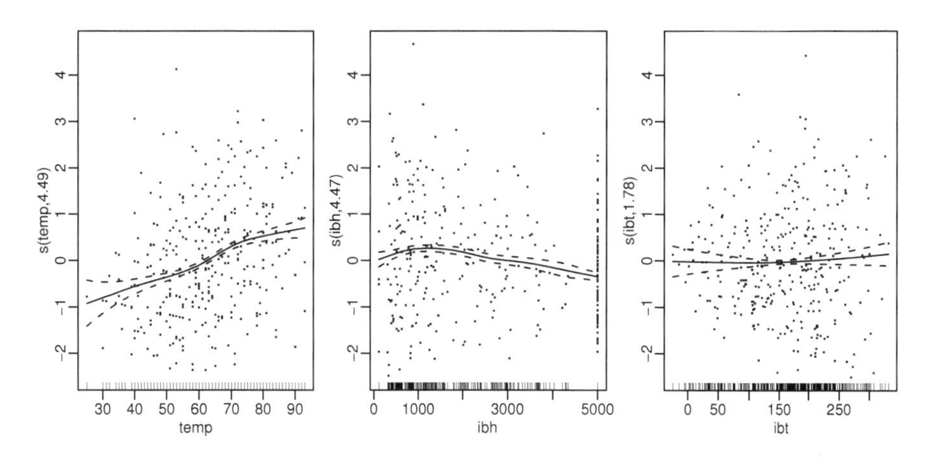

Figure 12.5 *Transformation on the predictors for the Poisson GAM.*

We see that the selected transformations are quite similar to those observed previously.

12.4 Alternating Conditional Expectations

In the additive model:

$$y = \alpha + \sum_{j=1}^{p} f_j(X_j) + \varepsilon$$

but in the transform both sides (TBS) model:

$$\theta(y) = \alpha + \sum_{j=1}^{p} f_j(X_j) + \varepsilon$$

For example, $y = e^{x_1+\sqrt{x_2}}$ cannot be modeled well by additive models, but can if we transform both sides: $\log y = x_1 + \sqrt{x_2}$. This fits within the transform-both-sides (TBS) model framework. A more complicated alternative approach would be nonlinear regression. One particular way of fitting TBS models is *alternating conditional expectation* (ACE) which is designed to minimize $\sum_i(\theta(y_i) - \sum f_j(x_{ij}))^2$. Distractingly, this can be trivially minimized by setting $\theta = f_j = 0$ for all j. To avoid this solution, we impose the restriction that the variance of $\theta(y)$ be one. The fitting proceeds using the following algorithm:

1. Initialize:

$$\theta(y) = \frac{y - \bar{y}}{SD(y)} \qquad f_j = \hat{\beta}_j x_j \qquad j = 1, \dots p$$

2. Cycle:

$$f_j \quad = \quad S(x_j, \theta(y) - \sum_{i \neq j} f_i(x_i))$$

$$\theta = S(y, \sum_j f_j(x_j))$$

Renormalize at the end of each cycle:

$$\theta(y) \leftarrow \frac{\theta(y) - \overline{\theta(y)}}{SD(\overline{\theta(y)})}$$

We repeat until convergence. ACE is comparable to the additive model, except now we allow transformation of the response as well. In principle, you can use any reasonable smoother S, but the original smoother used was the supersmoother. This cannot be easily changed in the R software implementation.

For our example, we start with the same three predictors in the `ozone` data:

```
> x <- ozone[,c("temp","ibh","ibt")]
> library(acepack)
> acefit <- ace(x,ozone$O3)
```

Note that the `ace` function interface is quite rudimentary as we must give it the X matrix explicitly. The function returns the components `ty` which contains $\theta(y)$ and `tx` which is a matrix whose columns contain the $f_j(x_j)$. We can get a sense of how well these transformations work by fitting a linear model that uses the transformed variables:

```
> summary(lm(acefit$ty ~ acefit$tx))
Coefficients:
              Estimate Std. Error t value Pr(>|t|)
(Intercept)    9.2e-18     0.0290 3.2e-16   1.0000
acefit$txtemp   0.9676     0.0509   19.01  < 2e-16
acefit$txibh    1.1801     0.1360    8.68  2.2e-16
acefit$txibt    1.3712     0.5123    2.68   0.0078

Residual standard error: 0.527 on 326 degrees of freedom
Multiple R-Squared: 0.726,      Adjusted R-squared: 0.723
F-statistic:  288 on 3 and 326 degrees of freedom,      p-value:    0
```

All three transformed predictors are strongly significant and the fit is superior to the original model. The R^2 for the comparable additive model was 0.703. So the additional transformation of the response did improve the fit. Now we examine the transforms on the response and the three predictors:

```
> plot(ozone$O3,acefit$ty,xlab="O3",
  ylab=expression(theta(O3)))
> plot(x[,1],acefit$tx[,1],xlab="temp",ylab="f(temp)")
> plot(x[,2],acefit$tx[,2],xlab="ibh",ylab="f(ibh)")
> plot(x[,3],acefit$tx[,3],xlab="ibt",ylab="f(ibt)")
```

See Figure 12.6. The transform on the response is close to, but not quite, linear. The transformations on `temp` and `ibh` are similar to those found by the additive model. The transformation for `ibt` looks implausibly rough in some parts.

Now let's see how we do on the full data:

```
> x <- ozone[,-1]
> acefit <- ace(x,ozone$O3)
```

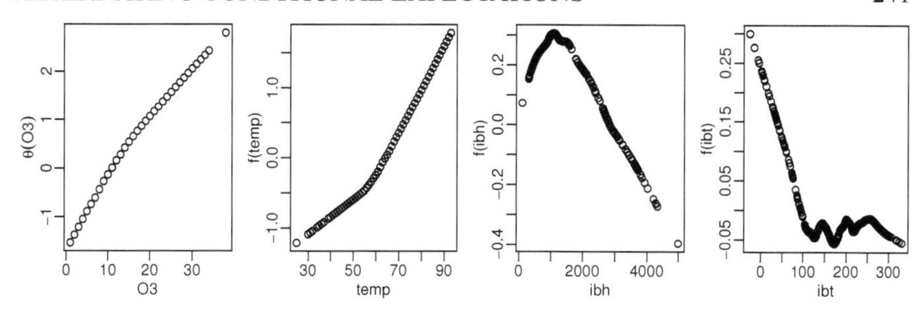

Figure 12.6 *ACE transformations: the first panel shows the transformation on the response while the remaining three show the transformations on the predictors.*

```
> summary(lm(acefit$ty ~ acefit$tx))
Coefficients:
                  Estimate Std. Error  t value Pr(>|t|)
(Intercept)        -5.8e-17     0.0225 -2.6e-15   1.0000
acefit$txvh          1.1715     0.3852     3.04   0.0026
acefit$txwind        1.0739     0.4047     2.65   0.0084
acefit$txhumidity    0.6515     0.2455     2.65   0.0084
acefit$txtemp        0.9163     0.1236     7.41  1.1e-12
acefit$txibh         1.3510     0.4370     3.09   0.0022
acefit$txdpg         1.3217     0.1672     7.91  4.4e-14
acefit$txibt         0.9256     0.1967     4.70  3.8e-06
acefit$txvis         1.3864     0.2303     6.02  4.8e-09
acefit$txdoy         1.2837     0.1097    11.70  < 2e-16

Residual standard error: 0.409 on 320 degrees of freedom
Multiple R-Squared: 0.838,      Adjusted R-squared: 0.833
F-statistic:  184 on 9 and 320 degrees of freedom,      p-value:    0
```

A very good fit, but we must be cautious. Notice that all the predictors are strongly significant. This might be a reflection of reality or it could just be that the ACE model is overfitting the data by using implausible transformations as seen on the `ibt` variable above.

ACE can be useful in searching for good transformations while building a linear model. We might examine the fitted transformations as seen in Figure 12.6 to suggest appropriate parametric forms. More caution is necessary if the model is to be used in its own right, because of the tendency to overfit.

An alternative view of ACE is to consider the problem of choosing θ and f_j's such that $\theta(Y)$ and $\sum_j f_j(X_j)$ are maximally correlated. ACE solves this problem. For this reason, ACE can be viewed as a correlation method rather than a regression method.

The canonical correlation method is an ancestor to ACE. Given two sets of random variables $X_1, \ldots X_m$ and $Y_1, \ldots Y_n$, we find unit vectors a and b such that:

$$corr(a^T X, b^T Y)$$

is maximized. One generalization of canonical correlation is to allow some the X's and Y's to be power transforms of the original variables; this results in a parametric form of ACE. For example:

```
> y <- cbind(ozone$O3,ozone$O3^2,sqrt(ozone$O3))
> x <- ozone[,c("temp","ibh","ibt")]
> cancor(x,y)
$cor
[1] 0.832346 0.217517 0.016908

$xcoef
              [,1]        [,2]         [,3]
[1,] -3.4951e-03  3.6335e-03 -6.7913e-03
[2,]  1.3667e-05 -5.2054e-05 -5.2243e-06
[3,]  1.6744e-04 -1.7384e-03  1.2436e-03

$ycoef
              [,1]        [,2]       [,3]
[1,] -0.00390830 -0.00539076 -0.1802230
[2,]  0.00009253 -0.00044172  0.0022167
[3,] -0.03928664  0.12068982  0.7948130
```

We see that it is possible to obtain a correlation of 0.832 by taking particular linear combinations of $O3$, $O3^2$ and $\sqrt{(O3)}$ with the three predictors. The other two orthogonal combinations are not of interest to us here. Remember that R^2 is the correlation squared in a simple linear model and $0.832^2 = 0.692$, so this is not a particularly competitive fit.

There are some oddities about ACE. For a single predictor, ACE is symmetric in X and Y, which is not the usual situation in regression. Furthermore, ACE does not necessarily reproduce the true model. Consider the population form of the problem and take $Y = X + \varepsilon$ and $\varepsilon \sim N(0,1)$ and $X \sim U(0,1)$, then $E(Y|X) = X$ but $E(X|Y) \neq Y$ which is not what one might expect, because f and θ will not both be identity transformations as the model might suggest.

12.5 Additivity and Variance Stabilization

Additivity and variance stabilization (AVAS) is another TBS model and is quite similar to ACE. We choose the f_j to optimize the fit, but we also want constant variance for the response:

$$var[\theta(Y)|\sum_{j=1}^{p} f_j(X_j)] = \text{constant}$$

So we choose the f_j's to get a good additive fit and choose the θ to get constant variance.

Here is how the method of fitting θ works: Suppose $Var(Y) \equiv V(Y)$ is not constant. We transform to constancy by:

$$\theta(t) = \int_0^t \frac{d\mu}{\sqrt{V(\mu)}}$$

We use data to estimate $V(y)$, then get θ. The purpose of the AVAS method is to obtain additivity and variance stabilization and not necessarily to produce the best possible fit. We demonstrate its application on the `ozone` data:

```
> avasfit <- avas(x,ozone$O3)
```

Plot the transformations selected:

```
> plot(ozone$O3,avasfit$ty,xlab="O3",ylab=expression(theta(O3)))
> plot(x[,1],avasfit$tx[,1],xlab="temp",ylab="f(temp)")
> plot(x[,2],avasfit$tx[,2],xlab="ibh",ylab="f(ibh)")
> plot(x[,3],avasfit$tx[,3],xlab="ibt",ylab="f(ibt)")
```

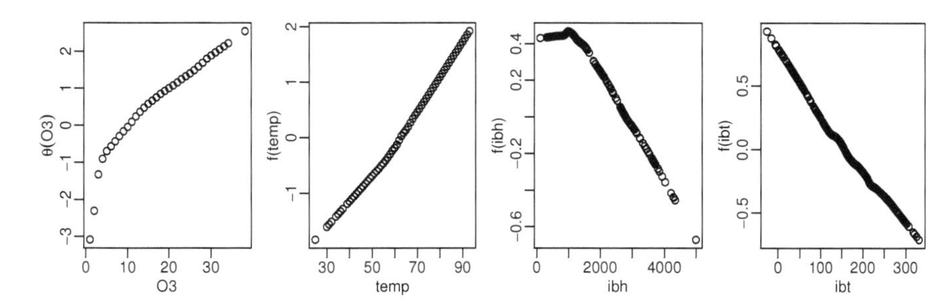

Figure 12.7 *AVAS transformations — the first panel shows the transformation on the response while the remaining three show the transformations on the predictors.*

See Figure 12.7. It would be convenient if the transformation on the response matched a simple functional form. We see if this is possible. We need to sort the response to get the line plots to work:

```
> i <- order(ozone$O3)
> plot(ozone$O3[i],avasfit$ty[i],type="l",xlab="O3",ylab=expression(theta(O3)))
> gs <- lm(avasfit$ty[i] ~ sqrt(ozone$O3[i]))
> lines(ozone$O3[i],gs$fit,lty=2)
> gl <- lm(avasfit$ty[i] ~ log(ozone$O3[i]))
> lines(ozone$O3[i],gl$fit,lty=5)
```

See the ;eft panel of Figure 12.8. We have shown the square-root fit as a dotted line and log fit as a dashed line. Neither one fits well across the whole range. Now look at the overall fit:

```
> lmod <- lm(avasfit$ty ~ avasfit$tx)
> summary(lmod)
Coefficients:
                 Estimate Std. Error t value Pr(>|t|)
(Intercept)      2.87e-07   3.10e-02 9.3e-06    1.000
avasfit$txtemp   9.02e-01   7.50e-02   12.02  < 2e-16
avasfit$txibh    7.98e-01   1.09e-01    7.33  1.8e-12
avasfit$txibt    5.69e-01   2.39e-01    2.38    0.018

Residual standard error: 0.563 on 326 degrees of freedom
Multiple R-Squared: 0.687,      Adjusted R-squared: 0.684
F-statistic:  238 on 3 and 326 degrees of freedom,      p-value:    0
```

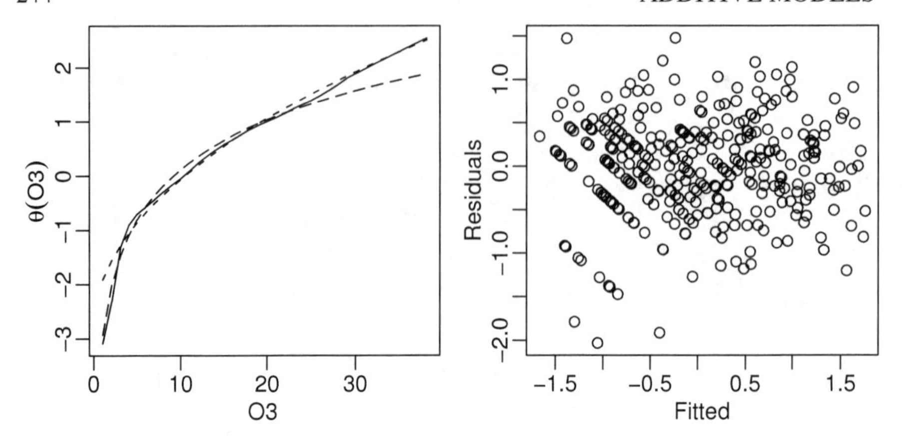

Figure 12.8 *The left panel checks for simple fits to the AVAS transformation on the response given by the solid line. The log fit is given by the dashed line while the square-root fit is given by the dotted line. The right panel shows the residuals vs. fitted values plot for the AVAS model.*

The fit is not so good, but check the diagnostics:

```
> plot(predict(lmod),residuals(lmod),xlab="Fitted",ylab="Residuals")
```

The plot is shown in the right panel of Figure 12.8.

AVAS does not optimize the fit; it trades some of the optimality in order to obtain constant variance. Whether this is a good trade depends on how much relative value you put on the accuracy of point predictions and accurate estimation of the standard error of prediction. In other words, is it more important to try to be right or to know how much you are wrong? The choice will depend on the application.

12.6 Generalized Additive Mixed Models

The generalized additive mixed model (GAMM) manages to combine the three major themes of this book. The response can be nonnormal from the exponential family of distributions. The error structure can allow for grouping and hierarchical arrangements in the data. Finally we can allow for smooth transformations of the response. We demonstrate this method on the epilepsy data from Section 10.2:

```
> data(epilepsy)
> egamm <- gamm(seizures ~ treat*expind+s(age),family=poisson,
  random=list(id=~1),data=epilepsy,subset=(id!=49))
> summary(egamm$gam)
Parametric coefficients:
                Estimate  std. err.    t ratio    Pr(>|t|)
 (Intercept)      3.1607     0.1435      22.02    <2e-16
       treat   -0.010368     0.2001   -0.05182    0.959
      expind     -1.2745    0.07574     -16.83    <2e-16
treat:expind    -0.30238     0.1126     -2.684    0.00769
```

```
Approximate significance of smooth terms:
            edf       chi.sq     p-value
s(age)     1.014     0.30698     0.586

R-sq.(adj) =  0.328  Scale est. = 2.5754    n = 290
```

We see that the age effect is not significant. Again the interaction effect is significant which shows, in this case, a beneficial effect for the drug. We would like to use an offset here for compatibility with the previous analysis.

12.7 Multivariate Adaptive Regression Splines

Multivariate adaptive regression splines (MARS) were introduced by Friedman (1991). We wish to find a model of the form:

$$\hat{f}(x) = \sum_{j=1}^{k} c_j B_j(x)$$

where the basis functions, $B_j(x)$, are formed from products of terms of the form $[\pm(x_i - t)]_+^q$. The $[\,]_+$ denotes taking the positive part. For $q = 1$, this is sometimes called a hockey-stick function and can be seen in the right panel of Figure 11.6. The $q = 1$ case is the most common choice and one might think this results in a piecewise linear fit, as in Section 11.2. However, the fit is more flexible than this. Consider the product of two hockey-stick functions in one dimension; this forms a quadratic shape. Furthermore, if we form the product of terms in two or more variables, we have an interaction term.

The model building proceeds iteratively. We start with no basis functions. We then search over all variables and possible knotpoints t to find the one basis function that produces the best fit to the data. We now repeat this process to find the next best basis function addition given that the first basis function is already included in the model. We might impose rules on what new basis functions are included. For example, we might disallow interactions or only allow two-way interactions. This will enhance interpretability, possibly at the cost of fit. The number of basis functions added determines the overall smoothness of the fit. We can use cross-validation to determine how many basis functions are enough.

When interactions are disallowed, the MARS approach will be a type of additive model. The MARS model building will be iterative, in contrast to the fit using the `gam` function of the `mgcv` package that fits and determines overall smoothness in one step. If a strictly additive model is all that is needed, the `mgcv` approach will typically be more effective. The MARS approach will have a relative advantage when interactions are considered, particularly when there are a larger number of variables. Here there will be a large number of potential interactions that cannot be simultaneously entertained. The iterative approach of MARS will be more appropriate here.

We apply the MARS method to the `ozone` dataset. The `mda` library needs to be installed and loaded:

```
> library(mda)
> data(ozone)
```

```
> a <- mars(ozone[,-1],ozone[,1])
```

The interface is quite rudimentary. The default choice allows only additive (first-order) predictors and chooses the model size using a GCV criterion. The basis functions can be used as predictors in a linear regression model:

```
> summary(lm(ozone[,1] ~ a$x[,-1]))
Coefficients:
              Estimate Std. Error t value Pr(>|t|)
(Intercept)   8.993451   0.804557  11.178  < 2e-16
a$x[, -1]1    0.246074   0.053263   4.620 5.58e-06
a$x[, -1]2   -0.002999   0.001075  -2.788 0.005614
a$x[, -1]3    0.047841   0.017223   2.778 0.005799
a$x[, -1]4   -0.113577   0.023158  -4.904 1.50e-06
a$x[, -1]5   -0.101047   0.014229  -7.101 8.12e-12
a$x[, -1]6    0.024818   0.005682   4.368 1.70e-05
a$x[, -1]7    0.017516   0.004713   3.716 0.000239
a$x[, -1]8   -0.091784   0.020821  -4.408 1.43e-05
a$x[, -1]9   -0.138902   0.023019  -6.034 4.44e-09
a$x[, -1]10 -0.553584   0.173751  -3.186 0.001585
a$x[, -1]11  0.029704   0.010743   2.765 0.006023

Residual standard error: 3.637 on 318 degrees of freedom
Multiple R-Squared: 0.8008,Adjusted R-squared: 0.7939
F-statistic: 116.2 on 11 and 318 DF,  p-value: < 2.2e-16
```

The fit is good in terms of R^2, but the model size is also larger. It is also an additive model, so we can reasonably compare it to the additive model presented at the end of Section 12.2. That model had an adjusted R^2 of 79.3% using 19.4 parameters.

Let's reduce the model size to that used for previous models. The parameter `nk` controls the maximum number of model terms:

```
> a <- mars(ozone[,-1],ozone[,1],nk=7)
> summary(lm(ozone[,1] ~ a$x[,-1]))
Coefficients:
              Estimate Std. Error t value Pr(>|t|)
(Intercept) 12.6634816  0.7513996  16.853  < 2e-16
a$x[, -1]1   0.4839477  0.0297353  16.275  < 2e-16
a$x[, -1]2  -0.0964835  0.0431022  -2.238   0.0259
a$x[, -1]3  -0.0014196  0.0001992  -7.126 6.80e-12
a$x[, -1]4  -0.0020999  0.0010856  -1.934   0.0539
a$x[, -1]5  -0.0124208  0.0027842  -4.461 1.13e-05
a$x[, -1]6  -0.1080424  0.0206664  -5.228 3.08e-07

Residual standard error: 4.158 on 323 degrees of freedom
Multiple R-Squared: 0.7355,Adjusted R-squared: 0.7306
F-statistic: 149.7 on 6 and 323 DF,  p-value: < 2.2e-16
```

This fit is worse, but remember we are disallowing any interaction terms. Now let's allow second-order (two-way) interaction terms. `nk` was chosen to get the same model size as before:

```
> a <- mars(ozone[,-1],ozone[,1],nk=10,degree=2)
> summary(lm(ozone[,1] ~ a$x[,-1]))
Coefficients:
              Estimate Std. Error t value Pr(>|t|)
(Intercept) 12.0906983  0.6478959  18.661  < 2e-16
a$x[, -1]1   0.5743487  0.0317558  18.086  < 2e-16
a$x[, -1]2  -0.1190568  0.0412742  -2.885  0.00418
a$x[, -1]3  -0.0011486  0.0001629  -7.050 1.09e-11
a$x[, -1]4  -0.0082510  0.0013801  -5.979 5.95e-09
a$x[, -1]5  -0.0128281  0.0026556  -4.831 2.11e-06
a$x[, -1]6  -0.1023343  0.0197303  -5.187 3.79e-07

Residual standard error: 3.969 on 323 degrees of freedom
Multiple R-Squared: 0.7591,Adjusted R-squared: 0.7546
F-statistic: 169.6 on 6 and 323 DF,  p-value: < 2.2e-16
```

This is a good fit. Compare this with an additive model approach. Since there are nine predictors, this would mean 36 possible two-way interaction terms. Such a model would be complex to estimate and interpret. In contrast, the MARS approach reduces the complexity by carefully selecting the interaction terms.

Now let's see how the terms enter into the model. We can examine the actual form of the basis functions. Start by examining the indicator matrix:

```
> a$factor[a$selected.terms,]
     vh wind humidity temp ibh dpg ibt vis doy
[1,]  0    0        0    0   0   0   0   0   0
[2,]  0    0        0    1   0   0   0   0   0
[3,]  0    0        0   -1   0   0   0   0   0
[4,]  0    0        0    0   1   0   0   0   0
[5,]  0    0       -1    1   0   0   0   0   0
[6,]  0    0        0    0   0   0   0   0   1
[7,]  0    0        0    0   0   0   0   0  -1
```

The first term, given by the first row, is the intercept and involves no variables. The sixth and seventh involve just doy. The "1" indicates a right hockey stick and the "-1" a left hockey stick. The ibh term just has the right hockey stick. Depicting the effect of doy and ibh just requires plotting the transformation as a function of the predictor:

```
> plot(ozone[,6],a$x[,4]*a$coef[4],xlab="ibh",ylab="Contribution of ibh")
> plot(ozone[,10],a$x[,7]*a$coef[7]+a$x[,6]*a$coef[6],xlab="Day",
  ylab="Contribution of day")
```

Temperature and humidity have an interaction so we must combine all terms involving these. Our approach is to compute the predicted value of the response over a grid of values where temperature and humidity are varied while holding the other predictors at their median values. The interaction is displayed as a contour plot and a three-dimensionl plot of the surface:

```
> humidity <- seq(10,100,len=20)
> temp <- seq(20,100,len=20)
> medians <- apply(ozone,2,median)
```

```
> pdf <- matrix(medians,nrow=400,ncol=10,byrow=T)
> pdf[,4] <- rep(humidity,20)
> pdf[,5] <- rep(temp,rep(20,20))
> pdf <- as.data.frame(pdf)
> names(pdf) <- names(medians)
> z <- predict(a,pdf[,-1])
> zm <- matrix(z,ncol=20,nrow=20)
> contour(humidity,temp,zm,xlab="Humidity",ylab="Temperature")
> persp(humidity,temp,zm,xlab="Humidity",ylab="Temperature",
  zlab="Ozone",theta=-30)
```

Now check the diagnostics:

```
> qqnorm(a$res,main="")
> plot(a$fit,a$res,xlab="Fitted",ylab="Residuals")
```

These plots show no problem with normality, but some indication of nonconstant variance. See the bottom two panels of Figure 12.9.

It is interesting to compare the MARS approach to the univariate version as demonstrated in Figure 11.7. There we used a moderate number of knots in just one dimension while MARS gets by with just a few knots in higher dimensions. The key is to choose the right knots. MARS can be favorably compared to linear regression: it has additional flexibility to find nonlinearity in the predictors in higher dimensions. MARS can also be favorably compared to the tree method discussed in the next chapter: it allows for continuous fits but still maintains good interpretability.

Further Reading: Hastie and Tibshirani (1990) provide the original overview of additive modeling, while Wood (2006) gives a more recent introduction. Gu (2002) presents another approach to the problem. Green and Silverman (1993) show the link to GLMs. Hastie, Tibshirani, and Friedman (2001) discuss additive models as part of larger review and compare them to competitive methods.

Exercises

1. The `fat` data gives percentage of body fat, age, weight, height, and 10 body circumference measurements, such as the abdomen, are recorded for 252 men. Body fat is estimated through an underwater weighing technique, but this is inconvenient to use widely. Develop an additive model that allows the estimation of body fat for men using only a scale and a measuring tape. Your model should predict %body fat according to Siri. You may not use Brozek's %body fat, density or fat free weight as predictors.

2. Find a good model for `volume` in terms of `girth` and `height` using the `trees` data. We might expect that `Volume = c * Height * Girth`^{2} suggesting a logarithmic transformation on all variables to achieve a linear model. What models do the ACE and AVAS procedures suggest for this data?

3. Refer to the `pima` dataset described in Question 3 of Chapter 2. First take care to deal with the clearly mistaken observations for some variables.

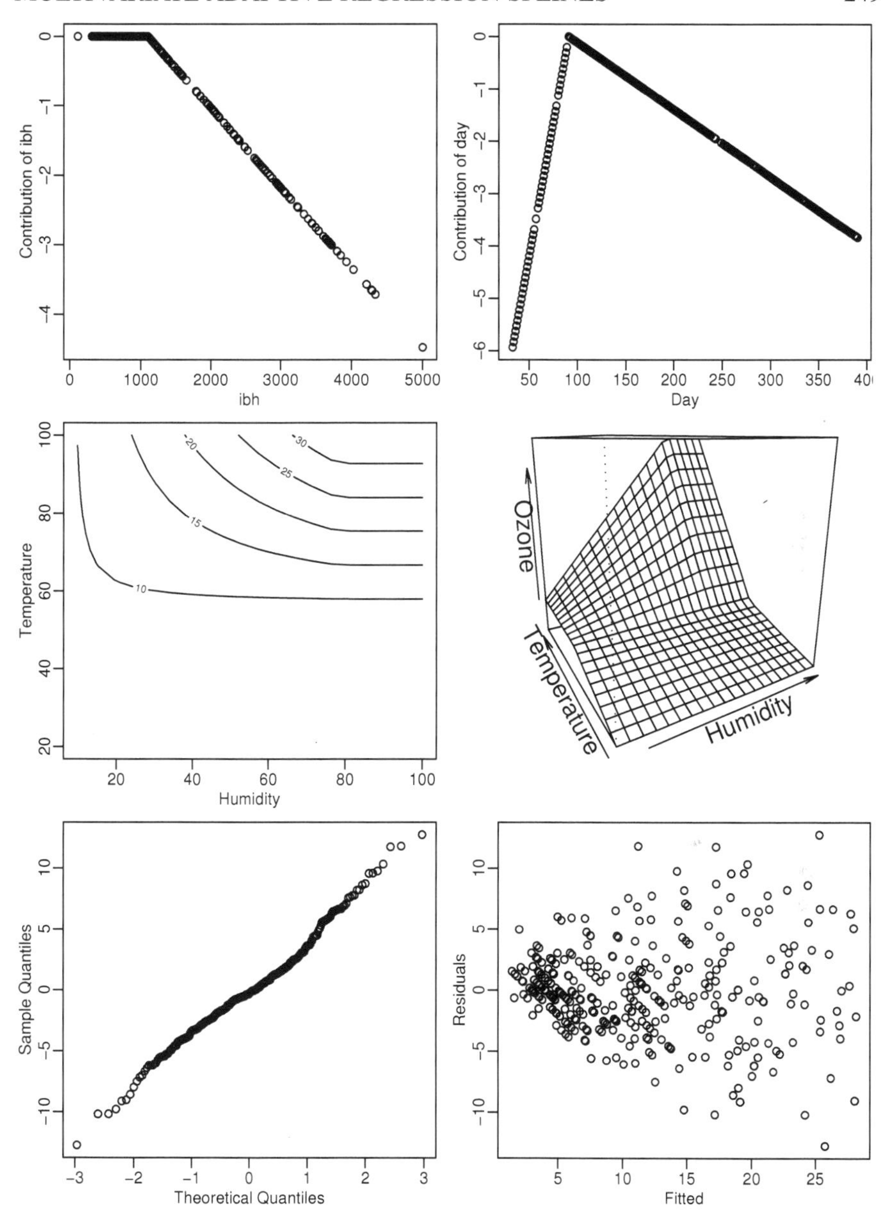

Figure 12.9 *Contribution of predictors in the MARS model and diagnostics.*

(a) Fit a generalized additive model with the result of the diabetes test as the response and all the other variables as predictors.

(b) Perform diagnostics on your model, reporting any potential violations.

(c) Predict the outcome for a woman with predictor values 1, 99, 64, 22, 76, 27, 0.25, 25 (same order as in dataset). How certain is this prediction?

(d) If you completed the logistic regression analysis of the data earlier, compare the two analyses.

4. The `dvisits` data comes from the Australian Health Survey of 1977–78 and consist of 5190 single adults where young and old have been oversampled.

(a) Build a generalized additive model with `doctorco` as the response and `sex`, `age`, `agesq`, `income`, `levyplus`, `freepoor`, `freerepa`, `illness`, `actdays`, `hscore`, `chcond1` and `chcond2` as possible predictor variables. Select an appropriate size for your model.

(b) Check the diagnostics.

(c) What sort of person would be predicted to visit the doctor the most under your selected model?

(d) For the last person in the dataset, compute the predicted probability distribution for their visits to the doctor, i.e., give the probability they visit 0, 1, 2 etc. times.

(e) If you have previously completed the analysis of this data in the exercises for Chapter 3, compare the results.

5. Use the additive model approach to reanalyze the `motorins` data introduced in Section 7.1. For compatibility with the previous analysis, restrict yourself to data from zone one. Try both a Gaussian additive model with a logged response and a gamma GAM for the untransformed response. Compare your analysis to the gamma GLM analysis in the text.

6. The `ethanol` dataset in the `lattice` package presents data from ethanol fuel burned in a single-cylinder engine. The emissions of nitrogen oxides should be considered as the response and engine compression and equivalence ratio as the predictors. Study the example plots given on the help page for `ethanol` that reveal the relationship between the variables.

- Apply the additive model approach to the data.
- Try the MARS approach.

Did either approach reveal the structure that seems apparent in the plots?

CHAPTER 13

Trees

13.1 Regression Trees

Regression trees are similar to additive models in that they represent a compromise between the linear model and the completely nonparametric approach. Tree methodology has roots in both the statistics and computer science literature. A precursor to current methodology was CHAID developed by Morgan and Sonquist (1963) although the book by Breiman, Friedman, Olshen, and Stone (1984) introduced the main ideas to statistics. Concurrently, tree methodology was developed in machine learning starting in the 1970s — see Quinlan (1993) for an overview.

Most statistical work starts from the specification of a model. The model says how we believe the data is generated and contains both a systematic and a random component. The model is not completely specified and so we use the data to select a particular model by either estimating parameters or perhaps by fitting functions, as in our recent nonparametric approaches. Clearly this strategy has been effective in a wide range of situations. However, the insistence on specifying a model, right from the start, does limit statistics. It is often difficult to specify a model, particularly for larger and more complex datasets. Furthermore, it is often impractical to develop inferential methods for more complex statistical models.

Tukey (1977) advocated exploratory data analysis (EDA) in his book. Graphical and descriptive statistics can sometimes make the message of the data very clear or at least suggest a suitable form for the model. However, EDA is not a complete solution and sometimes we need definite predictions or conclusions.

Regression trees are an example of a statistical method that is best described by the algorithm used in their construction. One can uncover the implicit model underlying regression trees, but the algorithm is the true starting point. Any method of analysis should ultimately be judged on whether it successfully predicts or explains something. Statistical models may achieve this, but algorithmically based methods are also competitive. The distinction between algorithm based and model-based methods is discussed in Breiman (2001b). In the computer science literature, tree methodology has been applied to decision tree problems where there is no stochastic structure and we simply want to build a rule for making the correct decision.

We use the recursive partitioning regression algorithm:

1. Consider all partitions of the region of the predictors into two regions where the division is parallel to one of the axes. In other words, we partition a single predictor by choosing a point along the range of that predictor to make the split. It does not matter exactly where we make the split between two adjacent points so there will be at most $(n-1)p$ partitions to consider.

2. For each partition, we take the mean of the response in that partition. We then compute:

$$RSS(partition) = RSS(part_1) + RSS(part_2)$$

We then choose the partition that minimizes the residual sum of squares (RSS). We do need to consider many partitions, but the computations on each partition are simple, so that fit can be accomplished without excessive effort.

3. We now subpartition the partitions in a recursive manner. We only allow partitions within existing partitions and not across them. This means that the partitioning can be represented using a tree. There is no restriction preventing us from splitting the same variables consecutively.

For categorical predictors, it is possible to split on the levels of the factor. For an ordered factor with L levels, there are only $L-1$ possible splits. For an unordered factor, there are $2^{L-1}-1$ possible splits. Although, this is a large number of possibilities as L grows, there is a way to limit the number that need to be considered. Notice that there is no point in monotonely transforming a quantitative predictor as this will have no effect on the partitioning algorithm. Transforming the response will make a difference because it will change the computation of the RSS.

Missing values can be handled quite easily by tree methods. When we construct the tree, we may encounter missing values for a predictor when we are considering a split on that variable. We may simply exclude such points from the computation provided we weight appropriately. This approach is suitable for data where the observations are missing in a noninformative manner. If we believe the fact of being missing express some information, we might choose to treat missingness as an additional level of a factor. For continuous predictors, we could discretize the data into ranges so that it becomes a factor and then add missingness as an additional level. When we wish to predict the response for a new value with missing values, we can drop the prediction down through the tree until the missing values prevent us from going further. We can then use the mean value for that internal node. An alternative is to use surrogate splits, described below.

Tree models are well suited to finding interactions. If we split on one variable and then split on another variable within the partitions of the first variable, we are finding an interaction between these two variables. As we construct further splits within splits, we are finding higher and higher order interactions. This may be a disadvantage as true high-order interactions are not common in reality. The MARS method discussed in Section 12.7 counteracts this by limiting the amount of interaction.

Trees are quite popular because the structure is easier for nontechnical people to understand. The term CART stands for Classification and Regression Trees and is also the name of a commercial software product.

We apply the regression tree methodology to study the relationship between atmospheric ozone concentration and meteorology in the Los Angeles Basin in 1976. A number of cases with missing variables have been removed for simplicity. The data were first presented by Breiman and Friedman (1985). We wish to predict the ozone level from the other predictors. We read in the data and summarize numerically and graphically:

```
> data(ozone)
> summary(ozone)
> pairs(ozone,pch=".")
```

The plots (not shown) reveal several nonlinear relationships indicating that a linear regression might not be appropriate without the addition of some transformations. Now fit a tree:

```
> library(rpart)
> (roz <- rpart(O3 ~ .,ozone))
n= 330

node), split, n, deviance, yval
      * denotes terminal node

 1) root 330 21115.00 11.7760
   2) temp< 67.5 214  4114.30  7.4252
     4) ibh>=3573.5 108   689.63  5.1481 *
     5) ibh< 3573.5 106  2294.10  9.7453
      10) dpg< -9.5 35   362.69  6.4571 *
      11) dpg>=-9.5 71  1366.50 11.3660
        22) ibt< 159 40   287.90  9.0500 *
        23) ibt>=159 31   587.10 14.3550 *
   3) temp>=67.5 116  5478.40 19.8020
     6) ibt< 226.5 55  1276.80 15.9450
      12) humidity< 59.5 10   167.60 10.8000 *
      13) humidity>=59.5 45   785.64 17.0890 *
     7) ibt>=226.5 61  2646.30 23.2790
      14) doy>=306.5 8   398.00 16.0000 *
      15) doy< 306.5 53  1760.50 24.3770
        30) vis>=55 36  1149.90 22.9440 *
        31) vis< 55 17   380.12 27.4120 *
```

We see that the first split (nodes 2 and 3) is on temperature, 214 observations have temperatures less than 67.5 with a mean response value of 7.4, whereas 116 observations have temperatures greater than 67.5 with a mean response value of 20. The total RSS has been reduced from 21,115 to $4114 + 5478 = 9592$. Although the relevant information can be gleaned from the text-based output, a graphical display is nicer as in Figure 13.1. In the first version of the plot, the depth of the branches is proportional to the reduction in error due to the split. The disadvantage is that the labels can be hard to read in lower parts of the tree where the reduction in error is much smaller. The second version of the plot uses a uniform spacing to allow more room for labeling:

```
> plot(roz,margin=.10)
> text(roz)
> plot(roz,compress=T,uniform=T,branch=0.4,margin=.10)
> text(roz)
```

We see that the first split on temperature produces a large reduction in the RSS. Some of the subsequent splits do not do much. The immediate message is that high

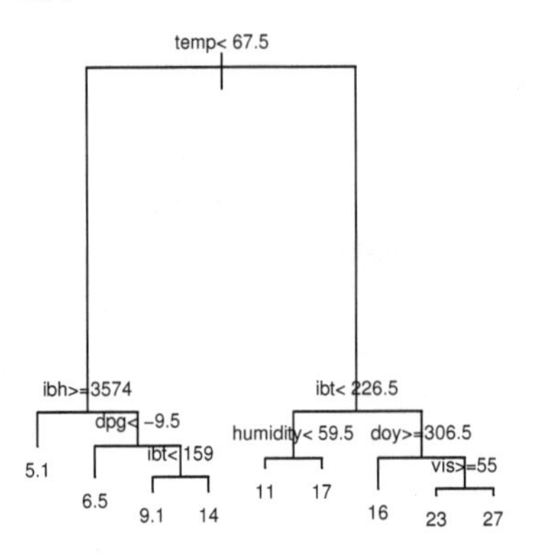

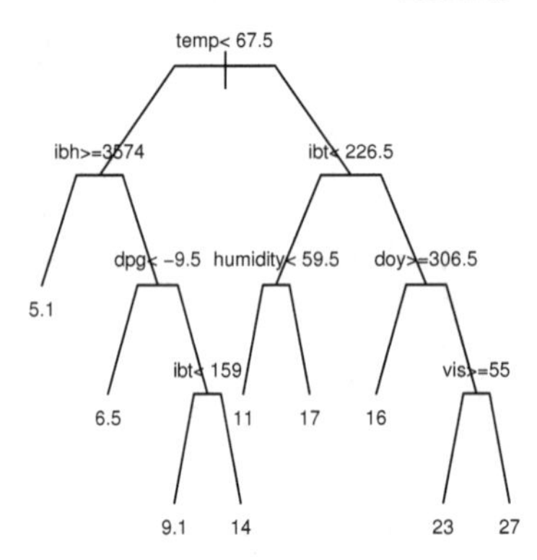

Figure 13.1 *Tree model for the* `ozone` *data. On the left, the depth of the branches is proportional to the improvement in fit. On the right, the depth is held constant to improve readability. If the logical condition at a node is true, follow the branch to the left.*

temperatures are associated with high ozone levels. A regression tree is a regression model, so diagnostics are called for:

```
> plot(predict(roz),residuals(roz),xlab="Fitted",ylab="Residuals")
> qqnorm(residuals(roz))
```

See Figure 13.2. There are no visible problems here. If nonconstant variance is observed, one might consider transforming the response. Trees are also somewhat sensitive to outliers as they are based on local means. Outliers may be observed in the QQ plot, but, as with linear models, they may conceal themselves and be influential on the fit. Suppose we wanted to predict the response for a new value — for example

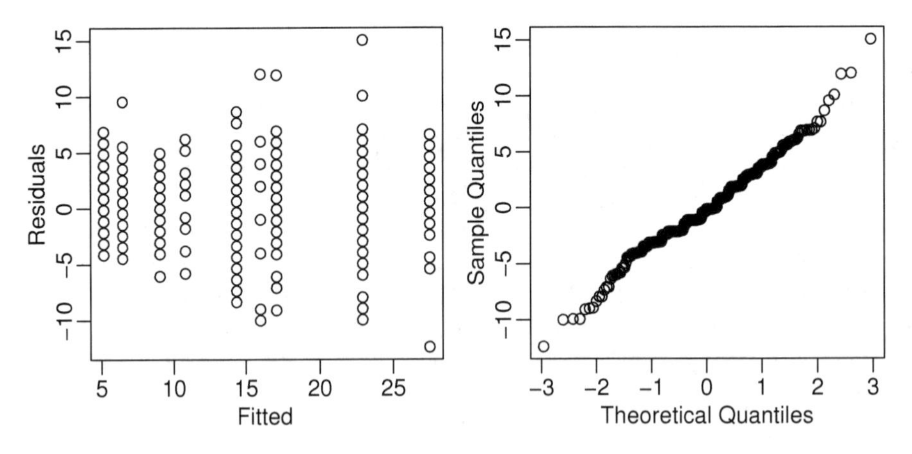

Figure 13.2 *Residuals and fitted values for the tree model of the Ozone data are shown in the left panel. A QQ plot of the residuals is shown in the right panel.*

the median value in the dataset:

```
> (x0 <- apply(ozone[,-1],2,median))
      vh     wind humidity     temp      ibh      dpg      ibt      vis
  5760.0      5.0     64.0     62.0   2112.5     24.0    167.5    120.0
     doy
   205.5
> predict(roz,data.frame(t(x0)))
     1
14.355
```

You should be able to verify this prediction by following the splits down through the tree shown in Figure 13.1.

13.2 Tree Pruning

The recursive partitioning algorithm describes how to grow the tree, but what is the optimal size for the tree? The default form of `rpart` does restrict the size of the tree, but some intervention is probably necessary to select the best tree size.

One possibility, called a *greedy strategy*, is to keep partitioning until the reduction in overall cost (RSS for this type of tree) is not reduced by more than ε. However, it is difficult to set ε in a sensible way. Furthermore, a greedy strategy may stop too soon. For example, consider data laid out in Table 13.1: Neither the horizontal nor

x_2	1	2
	2	1
		x_1

Table 13.1 *There are four data points arranged in a square. The number shows the value of y at that point.*

the vertical split will improve the fit at all. Both splits are required to get a better fit. However, this drawback is common to most tree-growing strategies as looking more than one step ahead greatly increases the number of splits that must be considered. Nevertheless, it does illustrate the point that the incremental improvements due to each expansion of the tree may not necessarily be always decreasing.

The observed RSS for a tree will be an underestimate of how well the tree will make predictions. This phenomenon is common to most models. One generic method of obtaining a better estimate of predictive ability is cross-validation (CV).

For a given tree, leave out one observation, recalculate the tree and use that tree to predict the left-out observation. For regression, this criterion would be:

$$\sum_{j=1}^{n} (y_j - \hat{f}_{(j)}(x_j))^2$$

where $\hat{f}_{(j)}(x_j)$ denotes the predicted value of the tree given the input x_j when case j is not used in the construction of the tree. For other types of tree, a different criterion would be used. For classification problems, it might be the deviance. CV is a more realistic estimate of how the tree will perform in practice. Leave-out-one cross-validation is computationally expensive for trees so usually k-fold cross-validation is used. The data is randomly divided into k roughly equal parts and the remainder is used to predict those left out. As well as being less expensive computationally than the full leave-out-one method, it may even work better. One drawback is that the partition is random so that repeating the method will give different numerical results.

However, there may be very many possible trees if we consider all subsets of a large tree; cross-validation would just be too expensive. We need a method to reduce the set of trees to be considered to just those that are worth considering. This is where cost-complexity pruning is useful. We define a cost-complexity function for trees:

$$CC(Tree) = \sum_{\text{terminal nodes: i}} \text{RSS}_{\text{i}} + \lambda(\text{number of terminal nodes})$$

If λ is large, then the tree that minimizes this cost will be small and vice versa. We can determine the best tree of any given size by growing a large tree and then pruning it back. Given a tree of size n, we can determine the best tree of size $n-1$ by considering all the possible ways of combining adjacent nodes. We pick the one that increases the fit criterion by the least amount. The strategy is akin to backward elimination in linear regression variable selection except that it can be shown that it generates the optimal sequence of trees of a given size.

We now use cross-validation to select from this sequence of trees. By default, `rpart` selects a tree size that may not be large enough to include all those trees we might want to consider. We force it to consider a larger tree and then examine the cross-validation criterion for all the subtrees. The parameter `cp` plays a similar role to the smoothing parameter in nonparametric regression and is defined as the ratio of λ to the RSS of the root tree (a tree with no branches). When we call `rpart` initially, it computes the whole sequence of trees and we merely need to use functions like `printcp` to examine the intermediate possibilities:

```
> roze <- rpart(O3 ~ .,ozone,cp=0.001)
> printcp(roze)

        CP nsplit rel error xerror   xstd
1  0.54570      0     1.000  1.015 0.0771
2  0.07366      1     0.454  0.486 0.0416
3  0.05354      2     0.381  0.411 0.0383
4  0.02676      3     0.327  0.385 0.0358
5  0.02328      4     0.300  0.387 0.0364
6  0.01532      6     0.254  0.377 0.0363
7  0.01091      7     0.239  0.372 0.0364
8  0.00707      8     0.228  0.376 0.0391
9  0.00599      9     0.221  0.375 0.0402
10 0.00497     12     0.203  0.372 0.0408
11 0.00319     17     0.179  0.380 0.0415
12 0.00222     19     0.172  0.378 0.0416
13 0.00144     23     0.164  0.382 0.0420
14 0.00113     24     0.162  0.378 0.0420
15 0.00100     26     0.160  0.379 0.0422
```

In this table, we see the value of the `cp` parameter, the number of splits in the tree, the RSS of the tree divided by the RSS of the null tree. `xerror` denotes the cross-validated error which is also scaled by the RSS of the null tree. Since the partition of the data into 10 parts is random, this CV error is also random, which makes the given standard error useful. The random division also means that if you repeat this command, you will not get exactly the same answer. We can select the size of the tree by minimizing the value of `xerror` and selecting the corresponding value of `CP`:

```
> rozr <- prune.rpart(roze,0.01091)
```

This selected tree turns out to be the same as the default choice in this instance. Another strategy for selecting the tree size is to select the smallest tree with a CV error within one standard error of the minimum — in this case, $0.372 + 0.036 = 0.408$. So we would take the two-split tree. We can illustrate this by plotting the CV error and a line showing one standard deviation above this value as shown in Figure 13.3:

```
> plotcp(roz)
```

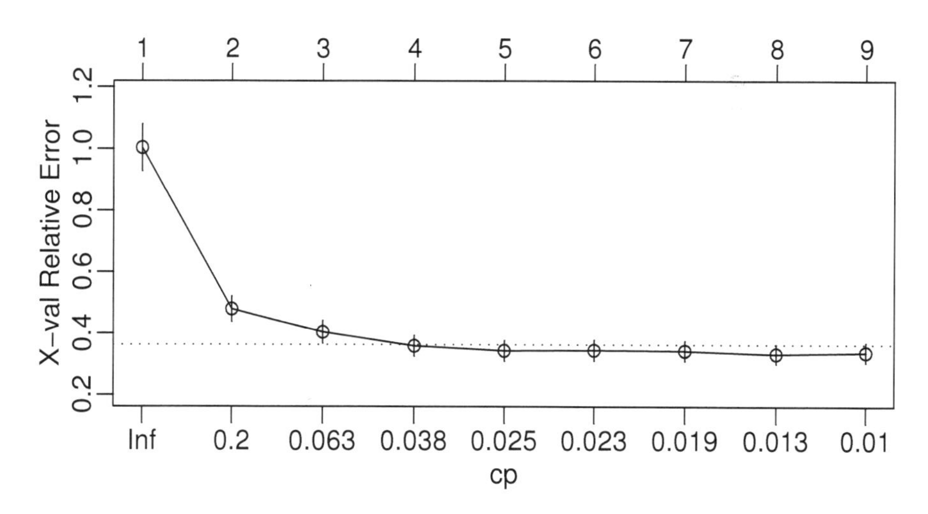

Figure 13.3 *Cross-validation plot for ozone tree model.*

You can get some fancier output by:

```
> post(roz,digits=3,filename="")
```

If you do not specify the filename, nothing will appear on-screen, but you will find a file called `roz.ps` in the directory from which you started R. See Figure 13.4: Let's compare the result to the earlier linear regression. We achieved an R^2 of about 70% using only six parameters in the previous chapter. We can select a tree with five splits and hence effectively six parameters and compare them:

```
> rozr <- prune.rpart(roz,0.0154)
> 1-sum(residuals(rozr)^2)/sum((ozone$O3-mean(ozone$O3))^2)
[1] 0.74603
```

We see that the tree model achieved a better fit than the equivalent linear model. Of course, it would be a mistake to generalize from this, but it is a good demonstration of the value of trees. A tree fit is piecewise constant over the regions defined by the

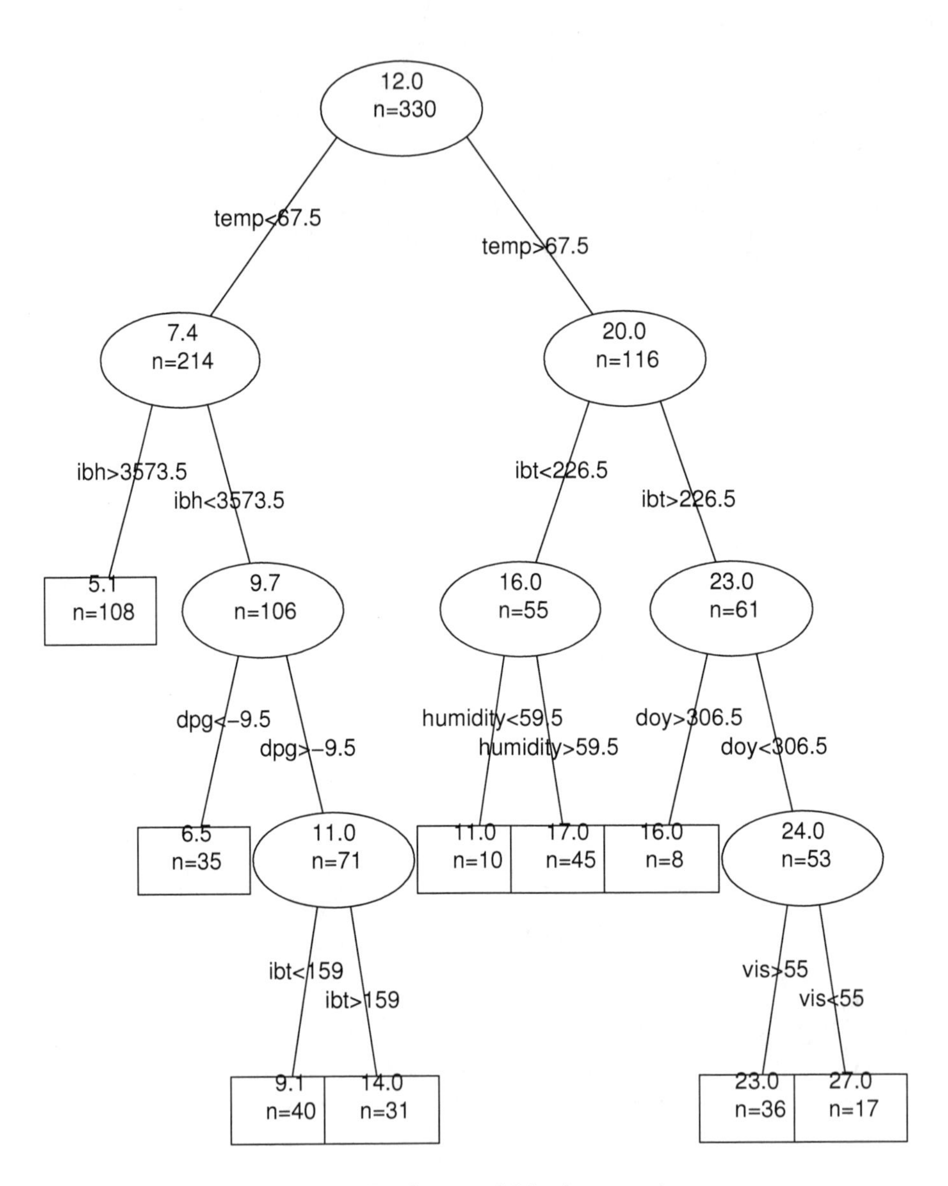

Figure 13.4 *Final tree model for the ozone data.*

partitions, so one might not expect a particularly good fit. However, we can see from this example that it can outperform linear regression.

13.3 Classification Trees

Trees can be used for several different types of response data. For the regression tree, we computed the mean within each partition. This is just the null model for a regression. We can extend the tree method to other types of response by fitting an appropriate null model on each partition. For example, we can extend the idea to binomial, multinomial, Poisson and survival data by using a deviance, instead of the RSS, as a criterion.

Classification trees work similarly to regression trees except the residual sum of squares is no longer a suitable criterion for splitting the nodes. The splits should divide the observations within a node so that the class types within a split are mostly of one kind (or failing that, just few kinds). We can measure the *purity* of the node with several possible measures. Let n_{ik} be the number of observations of type k within terminal node i and p_{ik} be the observed proportion of type k within node i. Let D_i be the measure for node i so that the total measure is $\sum D_i$. There are several choices for D_i:

1. Deviance:

$$D_i = -2\sum_k n_{ik} \log p_{ik}$$

2. Entropy:

$$D_i = -\sum_k p_{ik} \log p_{ik}$$

3. Gini index:

$$D_i = 1 - \sum_k p_{ik}^2$$

All these measures share the characteristic that they are minimized when all members of the node are of the same type. The `rpart` function uses the Gini index by default.

We illustrate the classification tree method in a problem involving the identification of the sex and species of an historical specimen of kangaroo. We have some training data consisting of 148 cases with the following variables: there are three possible species, *Giganteus*, *Melanops* and *Fuliginosus*, the sex of the animal and 18 skull measurements. The data were published in Andrews and Herzberg (1985). The historical specimen is from the Rijksmuseum van Natuurlijkee Historie in Leiden which had the following skull measurements in the same order as in the data:

```
1115 NA 748 182 NA NA 178 311 756 226 NA NA NA 48 1009 NA 204 593
```

We have a choice in how we model the response. One possibility is to form a six-level response representing all possible combinations of sex and species. Another approach is to form separate trees for identifying the sex and the species. We take the latter approach below, focusing on the the species. This choice is motivated by the belief that different features are likely to discriminate the sex and the species so that attempting to model them both in the same tree might result in a larger, more complex tree that might be less powerful than two smaller trees. Even so, it would be worth trying the first approach although we shall not do so here. We start by reading in and specifying the museum case:

```
> data(kanga)
> x0 <- c(1115,NA,748,182,NA,NA,178,311,756,226,NA,NA,NA,48,1009,NA,204,593)
```

We have missing values for the case to be classified. We have two options. We can build a tree model that will classify if there are missing values in the input or we can build a tree model that uses only variables that are observed. If we believe that the missing values were in some way informative, the first choice would be fine. In this particular case, that does not seem plausible, so the latter approach is preferred. However, if we want to build a model that could be used for future unspecified cases, then we would have to deal directly with the missing values. For this special purpose situation, where we want to classify one particular kangaroo, this is not a concern.

We exclude all variables that are missing in the test case. We drop sex since we will not be modeling it yet. We form a convenient data frame:

```
> kanga <- kanga[,c(T,F,!is.na(x0))]
> kanga[1:2,]
    species basilar.length palate.length palate.width squamosal.depth
1 giganteus           1312           882           NA             180
2 giganteus           1439           985          230             150
  lacrymal.width zygomatic.width orbital.width foramina.length
1            394             782           249              88
2            416             824           233             100
mandible.length  mandible.depth ramus.height
          10861             179          591
          11582             181          643
```

We still have missing values in the training set. We have a number of options:

1. Build a tree model that discretizes the predictors into factors and then treats missing values as another level of the factors. This might be appropriate if we think missing values are informative in some way. Information would be lost in the discretization. For this data, we have no reason to believe that the data is not missing at random and furthermore we have already decided to ignore the missing values in the test case.

2. Fill in or estimate the missing values and then build a tree. We could use missing data fill-in methods as used in other regression problems. This is not easy to implement and there are concerns about the bias caused by such methods.

3. The tree-fitting algorithm can handle missing values naturally. If a value for some case is not available, then it is simply excluded from the criterion. When we want to classify a new case with missing values, we follow the tree down until we reach a split which involves a missing value in our new case and take the majority verdict in that node. A more complicated approach is to allow a second-choice variable for splitting at a node called a *surrogate split*. Information on the surrogate splits may be obtained by using the `summary` command on the tree object.

4. Leave out the missing cases entirely.

We first check where the missing values occur:

```
> apply(kanga,2,function(x) sum(is.na(x)))
        species  basilar.length    palate.length    palate.width
              0               1                1              24
squamosal.depth  lacrymal.width  zygomatic.width   orbital.width
              1               0                1               0
foramina.length mandible.length   mandible.depth    ramus.height
              0              12                0               0
```

We observe that the majority of missing values occur in just two variables: mandible length and palate width. Suppose we throw out those variables and then remove the remaining missing cases. We compute the pairwise correlation of these variables with the other variables.

```
> round(cor(kanga[,-1],use="pairwise.complete.obs")[,c(3,9)],2)
                palate.width mandible.length
basilar.length          0.77            0.98
palate.length           0.81            0.98
palate.width            1.00            0.81
squamosal.depth         0.69            0.80
lacrymal.width          0.77            0.92
zygomatic.width         0.78            0.92
orbital.width           0.12            0.25
foramina.length         0.19            0.23
mandible.length         0.81            1.00
mandible.depth          0.62            0.85
ramus.height            0.73            0.94
```

We see that these two variables are highly correlated with other variables in the data. We claim that there is not much additional information in these two variables and we can reasonably discard them. We do this and then remove the remaining missing cases:

```
> newko <- na.omit(kanga[,-c(4,10)])
> dim(newko)
[1] 144  10
```

After excluding these two variables, we only lose four cases more by throwing out all the missing value cases. Alternatively, suppose we just throw out the missing value cases on the original data:

```
> dim(na.omit(kanga))
[1] 112  12
```

We would lose 36 cases by simply throwing out all the missing values. Removing a combination of variables and cases seems a better choice for this data.

We should also plot the data to see how the classes separate. An example of such a plot is:

```
> plot(foramina.length ~ zygomatic.width,data=newko,
  pch=substring(species,1,1))
```

We see in the left panel of Figure 13.5 that the classes do not separate well, at least for these two variables. We now fit a classification tree as follows: Because the response

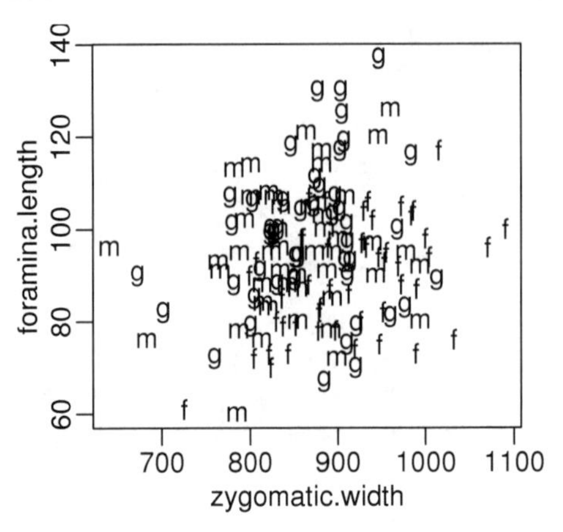

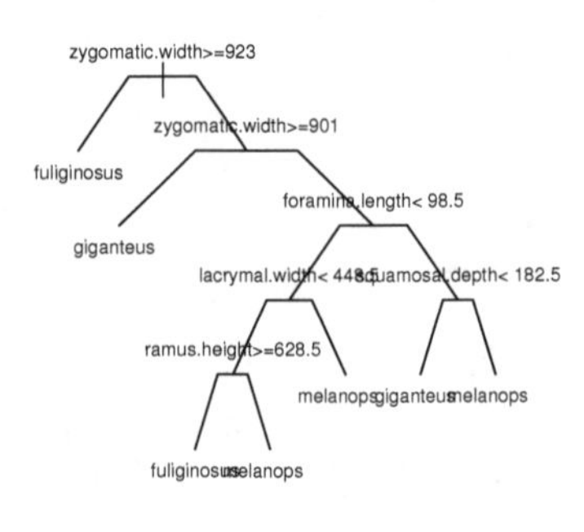

Figure 13.5 *Historical kangaroo tree model. The left panel shows the three species, m =melanops, g =giganteus and f =fuliginosus, as they vary with two of the measurements. The right panel shows the chosen tree.*

is a factor, classification rather than regression is automatically used. Gini's index is the default choice of criterion. Here we specify a smaller value of the complexity parameter `cp` than the default, so that larger trees are also considered:

```
> kt <- rpart(species ~ ., data=newko,cp=0.001)
> printcp(kt)
Root node error: 95/144 = 0.66

n= 144

      CP nsplit rel error xerror   xstd
1 0.1789      0     1.000  1.105 0.0561
2 0.1053      1     0.821  0.979 0.0604
3 0.0500      2     0.716  0.874 0.0624
4 0.0211      6     0.516  0.800 0.0631
5 0.0105      7     0.495  0.853 0.0627
6 0.0010      8     0.484  0.905 0.0620
```

The cross-validated error (expressed in relative terms in the `rel error` column) reaches a minimum for the six-split tree. We select this tree:

```
> ktp <- prune(kt,cp=0.0211)
> ktp
n= 144

node), split, n, loss, yval, (yprob)
      * denotes terminal node

 1) root 144 95 fuliginosus (0.340278 0.333333 0.326389)
  2) zygomatic.w>=923 37 13 fuliginosus (0.648649 0.162162 0.189189) *
  3) zygomatic.w< 923 107 65 giganteus (0.233645 0.392523 0.373832)
```

```
    6) zygomatic.w>=901 16  3 giganteus (0.125000 0.812500 0.062500) *
    7) zygomatic.w< 901 91 52 melanops (0.252747 0.318681 0.428571)
     14) foramina.l< 98.5 58 33 melanops (0.362069 0.206897 0.431034)
       28) lacrymal.w< 448.5 50 29 fuliginosus (0.420000 0.240000 0.340000)
        56) ramus.h>=628.5 33 14 fuliginosus (0.575758 0.181818 0.242424) *
        57) ramus.h< 628.5 17  8 melanops (0.117647 0.352941 0.529412) *
       29) lacrymal.w>=448.5 8  0 melanops (0.000000 0.000000 1.000000) *
     15) foramina.l>=98.5 33 16 giganteus (0.060606 0.515152 0.424242)
       30) squamosal.d< 182.5 26 10 giganteus (0.076923 0.615385 0.307692) *
       31) squamosal.d>=182.5 7  1 melanops (0.000000 0.142857 0.857143) *
> plot(ktp,compress=T,uniform=T,branch=0.4,margin=0.1)
> text(ktp)
```

This tree is not particularly successful as the relative error is estimated as 80% of just guessing the species. Some of the terminal nodes are quite pure, for example, #29 and #31, while others retain much uncertainty, for example, #56 and #57. We now compute the misclassification error:

```
> (tt <- table(actual=newko$species,predicted=predict(ktp,type="class")))
             predicted
actual         fuliginosus giganteus melanops
  fuliginosus          43         4        2
  giganteus            12        29        7
  melanops             15         9       23
> 1-sum(diag(tt))/sum(tt)
[1] 0.34028
```

We see that the error rate is 34%. We might hope to do better. We see that we can generally correctly identify *fuliginosus*, but we are more likely to be in error in distinguishing *melanops* and *giganteus*.

A look at the left panel of Figure 13.5 explains why we may have difficulty in classification. Any single measure will reflect mostly the overall size of the skull. For example, suppose we wanted to distinguish male human skulls from female human skulls. Most interesting measures will correlate strongly with size. If we just use one measure, then the rule will likely be: if the measure is small, then pick male; if it is large pick, female. This cannot be expected to work particularly well. There is something about the *relative* dimensions of the skulls that ought be more informative.

One possibility is to allow splits on linear combinations of variables. This is allowed in some classification tree software implementations. An alternative idea is to apply the method to the principal component scores rather than the raw data. Principal components (PC) seek out the main directions of variation in the data and might generate more effective predictors for classification in this example:

```
> pck <- princomp(newko[,-1])
> pcdf <- data.frame(species=newko$species,pck$scores)
> kt <- rpart(species ~ ., pcdf,cp=0.001)
> printcp(kt)
Root node error: 95/144 = 0.66

n= 144
```

```
     CP nsplit rel error xerror   xstd
1 0.4000      0     1.000  1.126 0.0552
2 0.1789      1     0.600  0.621 0.0621
3 0.0421      2     0.421  0.558 0.0609
4 0.0105      3     0.379  0.568 0.0612
5 0.0010      5     0.358  0.589 0.0616
```

We find a significantly smaller relative CV error (0.558). Before we can predict the test case, we need to do some work to remove the missing values, unused variables and apply the principal component transformation:

```
> nx0 <- x0[!is.na(x0)]
> nx0 <- nx0[-c(3,9)]
> nx0 <- (nx0-pck$center)/pck$scale
> nx0 %*% pck$loadings

       Comp.1  Comp.2  Comp.3 Comp.4 Comp.5 Comp.6  Comp.7
  [1,] 499.93 -74.834 -37.632 23.169 3.9564 16.584 -54.017

        Comp.8  Comp.9
  [1,] -35.995 -16.705
```

Our chosen tree is:

```
> ktp <- prune.rpart(kt,0.0421)
> ktp
n= 144

node), split, n, loss, yval, (yprob)
      * denotes terminal node

 1) root 144 95 fuliginosus (0.340278 0.333333 0.326389)
   2) Comp.2< -15.126 49  8 fuliginosus (0.836735 0.040816 0.122449) *
   3) Comp.2>=-15.126 95 49 giganteus (0.084211 0.484211 0.431579)
     6) Comp.4>=-9.513 63 24 giganteus (0.111111 0.619048 0.269841)
      12) Comp.3>=-18.996 55 17 giganteus (0.090909 0.690909 0.218182) *
      13) Comp.3< -18.996 8  3 melanops (0.250000 0.125000 0.625000) *
     7) Comp.4< -9.513 32  8 melanops (0.031250 0.218750 0.750000) *
```

It is interesting that the first PC is not used. Typically, the first PC represents an overall average or total size. Other PCs represent contrasts between variables which would describe shape features in this case. We see that the test case is classified as *fuliginosus*, which agrees with the experts. We can also compute the error rate as before:

```
> (tt <- table(newko$species,predict(ktp,type="class")))

              fuliginosus giganteus melanops
  fuliginosus 41           5         3
  giganteus    2          38         8
  melanops     6          12        29
> 1-sum(diag(tt))/sum(tt)
[1] 0.25
```

We see that the error rate has been reduced to 25%. It would be worth considering other combinations of predictors in an attempt to reduce the error rate further.

Further Reading: Breiman, Friedman, Olshen, and Stone (1984) is the classic book on trees. Ripley (1996) and Hastie, Tibshirani, and Friedman (2001) also discuss trees and compare them to other methods. See also the random forests method described in Breiman (2001a) that results in more stable assessments of the effects of predictors.

Exercises

1. Four hundred three African Americans were interviewed in a study to understand the prevalence of obesity, diabetes, and other cardiovascular risk factors in central Virginia. Data is presented in `diabetes`. Build a regression tree-based model for predicting glycosolated hemoglobin in terms of the other relevant variables. Interpret your model. Use the model to predict the glycosolated hemoglobin for a subject with:

```
  id chol stab.glu hdl ratio   location age gender
1004  213       72  58   3.3 Buckingham  56 female
 height weight  frame bp.1s bp.1d bp.2s bp.2d waist hip
     64    131 medium   108    55    NA    NA    30  40
   time.ppn
        720
```

   Glycosolated hemoglobin greater than 7.0 is usually taken as a positive diagnosis of diabetes. Now build a classification tree for the diagnosis of diabetes and compare your model to the corresponding regression tree.

2. Refer to the `pima` dataset described in Question 3 of Chapter 2. First take care to deal with the clearly mistaken observations for some variables.

   (a) Fit a tree model with the result of the diabetes test as the response and all the other variables as predictors.

   (b) Perform diagnostics on your model, reporting any potential violations.

   (c) Predict the outcome for a woman with predictor values 1, 99, 64, 22, 76, 27, 0.25, 25 (same order as in dataset). How certain is this prediction?

   (d) If you completed the logistic regression analysis of the data earlier, compare the two analyses.

3. The dataset `wbcd` is described in Question 2 of Chapter 2.

   (a) Fit a tree model with `Class` as the response and the other nine variables as predictors.

   (b) Use the model to predict the outcome for a new patient with predictor variables 1, 1, 3, 2, 1, 1, 4, 1, 1 (same order as above).

   (c) Suppose that a cancer is classified as benign if $p > 0.5$ and malignant if $p < 0.5$. Compute the number of errors of both types that will be made if this method is applied to the current data with the reduced model.

(d) Suppose we change the cutoff to 0.9 so that $p < 0.9$ is classified as malignant and $p > 0.9$ as benign. Compute the number of errors in this case. Discuss the issues in determining the cutoff.

(e) It is usually misleading to use the same data to fit a model and test its predictive ability. To investigate this, split the data into two parts and assign every third observation to a test set and the remaining two thirds of the data to a training set. Use the training set to determine the model and the test set to assess its predictive performance. Compare the outcome to the previously obtained results.

(f) If you completed the logistic regression analysis of the data earlier, compare the two analyses.

4. The dataset `uswages` is drawn as a sample from the Current Population Survey in 1988.

   (a) Build a tree regression model to predict `wage`.

   (b) Check the diagnostics of your model.

   (c) Use your model to predict the wage of a subject with predictor characteristics 12, 33, 0, 1, 0, 0, 0, 1, 0 where the values occur in the same order as in the data frame.

   (d) Conduct a quick linear model analysis and compare the results with the tree model. In particular, what do the two models say about the relationship between the predictors and the response?

5. The `dvisits` data comes from the Australian Health Survey of 1977–78 and consist of 5190 single adults where young and old have been oversampled.

   (a) Build a Poisson tree model with `doctorco` as the response and `sex`, `age`, `agesq`, `income`, `levyplus`, `freepoor`, `freerepa`, `illness`, `actdays`, `hscore`, `chcond1` and `chcond2` as possible predictor variables. Consult the `rpart` documentation for how to specify a Poisson response.

   (b) Check the diagnostics.

   (c) What sort of person would be predicted to visit the doctor the most under your selected model?

   (d) For the last person in the dataset, compute the predicted probability distribution for their visits to the doctor, i.e., give the probability they visit 0, 1, 2 etc. times.

   (e) If you have previously completed the analysis of this data in the exercises for Chapter 3, compare the results.

CHAPTER 14

Neural Networks

Neural networks (NN) were originally developed as an attempt to emulate the human brain. The brain has about 1.5×10^{10} neurons each with 10 to 10^4 connections called synapses. The speed of messages between neurons is about 100 m/sec which is much slower than CPU speed. Given that our fastest reaction time is around 100 ms and neuron computation time is 1–10 ms, then the number of steps must be less than 100. This is inconceivably small for a sequential computation, even in machine code; therefore, the brain must be computing in parallel.

The original idea behind neural networks was to use a computer-based model of the human brain to perform complex tasks. We can recognize people in fractions of a second, but this task is difficult for computers. So why not make software more like the human brain?

Despite the promise, there are some drawbacks. The brain model of connected neurons, first suggested by McCulloch and Pitts (1943), is too simplistic given more recent research. There are also more controversial philosophical questions about how any algorithmic computation can mimic some of the functions of the brain. This is discussed in Penrose (1989). For these and other reasons, the methodology is more properly called *artificial* neural nets. As with artificial intelligence, the promise of NNs is not matched by the reality of their performance. Arnold Schwarzenegger's brain in *The Terminator* was powered by a neural net, but the true accomplishments of NNs are much more modest. Nevertheless, they can be useful as we shall see.

NNs are used for various purposes. They can be used as biological models, which was the original motivation. They can also be used as a hardware implementation for adaptive control. But the area of application we are interested in is data analysis. There are NN models that rival the regression, classification and clustering methods normally used by statisticians.

A perceptron is a model of a neuron and is the basic building block of a neural network as depicted in Figure 14.1. The output x_o is determined from inputs x_i:

$$x_o = f_o\Big(\sum_{inputs:i} w_i x_i\Big)$$

where f_o is called the *activation function*. Standard choices include the identity, logistic and indicator functions. The w_i are *weights*. The NN *learns* the weights from the data. A statistician would prefer to say that the NN estimates the parameters from the data. Thus NN terminology differs from statistical usage in ways that can be confusing.

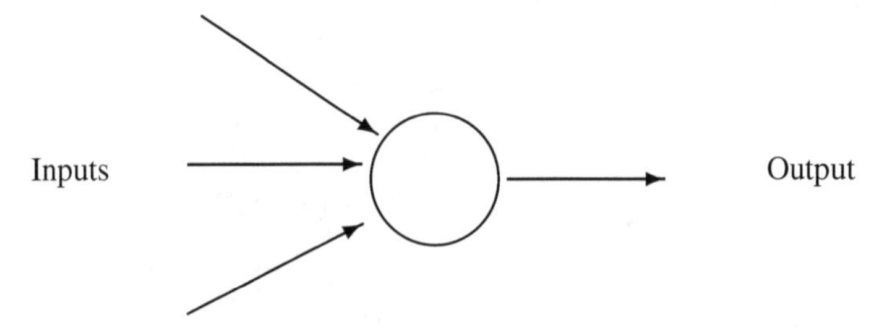

Figure 14.1 *A perceptron.*

14.1 Statistical Models as NNs

Three common statistical models are analogous to the single perceptron NN. For multiple linear regression:

$$y = \sum_i w_i x_i$$

So here f_o is the identity function. We can define $x_1 \equiv 1$ to get an intercept term. The NN alternative is to attach a weight, called a *bias*, to each neuron:

$$f(x) = x + \theta$$

A statistician would call the bias θ an intercept.

Logistic regression also fits easily within this framework if we define f_o as the logistic function. Of course, such an NN is not exactly equivalent to the corresponding statistical model unless that NN is fit in a very particular way.

Linear discriminant analysis is used to classify a binary response. Suppose there are two groups encoded by $y = 0$ or $y = 1$. In this case f_0 is the indicator function. Again the NN and and statistical approach are not exactly equivalent unless the same fitting procedures are used.

Other common statistical models can be approximated by adding more neurons Multivariate multiple linear regression with a bivariate response is $Y = X\beta + \varepsilon$ where Y, X, β, ε are all matrices and is depicted as an NN in Figure 14.2: Polynomial regression can be mimicked by using different activation function and more than one layer of neurons as seen in the second panel of Figure 14.2.

14.2 Feed-Forward Neural Network with One Hidden Layer

The feed-forward neural network with one hidden layer is the most common choice for regression-like modeling applications. It takes the form:

$$y_o = \phi_o(\sum_h w_{ho} \phi_h(\sum_i w_{ih} x_i))$$

The activation functions for the hidden layer, ϕ_h, are almost always logistic. If identity functions are used for the hidden layer and for the output, the resulting NN is quite similar to the partial least squares approach of Wold, Ruhe, Wold, and Dunn (1984). We will set one of our inputs to be constant at one so as to allow for an intercept/bias term. The choice of output activation function depends on the nature of

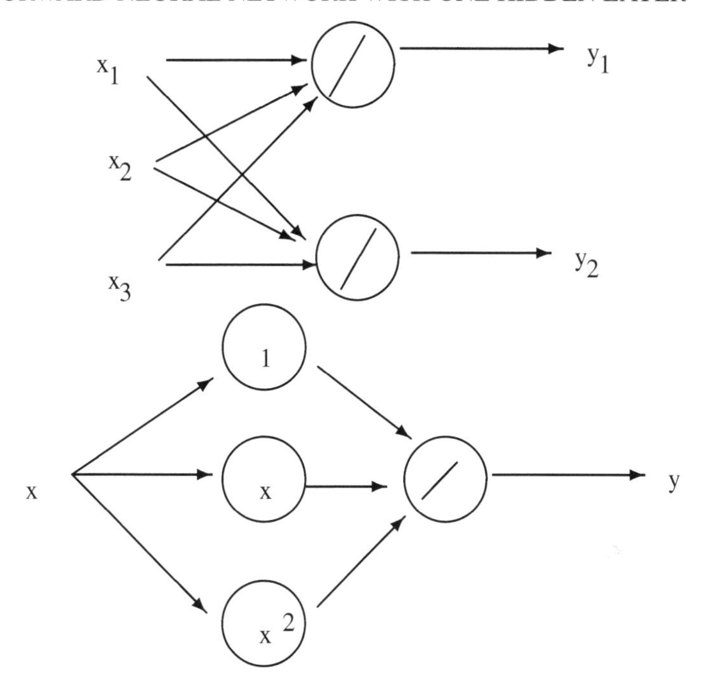

Figure 14.2 *NN equivalents of multivariate linear regression (shown on the top) and polynomial regression (shown on the bottom).*

the response. For continuous unrestricted output, an identity function is appropriate while for response bounded between zero and one, such as a binomial proportion, a logistic function should be used. We show the feed-forward NN in Figure 14.3.

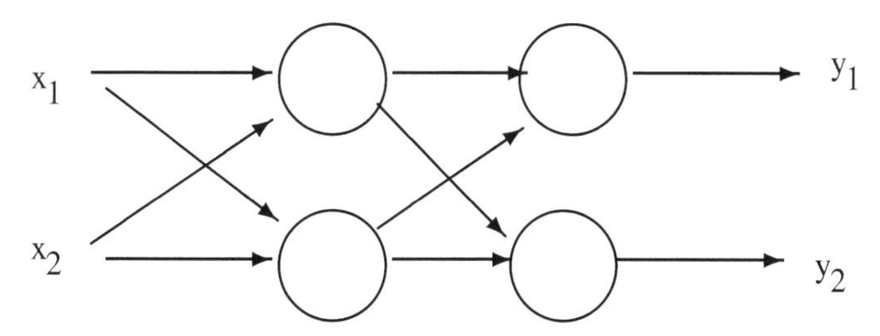

Figure 14.3 *Feed-forward neural network with one hidden layer.*

Sometimes a direct connection between the inputs and outputs is added called a *skip-layer* connection. More complexity can be added using more layers or feedbacks although this not always beneficial to the practical performance of the NN.

NNs can be elaborated to perform as universal approximators. This has been shown by authors such as Hornik, Stinchcombe, and White (1989). However, these results are of little practical value when confronted with a finite amount of data subject to noise. We must use the data estimate the parameters of the model or, in NN-speak, use the data to train the network. The weights w are chosen to minimize a criterion, such as:

$$E = \sum (y - \hat{y})^2$$

where y is the observed output and $\hat{y}$ is the predicted output. A different criterion would be more suitable for categorical responses.

NN researchers have developed different methods of estimation motivated by brain models of learning. These methods have generally not compared well with the numerical analysis-based approaches which are generally faster and more reliable. Nevertheless, in all but the most simple NNs, the criterion is a complicated function of the parameters. The function often has many local minima making it difficult to find the true minimum. A statistician, with no pretensions to mimicking brain functions, would be inclined to use standard methods of numerical analysis such as quasi-Newton methods, conjugate gradients or simulated annealing. The `nnet` function in R uses the BFGS method, as described in Fletcher (1987).

14.3 NN Application

We apply the NN method to the `ozone` data analyzed in previous chapters. The *nnet* package, due to Venables and Ripley (2002), must be loaded first:

```
> library(nnet)
> data(ozone)
```

We start with just three variables for simplicity of exposition as in previous analyses. We fit a feed-forward NN with one hidden layer containing two units with a linear output unit:

```
> nnmdl <- nnet(O3 ~ temp + ibh + ibt, ozone, size=2, linout=T)
# weights:  11
initial  value 65447.874069
final  value 21115.406061
converged
```

The RSS of 21,115 is equal to $\sum_i (y_i - \bar{y})^2$, so the fit is not any better than the null model. If you repeat this, your result may differ slightly because of the random starting point of the algorithm, but you will likely get a similar result. The problem lies with the initial selection of weights. It is hard to do this well when the variables have very different scales. The solution is to rescale the data to have zero mean and unit variance:

```
> sx <- scale(ozone)
```

Because a random starting point is used, the algorithm will not necessarily converge to the same solution if the fitting is repeated. Now we try refitting the model. We repeat this 100 times because of the random starting point. Here we find the best fit of the 100 attempts:

```
> bestrss <- 10000
> for(i in 1:100){
  nnmdl <- nnet(O3 ~ temp + ibh + ibt, sx, size=2, linout=T, trace=F)
  cat(nnmdl$value,"\n")
```

```
  if(nnmdl$value < bestrss){
  bestnn <- nnmdl
  bestrss <- nnmdl$value
  }}
> bestnn$value
[1] 88.031
```

The criterion function has 11 parameters or weights and has multiple minima. The problem is that we can never really know whether we have found the true minimum. All we can do is keep trying and stop if we do not find anything better after some number of attempts. The best strategy is not clear, although one can do better than the simple approach we have used above. Examine the estimated weights:

```
> summary(bestnn)
a 3-2-1 network with 11 weights
options were - linear output units
 b->h1 i1->h1 i2->h1 i3->h1
  1.12  -0.98   0.84   0.29
 b->h2 i1->h2 i2->h2 i3->h2
137.89 -74.74 240.66 137.89
  b->o  h1->o  h2->o
  2.59  -4.41   0.67
```

The notation `i2->h1`, for example, refers to the link between the second input variable and the first hidden neuron. `b` refers to the bias, which takes a constant value of one. We see that there is one skip-layer connection, `b->o`, from the bias to the output.

NNs have some drawbacks relative to competing statistical models. The parameters of an NN are uninterpretable whereas they often have some meaning in statistical models. Furthermore, NNs are not based on a probability model that expresses the structure and variation. As a consequence, there are no standard errors. It is possible to graft some statistical inference onto this NN model, but it is not easy. The bootstrap is a possible way of implementing this. The R^2 for the fit is:

```
> 1-88.03/sum((sx[,1]-mean(sx[,1]))^2)
[1] 0.73243
```

which is very similar to the additive model fit for these predictors.

Although the NN weights may be difficult to interpret, we can get some sense of the effect of the predictors by observing the marginal effect of changes in one or more predictor as other predictors are held fixed. Here, we vary each predictor individually while keeping the other predictors fixed at their mean values. Because the data has been centered and scaled for the NN fitting, we need to restore the original scales. The fits are shown in Figure 14.4:

```
> ozmeans <- attributes(sx)$"scaled:center"
> ozscales <- attributes(sx)$"scaled:scale"
> xx <- expand.grid(temp=seq(-3,3,0.1),ibh=0,ibt=0)
> plot(xx$temp*ozscales['temp']+ozmeans['temp'],
  predict(bestnn,new=xx)*ozscales['O3']+ozmeans['O3'],xlab="Temp",ylab="O3")
> xx <- expand.grid(temp=0,ibh=seq(-3,3,0.1),ibt=0)
> plot(xx$ibh*ozscales['ibh']+ozmeans['ibh'],
  predict(bestnn,new=xx)*ozscales['O3']+ozmeans['O3'],xlab="IBH",ylab="O3")
```

```
> xx <- expand.grid(temp=0,ibh=0,ibt=seq(-3,3,0.1))
> plot(xx$ibt*ozscales['ibt']+ozmeans['ibt'],
  predict(bestnn,new=xx)*ozscales['O3']+ozmeans['O3'],xlab="IBT",ylab="O3")
```

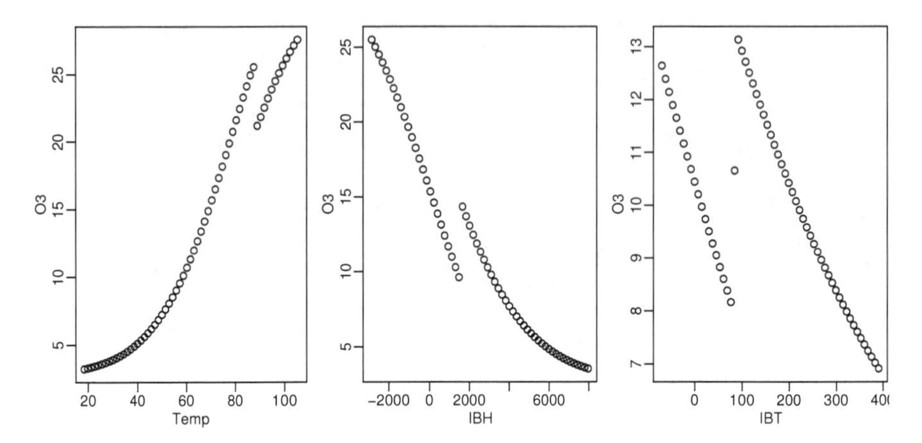

Figure 14.4 *Marginal effects of predictors for the NN fit. Other predictors are held fixed at their mean values.*

We see some surprising discontinuities in the plots which do not seem consistent with what we might expect for the effect of these predictors. If we examine the weights for this NN above, we see several large values. Consider that all the variables have been scaled to mean zero and variance one. Products formed using the large weights will vary substantially. The situation is analogous to the collinearity problem in linear regression where unreasonably large regression coefficients are often seen. The NN is choosing extreme weights in order to optimize the fit, but the predictions will be unstable, especially for extrapolations.

We can use a penalty function, as with smoothing splines, to obtain a more stable fit. Instead of minimizing E, we minimize:

$$E + \lambda \sum_i w_i^2$$

In NN terms, this is known as *weight decay*. The idea is similar to ridge regression. Let's try $\lambda = 0.001$ for 100 NN model fits:

```
> bestrss <- 10000
> for(i in 1:100){
  nnmdl <- nnet(O3 ~ temp + ibh + ibt, sx, size=2,linout=T,
  decay=0.001,trace=F)
  cat(nnmdl$value,"\n")
  if(nnmdl$value < bestrss){
  bestnn <- nnmdl
  bestrss <- nnmdl$value
  }}
> bestnn$value
[1] 92.055
```

The value of the best RSS is somewhat larger than before. We expect this since weight decay sacrifices some fit to the current data to obtain a more stable result. We repeat the assessment of the marginal effects as before and display the results in Figure 14.5:

```
> xx <- expand.grid(temp=seq(-3,3,0.1),ibh=0,ibt=0)
> plot(xx$temp*ozscales['temp']+ozmeans['temp'],
  predict(bestnn,new=xx)*ozscales['O3']+ozmeans['O3'],xlab="Temp",ylab="O3")
> xx <- expand.grid(temp=0,ibh=seq(-3,3,0.1),ibt=0)
> plot(xx$ibh*ozscales['ibh']+ozmeans['ibh'],
  predict(bestnn,new=xx)*ozscales['O3']+ozmeans['O3'],xlab="IBH",ylab="O3")
> xx <- expand.grid(temp=0,ibh=0,ibt=seq(-3,3,0.1))
> plot(xx$ibt*ozscales['ibt']+ozmeans['ibt'],
  predict(bestnn,new=xx)*ozscales['O3']+ozmeans['O3'],xlab="IBT",ylab="O3")
```

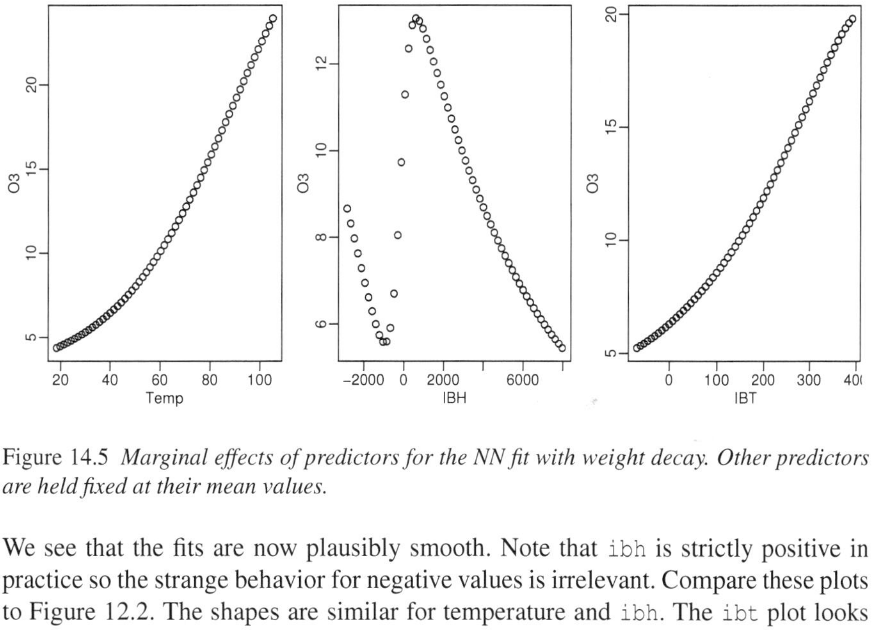

Figure 14.5 *Marginal effects of predictors for the NN fit with weight decay. Other predictors are held fixed at their mean values.*

We see that the fits are now plausibly smooth. Note that ibh is strictly positive in practice so the strange behavior for negative values is irrelevant. Compare these plots to Figure 12.2. The shapes are similar for temperature and ibh. The ibt plot looks quite different although we have no way to assess the significance of any of the terms in the NN fit.

NNs have interactions built in so one should also look at these. We could produce analogous plots to those in Figure 14.5 by varying two predictors at a time.

Now let's look at the full dataset. We use four hidden units because there are now more inputs.

```
> bestrss <- 10000
> for(i in 1:100){
   nnmdl <- nnet(O3 ~ ., sx, size=4, linout=T,trace=F)
   cat(nnmdl$value,"\n")
   if(nnmdl$value < bestrss){
    bestnn <- nnmdl
    bestrss <- nnmdl$value
```

```
 }}
> 1-bestnn$value/sum((sx[,1]-mean(sx[,1]))^2)
[1] 0.85063
```

The fit is good and there may be better minimum than we have found and increasing the number of hidden units would always improve the fit. The fit can be compared to those in previous chapters. The R^2 for the linear and tree model fits was substantially smaller, but these approaches place a premium on simplicity and interpretability. The fit for the corresponding additive model was better, but not quite as good as the NN. But the additive model also has the interpretability that the NN lacks. Finally, the MARS model fit better and was also interpretable.

Of course, it would be rash to draw firm conclusions from just one dataset. Furthermore, the value of the modeling approaches needs to be judged within the context of the particular problem. If explanation is the main goal of the data analysis, NNs are not a good choice. If prediction is the objective, we cannot judge just by the fit to the data we have now. It is more important how the model performs on future observations. We do not have fresh data here as we have used it all to fit the data. Some studies have withheld data for use in testing the prediction performance of the models considered. NNs have been generally competitive in these studies but by no means dominant.

14.4 Conclusion

NNs, as presented here, are a controlled flexible class of nonlinear regression models. By adding more hidden units we can control the complexity of the model in a measured way from relatively simple models up to models suitable for large datasets with complex structure. NNs are also attractive because they require less expertise to use successfully compared to statistical models. Nevertheless the user must still pay attention to basic statistical issues involving transformation and scaling of the data and outliers and influential points. See Faraway and Chatfield (1998) for an example of the application of neural networks and how they compare with statistical methods.

NNs are generally good for prediction but bad for understanding. The NN weights are almost uninterpretable. Although one can gain some insight from plotting the marginal effect of predictors, the NN inevitably introduces complex interactions that often do not reflect reality. Furthermore, without careful control, the NN can easily overfit the data resulting in overoptimistic predictions.

NNs are quite effective for large complex datasets compared to statistical methods where the burden of developing an appropriate sampling model can sometimes slow or even block progress. NNs do lack good statistical theory for inference, diagnostics and model selection. Of course, they were not developed with these statistical considerations in mind, but experience shows that such issues are often important.

NNs can outperform their statistical competitors for some problems provided they are carefully used. However, one should not be fooled by the evocative name, as NNs are just another tool in the box.

Further Reading: See Ripley (1996), Bishop (1995), Haykin (1998), Neal (1996) and Hertz, Krogh, and Palmer (1991) for more on NNs.

Exercises

This book has covered a wide range of statistical methods. For some datasets, only one method may be applicable, but for some others, we have a wide variety of choices. For continuous, binary or even count response data with some number of predictors, we can choose between:

1. Generalized linear models
2. Generalized additive models and associated methods
3. Trees
4. Neural networks

These methods have their strengths and weaknesses. It is not possible to claim that any one method dominates. Even so, we might hope to intuit what method might work well for a particular dataset. One way to gain an understanding of the relative value of these datasets is by the use of case studies. We want to make two sorts of comparisons:

Quantitative Does one method fit or predict better than another? Unfortunately, adding complexity to a model does tend to make it fit the current data better, but is no reliable indication of whether it will predict future data well. One way to avoid this problem is to reserve a portion of the data for testing your chosen model. A random subset of the data is selected and put to one side. The remaining data is used to select and fit the model. The model is then used to predict the response in the test set.

Qualitative Sometimes the purpose of a statistical analysis is to gain an understanding of the relationship between the variables. A good numerical fit to the data is then not helpful unless it can be intuitively understood. Judging the success of a method in this sense is necessarily more subjective.

Several of the datasets we have already used in previous exercises are suitable for case studies comparing the performance of the competing methods. We recommend the following datasets for this purpose: `diabetes`, `pima`, `wbcd`, `uswages`, `dvisits`, `fat` and `motorins`.

APPENDIX A

Likelihood Theory

This appendix is just an overview of the likelihood theory used in this book. For greater detail or a more gentle introduction, the reader is advised to consult a book on theoretical statistics such as Cox and Hinkley (1974), Bickel and Doksum (1977) or Rice (1998).

A.1 Maximum Likelihood

Consider n independent discrete random variables, $Y_1, \ldots, Y_n$, with probability distribution function $f(y|\theta)$ where θ is the, possibly vector-valued, parameter. Suppose we observe $\mathbf{y} = (y_1, \ldots, y_n)^T$, then we define the likelihood as:

$$P(\mathbf{Y} = \mathbf{y}) = \prod_{i=1}^{n} f(y_i|\theta) = L(\theta|y)$$

So the likelihood is a function of the parameter(s) given the data and is the probability of the observed data given a specified value of the parameter(s).

For continuous random variables, $Y_1, \ldots, Y_n$ with probability density function $f(y|\theta)$, we recognize that, in practice, we can only measure or observe data with limited precision. We may record y_i, but this effectively indicates an observation in the range $[y_i^l, y_i^u]$ so that:

$$P(Y_i = y_i) = P(y_i^l \le y_i \le y_i^u) = \int_{y_i^l}^{y_i^l} f(u|\theta)du \approx f(y_i|\theta)\delta_i$$

where $\delta_i = y_i^u - y_i^l$. We can now write the likelihood as:

$$L(\theta|y) \approx \prod_{i=1}^{n} f(y_i|\theta) \prod_{i=1}^{n} \delta_i$$

Now provided that δ_i is relatively small and does not depend on θ, we may ignore it and the likelihood is the same as in the discrete case.

As an example, suppose that Y is binomially distributed $B(n, p)$. The likelihood is:

$$L(p|y) = \binom{n}{y} p^y (1-p)^{n-y}$$

The *maximum likelihood* estimate (MLE) is the value of the parameter(s) that gives the largest probability to the observed data, or in other words, maximizes the likelihood function. The value at which the maximum occurs, $\hat{\theta}$, is the maximum likelihood estimate. In most cases, it is easier to maximize the log of likelihood function,

$l(\theta|y) = \log L(\theta|y)$. Since log is a monotone increasing function, the maximum occurs at the same $\hat{\theta}$.

In a few cases, we can find an exact analytical solution for $\hat{\theta}$. For the binomial, we have the log-likelihood:

$$l(p|y) = \log \binom{n}{y} + y \log p + (n-y)\log(1-p)$$

The *score function*, $u(\theta)$, is the derivative of the log-likelihood with respect to the parameters. For this example, we have:

$$u(p) = \frac{dl(p|y)}{dp} = \frac{y}{p} - \frac{n-y}{1-p}$$

We can find the maximum likelihood estimate $\hat{p}$ by solving $u(p) = 0$. We get $\hat{p} = y/n$. We should also verify that this stationary point actually represents a maximum.

Usually we want more than an estimate; some measure of the uncertainty in the estimate is valuable. This can be obtained via the Fisher information which is:

$$I(\theta) = \text{var } u(\theta) = -E\frac{\partial^2 l(\theta)}{\partial\theta\partial\theta^T}$$

If there is more than one parameter, $I(\theta)$ will be a matrix. The information at $\hat{\theta}$ is the second derivative at the maximum. Large values indicate high curvature so that the maximum is well defined and even close alternatives will have much lower likelihood. This would indicate a high level of confidence in the estimate. One can show that the variance of $\hat{\theta}$ can be estimated by:

$$\hat{\text{var}}\ (\hat{\theta}) = I^{-1}(\hat{\theta})$$

under mild conditions. Sometimes it is difficult to compute the expected value of the matrix of second derivatives. As an alternative, the observed, rather than expected, value at $\hat{\theta}$ may be used instead. For the binomial example this gives:

$$\hat{\text{var}}\ \hat{p} = \hat{p}(1-\hat{p})/n$$

We illustrate these concepts by plotting the log-likelihood for two binomial datasets: one where $n = 25, y = 10$ and another where $n = 50, y = 20$. We construct the log-likelihood function:

```
> loglik <- function(p,y,n) lchoose(n,y) + y*log(p) + (n-y)*log(1-p)
```

For ease of presentation, we normalize by subtracting the log-likelihood at the maximum likelihood estimate:

```
> nloglik <- function(p,y,n) loglik(p,y,n) - loglik(y/n,y,n)
```

Now plot the two log-likelihoods, as seen in Figure A.1:

```
> pr <- seq(0.05,0.95,by=0.01)
> matplot(pr,cbind(nloglik(pr,10,25),nloglik(pr,20,50)),type="l",
  xlab="p",ylab="log-likelihood")
```

We see that the maximum occurs at $p = 0.4$ in each case at a value of zero because of the normalization. For the larger sample, we see greater curvature and hence more information.

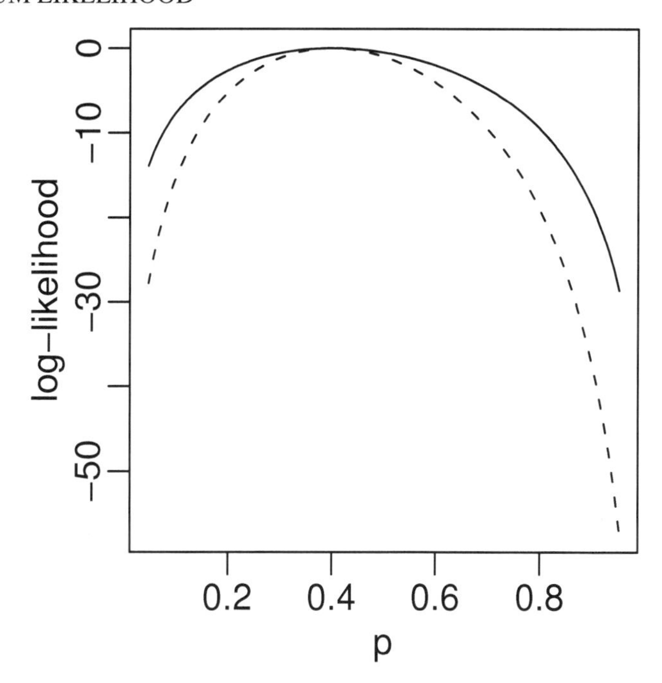

Figure A.1 *Normalized binomial log-likelihood for $n = 25, y = 10$ shown with a solid line and $n = 50, y = 20$ shown with a dotted line.*

Examples where likelihood can be maximized explicitly are confined to simple cases. Typically, numerical optimization is necessary. The *Newton–Raphson* method is the most well-known technique. Let θ_0 be an initial guess at θ, then we update using:

$$\theta_1 = \theta_0 - H^{-1}(\theta_0)u(\theta_0)$$

where H is the Hessian matrix of second derivatives:

$$H(\theta) = \frac{\partial^2 l(\theta)}{\partial\theta\partial\theta^T}$$

We iterate this method, putting θ_1 in place of θ_0 and so on, until the procedure (hopefully) converges. This method works well provided the log-likelihood is smooth and convex around the maximum and that the initial value is reasonably close. In less well-behaved cases, several things can go wrong:

- The likelihood has multiple maxima. The maximum that Newton–Raphson finds will depend on the choice of initial estimate. If you are aware that multiple maxima may exist, it is advisable to try multiple starting values to search for the overall maximum. The number and choice of these starting values is problematic. Such problems are common in fitting neural networks, but rare for generalized linear models.
- The maximum likelihood may occur at the boundary of the parameter space. This means that perhaps $u(\hat{\theta}) \neq 0$, which will confuse the Newton–Raphson method.

Mixed effect models have several variance parameters. In some cases, these are maximized at zero, which causes difficulties in the numerical optimization.

- The likelihood has a large number of parameters and is quite flat in the neighborhood of the maximum. The Newton–Raphson method may take a long time to converge.

The Fisher scoring method replaces H with $-I$ and sometimes gives superior results. This method is used in fitting GLMs and is equivalent to iteratively reweighted least squares.

A minimization function that uses a Newton-type method is available in R. We demonstrate its use for likelihood maximization. Note that we need to minimize $-l$ because nlm minimizes, not maximizes:

```
> f <- function(x) -loglik(x,10,25)
> mm <- nlm(f,0.5,hessian=T)
```

We use a starting value of 0.5 and find the optimum at:

```
> mm$estimate
[1] 0.4
```

The inverse of the Hessian at the optimum is equal to the standard estimate of the variance:

```
> c(1/mm$hessian,0.4*(1-0.4)/25)
[1] 0.0096016 0.0096000
```

Of course, this calculation is not necessary for the binomial, but it is useful for cases where exact calculation is not possible.

A.2 Hypothesis Testing

Consider two nested models, a larger model Ω and a smaller model ω. Let $\hat{\theta}_\Omega$ be the maximum likelihood estimate under the larger model, while $\hat{\theta}_\omega$ be the corresponding value when θ is restricted to the range proscribed by the smaller model. The *likelihood ratio test statistic* is:

$$2\log(L(\hat{\theta}_\omega)/L(\hat{\theta}_\Omega)) = -2(l(\hat{\theta}_\omega) - l(\hat{\theta}_\Omega))$$

Under some regularity conditions, this statistic is asymptotically distributed χ^2 with degrees of freedom equal to the difference in the number of identifiable parameters in the two models. The approximation may not be good for small samples and may fail entirely if the regularity conditions are broken. For example, if the smaller model places some parameters on the boundary of the parameter space, the χ^2 may not be valid. This can happen in mixed effects models when testing whether a particular variance component is zero.

The *Wald test* may be used to test hypotheses of the form $H_0 : \theta = \theta_0$ and the test statistic takes the form:

$$(\hat{\theta} - \theta_0)^T I(\hat{\theta})(\hat{\theta} - \theta_0)$$

Under the null, the test statistic has approximately a χ^2 distribution with degrees of freedom equal to the number of parameters being tested. Quite often, one does not wish to test all the parameters and the Wald test is confined to a subset. In particular, if we test only one parameter, $H_0 : \theta_i = \theta_{i0}$, the square root of the Wald test statistic is simply:

$$z = \frac{\hat{\theta}_i - \theta_{i0}}{se(\hat{\theta}_i)}$$

This is asymptotically normal. For a Gaussian linear model, these are the t-statistics and have an exact t-distribution, but for generalized linear and other models, the normal approximation must suffice.

The *score test* of the hypothesis $H_0 : \theta = \theta_0$ uses the statistic:

$$u(\theta_0)^T I^{-1}(\theta_0) u(\theta_0)$$

and is asymptotically χ^2 distributed with degrees of freedom equal to the number of parameters being tested.

There is no uniform advantage to any of these three tests. The score test does not require finding the maximum likelihood estimate, while the likelihood ratio test needs this computation to be done for both models. The Wald test needs just one maximum likelihood estimate. However, although the likelihood ratio test requires more information, the extra work is often rewarded. Although the likelihood ratio test is not always the best, it has been shown to be superior in a wide range of situations. Unless one has indications to the contrary or the computation is too burdensome, the likelihood ratio test is the recommended choice.

These test methods can be inverted to produce confidence intervals. To compute a $100(1-\alpha)\%$ confidence interval for θ, we calculate the range of hypothesized θ_0 such that $H_0 : \theta_0 = 0$ would not be rejected at the α level. The computation is simple for the single-parameter Wald test where the confidence interval for θ_i is:

$$\hat{\theta}_i \pm z^{1-\alpha/2} se(\hat{\theta}_i)$$

where z is the appropriate quantile of the normal distribution. The computation is trickier for the likelihood ratio test. If we are interested in a confidence interval for a single parameter θ_i, we will need to compute the log-likelihood for a range of θ_i with the other θ set to the maximizing values. This is known as the *profile likelihood* for θ_i. Once this is computed as $l_i(\theta_i|y)$, the confidence interval is:

$$\{\theta_i : 2(l(\hat{\theta}_i|y) - l(\theta_i|y)) < \chi_1^{1-\alpha}\}$$

As an example, this type of calculation is used in the computation of the confidence interval for the transformation parameter used in the Box–Cox method.

We can illustrate this by considering a binomial dataset where $n = 100$ and $y = 40$. We plot the normalized log-likelihood in Figure A.2 where we have drawn a horizontal line at half the distance of the 0.95 quantile of χ_1^2 below the maximum:

```
> pr <- seq(0.25,0.55,by=0.01)
> plot(pr,nloglik(pr,40,100),type="l",xlab="p",ylab="log-likelihood")
> abline(h=-qchisq(0.95,1)/2)
```

All p that have a likelihood above the line are contained within a 95% confidence interval for p. We can compute the range by solving for the points of intersection:

```
> g <- function(x) nloglik(x,40,100)+qchisq(0.95,1)/2
> uniroot(g,c(0.45,0.55))$root
```

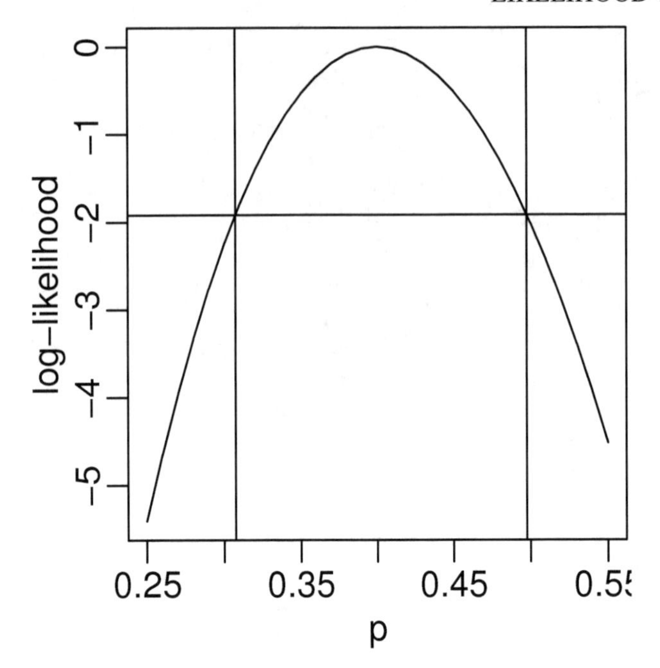

Figure A.2 *Likelihood ratio test-based confidence intervals for binomial p.*

```
[1] 0.49765
> uniroot(g,c(0.25,0.35))$root
[1] 0.30743
> abline(v=c(0.49765,0.30743))
```

The confidence interval is $(0.307, 0.498)$ as is indicated by the vertical lines on the plot. We can compute the Wald test-based interval as:

```
> se <- sqrt(0.4*(1-0.4)/100)
> cv <- qnorm(0.975)
> c(0.4-cv*se,0.4+cv*se)
[1] 0.30398 0.49602
```

which is very similar, but not identical, to the LRT-based intervals.

Suppose we are interested in the hypothesis, $H_0 : p = 0.5$. The LRT and p-value are:

```
> (lrstat <- 2*(loglik(0.4,40,100)-loglik(0.5,40,100)))
[1] 4.0271
> pchisq(lrstat,1,lower=F)
[1] 0.044775
```

So the null is barely rejected at the 5% level. The Wald test gives:

```
> (z <- (0.5-0.4)/se)
[1] 2.0412
> 2*pnorm(z,lower=F)
[1] 0.041227
```

Again, not very different from the LRT. The score test takes more effort to compute. The observed information is:

$$\frac{-d^2 l(p|y)}{dp^2} = \frac{y}{p^2} + \frac{n-y}{(1-p)^2}$$

We compute the score and information at $p = 0.5$ and then form the test and get the p-value:

```
> (sc <- 40/0.5-(100-40)/(1-0.5))
[1] -40
> (obsinf <- 40/0.5^2+(100-40)/(1-0.5)^2)
[1] 400
> (score.test <- 40*40/400)
[1] 4
> pchisq(4,1,lower=F)
[1] 0.0455
```

The outcome is again slightly different from the previous two tests. Asymptotically, the three tests agree. We have a moderate size sample in the example, so there is little difference. More substantial differences could be expected for smaller sample sizes.

APPENDIX B

R Information

R may be obtained from the R project website at www.r-project.org.

This book uses some functions and data that are not part of base R. You may wish to download these extras from the R website. The additional packages used are:

faraway, acepack, mda, brglm, sm, wavethresh, SuppDists, lme4, gam

MASS, rpart, mgcv and nnet are part of the "recommended" R installation; you will have these already unless you choose a nonstandard installation. Use the command:

```
> library()
```

within R to see what packages you have. Under Windows, to install the additional packages, choose the "Install packages from CRAN" menu option. You must have a network connection for this to work — if you are working offline, you may use the "Install packages from local zip file" menu option provided you have already obtained the necessary packages. Under other operating systems, such as Macintosh or Linux, the installation procedure differs. Consult the R website for details.

I have collected the data and functions that I have used in this book as an R package called faraway that you may obtain from CRAN. The book website is at people.bath.ac.uk/jjf23/ELM. The functions defined are:

```
halfnorm        Half normal plot
logit           logit transformation
ilogit          inverse logit transformation
```

Where add-on packages are needed in the text, you will find the appropriate library command. However, I have assumed that the faraway library is always loaded. You can add a line reading library(faraway) to your Rprofile file if you expect to use this package in every session. Otherwise, you will need to remember to type it each time.

I set the following options to achieve the output seen in this book:

```
> options(digits=5,show.signif.stars=FALSE)
```

The digits=5 reduces the number of digits shown when printing numbers from the default of seven. Note that this does not reduce the precision with which these numbers are internally stored. One might take this further as anything more than two or three significant digits in a displayed table is usually unnecessary and more importantly, distracting. I have also edited the output in the text to remove extraneous output or to improve the formatting.

The code and output shown in this book were generated under R version 2.2.0. R is regularly updated and improved, so more recent versions may show some differences in the output.

Getting Started with R: R requires some effort to learn. Such effort will be repaid with increased productivity. Free introductory guides to R may be obtained from the R project website at `www.r-project.org`. Introductory books have been written by Dalgaard (2002), Verzani (2004) and Maindonald and Braun (2003). Venables and Ripley (2002) also have an introduction to R along with more advanced material. Fox (2002) is intended as a companion to a standard regression text. You may also find Becker, Chambers, and Wilks (1998) and Chambers and Hastie (1991) to be useful references to the S language. Ripley and Venables (2000) wrote a more advanced text on programming in S or R.

While running R you can get help about a particular command; for example, if you want help about the `boxplot` command, just type `help(boxplot)`. If you do not know what the name of the command that you want to use, then type:

```
> help.start()
```

and then browse. You will be able to pick up the language from the examples in the text and from the help pages.

Bibliography

Agresti, A. (1984). *Analysis of Ordinal Categorical Data*. New York: Wiley.

Agresti, A. (2002). *Categorical Data Analysis* (2 ed.). New York: John Wiley.

Allison, T. and D. Cicchetti (1976). Sleep in mammals: Ecological and constitutional correlates. *Science 194*, 732–734.

Andrews, D. and A. Herzberg (1985). *Data : A Collection of Problems from Many Fields for the Student and Research Worker*. New York: Springer-Verlag.

Appleton, D., J. French, and M. Vanderpump (1996). Ignoring a covariate: An example of Simpson's paradox. *American Statistician 50*, 340–341.

Bates, D. (2005, May). Fitting linear mixed models in R. *R News 5*(1), 27–30.

Becker, R., J. Chambers, and A. Wilks (1998). *The New S Language: A Programming Environment for Data Analysis and Graphics* (revised ed.). Boca Raton, FL: CRC Press.

Bellman, R. (1961). *Adaptive Control Processes: A Guided Tour*. Princeton, NJ: Princeton University Press.

Bergman, B. and A. Hynen (1997). Dispersion effects from unreplicated designs in the 2^{k-p} series. *Technometrics 39*, 191–198.

Bickel, P. and K. Doksum (1977). *Mathematical Statistics: Basic Ideas and Selected Topics*. San Francisco: Holden Day.

Bishop, C. (1995). *Neural Networks for Pattern Recognition*. Oxford: Clarendon Press.

Bishop, Y., S. Fienberg, and P. Holland (1975). *Discrete Multivariate Analysis: Theory and Practice*. Cambridge, MA: MIT Press.

Blasius, J. and M. Greenacre (1998). *Visualization of Categorical Data*. San Diego: Academic Press.

Bliss, C. I. (1935). The calculation of the dose-mortality curve. *Annals of Applied Biology 22*, 134–167.

Bliss, C. I. (1967). *Statistics in Biology*. New York: McGraw Hill.

Bowman, A. and A. Azzalini (1997). *Applied Smoothing Techniques for Data Analysis: The Kernel Approach with S-Plus Illustrations*. Oxford: Oxford University Press.

Box, G. and R. Meyer (1986). Dispersion effects from fractional designs. *Technometrics 28*, 19–27.

Box, G. P., S. Bisgaard, and C. Fung (1988). An explanation and critique of Taguchi's contributions to quality engineering. *Quality and Reliability Engineering International 4*, 123–131.

Box, G. P., W. G. Hunter, and J. S. Hunter (1978). *Statistics for Experimenters*. New York: Wiley.

Breiman, L. (2001a). Random forests. *Machine Learning 45*, 5–32.

Breiman, L. (2001b). Statistical modeling: The two cultures (with comments and a rejoinder by the author). *Statistical Science 16*, 199–231.

Breiman, L., J. Friedman, R. Olshen, and C. Stone (1984). *Classification and Regression Trees*. Boca Raton, FL: Chapman & Hall.

Breiman, L. and J. H. Friedman (1985). Estimating optimal transformations for multiple regression and correlation. *Journal of the American Statistical Association 80*, 580–598.

Breslow, N. (1982). Covariance adjustment of relative-risk estimates in matched studies. *Biometrics 38*, 661–672.

Breslow, N. E. and D. G. Clayton (1993). Approximate inference in generalized linear mixed models. *Journal of the American Statistical Association 88*, 9–25.

Cameron, A. and P. Trivedi (1998). *Regression Analysis of Count Data*. Cambridge: Cambridge University Press.

Chambers, J. and T. Hastie (1991). *Statistical Models in S*. London: Chapman & Hall.

Christensen, R. (1997). *Log-Linear Models and Logistic Regression* (2 ed.). New York: Springer.

Cleveland, W. (1979). Robust locally weighted regression and smoothing scatterplots. *Journal of the American Statistical Association 74*, 829–836.

Clogg, C. and E. Shihadeh (1994). *Statistical Models for Ordinal Variables*. Thousands Oaks, CA: Sage.

Cochran, W. (1954). Some methods of strengthening the common χ^2 tests. *Biometrics 10*, 417–451.

Collett, D. (2003). *Modelling Binary Data* (2 ed.). London: Chapman & Hall.

Comizzoli, R. B., J. M. Landwehr, and J. D. Sinclair (1990). Robust materials and processes: Key to reliability. *AT&T Technical Journal 69*(6), 113–128.

Cox, D. (1970). *Analysis of Binary Data*. London: Spottiswoode, Ballantyne and Co.

Cox, D. and D. Hinkley (1974). *Theoretical Statistics*. London: Chapman & Hall/CRC.

Crainiceanu, C. and D. Ruppert (2004). Likelihood ratio tests in linear mixed models with one variance component. *Journal of the Royal Statistical Society, Series B 66*, 165–185.

Crowder, M. (1978). Beta-binomial anova for proportions. *Applied Statistics 27*, 34–37.

Crowder, M. J. and D. J. Hand (1990). *Analysis of Repeated Measures*. London: Chapman & Hall.

Dalal, S., E. Fowlkes, and B. Hoadley (1989). Risk analysis of the space shuttle: Pre-Challenger prediction of failure. *Journal of the American Statistical Association 84*, 945–957.

Dalgaard, P. (2002). *Introductory Statistics with R*. New York: Springer.

Daubechies, I. (1991). *Ten Lectures on Wavelets*. Philadelphia: SIAM.

Davies, O. (1954). *The Design and Analysis of Industrial Experiments*. New York: Wiley.

Dey, D., S. Ghosh, and B. Mallick (2000). *Generalized Linear Models : A Bayesian Perspective*. New York: Marcel Dekker.

Diggle, P. J., P. Heagerty, K. Y. Liang, and S. L. Zeger (2002). *Analysis of Longitudinal Data* (2 ed.). Oxford: Oxford University Press.

Dobson, A. (1990). *An Introduction to Generalized Linear Models*. London: Chapman and Hall.

Draper, N. and H. Smith (1998). *Applied Regression Analysis* (3rd ed.). New York: Wiley.

Eubank, R. (1988). *Spline Smoothing and Nonparametric Regression*. New York: Marcel Dekker.

Fahrmeir, L. and G. Tutz (2001). *Multivariate Statistical Modelling Based on Generalized Linear Models*. New York: Springer.

Faraway, J. (2004). *Linear Models with R*. Boca Raton, FL: Chapman & Hall/CRC.

Faraway, J. and C. Chatfield (1998). Time series forecasting with neural networks: A case study. *Applied Statistics 47*, 231–250.

Firth, D. (1993). Bias reduction of maximum likelihood estimates. *Biometrika 80*, 27–38.

Fitzmaurice, G., N. Laird, and J. Ware (2004). *Applied Longitudinal Analysis*. Hoboken, NJ: Wiley-Interscience.

Fletcher, R. (1987). *Practical Methods of Optimization* (2 ed.). Chichester, UK: John Wiley.

Fox, J. (2002). *An R and S-plus Companion to Applied Regression*. Thousand Oaks, CA: Sage.

Frees, E. (2004). *Longitudinal and Panel Data: Analysis and Applications in the Social Sciences*. Cambridge: Cambridge University Press.

Friedman, J. (1991). Multivariate adaptive regression splines (with discussion). *Annals of Statistics 19*, 1–141.

Frome, E. and R. DuFrain (1986). Maximum likelihood estimation for cytogenic dose-response curves. *Biometrics 42*, 73–84.

Gelman, A. (2005). Analysis of variance — why it is more important than ever (with discussion). *Annals of Statistics 33*, 1–53.

Gelman, A., J. Carlin, H. Stern, and D. Rubin (2003). *Bayesian Data Analysis* (2 ed.). Chapman & Hall/CRC.

Gill, J. (2001). *Generalized Linear Models: A Unified Approach*. Thousand Oaks, CA: Sage.

Goldstein, H. (1995). *Multilevel Statistical Models* (2 ed.). London: Arnold.

Green, P. and B. Silverman (1993). *Nonparametric Regression and Generalized Linear Models: A Roughness Penalty Approach*. London: Chapman & Hall.

Gu, C. (2002). *Smoothing Spline ANOVA Models*. New York: Springer Verlag.

Haberman, S. (1977). *The Analysis of Frequency Data*. Chicago, IL: University of Chicago Press.

Hall, S. (1994). Analysis of defectivity of semiconductor wafers by contigency table. *Proceedings of the Institute of Environmental Sciences 1*, 177–183.

Hallin, M. and J.-F. Ingenbleek (1983). The Swedish automobile portfolio in 1977. A statistical study. *Scandinavian Actuarial Journal 83*, 49–64.

Hand, D. (1981). *Discrimination and Classification*. Chichester, UK: Wiley.

Hardin, J. and J. Hilbe (2003). *Generalized Estimating Equations*. Boca Raton, FL: Chapman & Hall/CRC Press.

Härdle, W. (1991). *Smoothing Techniques with Implementation in S*. New York: Springer.

Harrell, F. (2001). *Regression Modelling Strategies*. New York: Springer Verlag.

Hart, J. (1997). *Nonparametric Smoothing and Lack-of-Fit Tests*. New York: Springer.

Hartigan, J. and B. Kleiner (1981). Mosaics for contingency tables. In W. Eddy (Ed.), *Computer Science and Statistics: Proceedings of the 13th Symposium on the Interface*, pp. 268–273. Springer Verlag.

Hastie, T. and R. Tibshirani (1990). *Generalized Additive Models*. London: Chapman & Hall.

Hastie, T., R. Tibshirani, and J. Friedman (2001). *The Elements of Statistical Learning: Data Mining, Inference, and Prediction*. New York: Springer.

Hauck, W. and A. Donner (1977). Wald's test as applied to hypotheses in logit analysis. *Journal of the American Statistical Association 72*, 851–853.

Haykin, S. (1998). *Neural Networks: A Comprehensive Foundation* (2 ed.). Prentice Hall.

Hertz, J., A. Krogh, and R. Palmer (1991). *Introduction to the Theory of Neural Computation*. Redwood City, CA: Addison–Wesley.

Hill, M. S. (1992). *The Panel Study of Income Dynamics: A User's Guide*. Newbury Park, CA: Sage.

Hinde, J. and C. Demetrio (1988). Overdispersion: Models and estimation. *Computational Statistics and Data Analysis 27*, 151–170.

Hornik, K., M. Stinchcombe, and H. White (1989). Multilayer feedforward networks are universal approximators. *Neural Networks 2*, 359–366.

Hosmer, D. and S. Lemeshow (2000). *Applied Logistic Regression* (2 ed.). New York: Wiley.

Johnson, M. P. and P. H. Raven (1973). Species number and endemism: The Galápagos archipelago revisited. *Science 179*, 893–895.

Kleinbaum, D. G. and M. Klein (2002). *Logistic Regression: A Self-Learning Text*. New York: Springer.

Lawless, J. (1987). Negative binomial and mixed poisson regression. *Canadian Journal of Statistics 15*, 209–225.

Le, C. T. (1998). *Applied Categorical Data Analysis*. Newbury Park, CA: Wiley.

Leonard, T. (2000). *A Course in Categorical Data Analysis*. Boca Raton, FL: Chapman & Hall/CRC Press.

Lindsey, J. K. (1997). *Applying Generalized Linear Models*. New York: Springer.

Lindsey, J. K. (1999). *Models for Repeated Measurements* (2 ed.). Oxford: Oxford University Press.

Loader, C. (1999). *Local Regression and Likelihood*. New York: Springer.

Lowe, C., C. Roberts, and S. Lloyd (1971). Malformations of the central nervous system and softness of local water supplies. *British Medical Journal 15*, 357–361.

Maindonald, J. and J. Braun (2003). *Data Analysis and Graphics Using R*. Cambridge, UK: Cambridge University Press.

Manly, B. (1978). Regression models for proportions with extraneous variance. *Biometrie-Praximetrie 18*, 1–18.

Mantel, N. and W. Haenszel (1959). Statistical aspects of the analysis of data from retrospective studies of disease. *Journal of the National Cancer Institute 22*, 719–748.

Margolese, M. (1970). Homosexuality: A new endocrine correlate. *Hormones and Behavior 1*, 151–155.

McCullagh, P. (1983). Quasi-likelihood functions. *Annals of Statistics 11*, 59–67.

McCullagh, P. and J. Nelder (1989). *Generalized Linear Models* (2 ed.). London: Chapman & Hall.

McCulloch, C. and S. Searle (2002). *Generalized, Linear, and Mixed Models*. New York: Wiley.

McCulloch, W. and W. Pitts (1943). A logical calculus of ideas immanent in neural activity. *Bulletin of Mathematical Biophysics 5*, 115–133.

Mehta, C. and N. Patel (1995). Exact logistic regression: theory and examples. *Statistics in Medicine 14*, 2143–2160.

Menard, S. (2002). *Applied Logistic Regression Analysis* (2 ed.). Thousands Oaks, CA: Sage.

Meyer, M. (2002). Uncounted votes: Does voting equipment matter? *Chance 15*(4), 33–38.

Milliken, G. A. and D. E. Johnson (1992). *Analysis of Messy Data*, Volume 1. New York: Van Nostrand Reinhold.

Morgan, J. and J. Sonquist (1963). Problems in the analysis of survey data, and a proposal. *Journal of the American Statistical Association 58*, 415–434.

Mortimore, P., P. Sammons, L. Stoll, D. Lewis, and R. Ecob (1988). *School Matters*. Wells, UK: Open Books.

Myers, R. and D. Montgomery (1997). A tutorial on generalized linear models. *Journal of Quality Technology 29*, 274–291.

Myers, R., D. Montgomery, and G. Vining (2002). *Generalized Linear Models: With Applications in Engineering and the Sciences*. New York: Wiley.

Nagelkerke, N. (1991). A note on a general definition of the coefficient of determination. *Biometrika 78*, 691–692.

Neal, R. (1996). *Bayesian Learning for Neural Networks*. New York: Springer–Verlag.

Nelder, J., Y. Lee, B. Bergman, A. Hynen, A. Huele, and J. Engel (1998). Letter to editor: Joint modeling of mean and dispersion. *Technometrics 40*, 168–175.

Nelder, J. and R. Wedderburn (1972). Generalized linear models. *Journal of the Royal Statistical Society, Series A 132*, 370–384.

Payne, C. (1987). *The GLIM System Release 3.77 Manual* (2 ed.). Oxford: Numerical Algorithms Group.

Penrose, R. (1989). *The Emperor's New Mind: Concerning Computers, Minds, and the Laws of Physics*. Oxford: Oxford University Press.

Pignatiello, J. J. and J. S. Ramberg (1985). Contribution to discussion of offline quality control, parameter design and the Taguchi method. *Journal of Quality Technology 17*, 198–206.

Pinheiro, J. C. and D. M. Bates (2000). *Mixed-Effects Models in S and S-PLUS*. New York: Springer.

Powers, D. and Y. Xie (2000). *Statistical Methods for Categorical Data Analysis*. San Diego, CA: Academic Press.

Pregibon, D. (1981). Logistic regression diagnostics. *Annals of Statistics 9*, 705–724.

Purott, R. and E. Reeder (1976). The effect of changes in dose rate on the yield of chromosome aberrations in human lymphocytes exposed to gamma radiation. *Mutation Research 35*, 437–444.

Quinlan, J. (1993). *C4.5: Programs for Machine Learning*. San Mateo, CA: Morgan Kaufman.

Raudenbush, S. and A. Bryk (2002). *Hierarchical Linear Models: Applications and Data Analysis Methods* (2 ed.). Thousand Oaks, CA: Sage.

Rice, J. (1998). *Mathematical Statistics and Data Analysis*. Monterey, CA: Brooks Cole.

Ripley, B. (1996). *Pattern Recognition and Neural Networks*. Cambridge, UK: Cambridge University Press.

Ripley, B. and W. Venables (2000). *S Programming*. New York: Springer-Verlag.

Rosenstone, S. J., D. R. Kinder, and W. E. Miller (1997). *American National Election Study*. Ann Arbor, MI: Inter-university Consortium for Political and Social Research.

Santner, T. and D. Duffy (1989). *The Statistical Analysis of Discrete Data*. New York: Springer.

Scheffé, H. (1959). *The Analysis of Variance*. New York: Wiley.

Searle, S., G. Casella, and C. McCulloch (1992). *Variance Components*. New York: Wiley.

Seshadri, V. (1993). *The Inverse Gaussian Distribution*. Oxford: Clarendon.

Sheldon, F. (1960). Statistical techniques applied to production situations. *Industrial and Engineering Chemistry 52*, 507–509.

Simonoff, J. (1996). *Smoothing Methods in Statistics*. New York: Springer.

Simonoff, J. (2003). *Analyzing Categorical Data*. New York: Springer.

Simpson, E. (1951). The interpretation of interaction in contingency tables. *Journal of the Royal Statistical Society, Series B 13*, 238–241.

Sinha, S. (2004). Robust analysis of generalized linear mixed models. *JASA 99*, 451–460.

Smyth, G., F. Huele, and A. Verbyla (2001). Exact and approximate reml for heteroscedastic regression. *Statistical Modelling: An International Journal 1*, 161–175.

Snedecor, G. and W. Cochran (1989). *Statistical Methods* (8 ed.). Ames, IA: Iowa State University Press.

Snee, R. (1974). Graphical display of two-way contingency tables. *American Statistician 28*, 9–12.

Steele, R. (1998). *Effect of surface and vision on balance*. Ph. D. thesis, Department of Physiotherapy, University of Queensland.

Stone, C. (1985). Additive regression and other nonparametric models. *Annals of Statistics 13*, 689–705.

Stram, D. and J. Lee (1994). Variance components testing in the longitudinal mixed-effects model. *Biometrics 50*, 1171–1179.

Stuart, A. (1955). A test for homogeneity of the marginal distributions in a two-way classification. *Biometrika 42*, 412–416.

Thall, P. F. and S. C. Vail (1990). Some covariance models for longitudinal count data with overdispersion. *Biometrics 46*, 657–671.

Tukey, J. (1977). *Exploratory Data Analysis*. New York: Addison Wesley.

Venables, W. and B. Ripley (2002). *Modern Applied Statistics with S* (4 ed.). New York: Springer.

Verbeke, G. and G. Molenberghs (2000). *Linear Mixed Models for Longitudinal Data*. New York: Springer.

Verzani, J. (2004). *Using R for Introductory Statistics*. Boca Raton, FL: Chapman & Hall/CRC.

Wahba, G. (1990). *Spline Models for Observational Data*. Philadelphia: SIAM.

Wand, M. and M. Jones (1995). *Kernel Smoothing*. London: Chapman & Hall.

Wedderburn, R. W. M. (1974). Quasilikelihood functions, generalized linear models and the Gauss–Newton method. *Biometrika 61*, 439–447.

Weisberg, S. (2005). *Applied Linear Regression* (3 ed.). New York: Wiley.

Whitmore, G. (1986). Inverse Gaussian ratio estimation. *Applied Statistics 35*, 8–15.

Wilkinson, G. and C. Rogers (1973). Symbolic description of factorial models for the analysis of variance. *Applied Statistics 22*, 392–399.

Williams, D. (1982). Extra-binomial variation in logistic linear models. *Applied Statistics 31*, 144–148.

Williams, D. (1987). Generalized linear model diagnostics using the deviance and single case deletions. *Applied Statistics 36*, 181–191.

Wold, S., A. Ruhe, H. Wold, and W. Dunn (1984). The collinearity problem in linear regression: The partial least squares (pls) approach to generalized inverses. *SIAM Journal on Scientific and Statistical Computing 5*, 735–743.

Wood, S. (2000). Modelling and smoothing parameter estimation with multiple quadratic penalties. *Journal of the Royal Statistal Society, Series B 62*, 413–428.

Wood, S. (2006). *An Introduction to Generalized Additive Models with R*. Boca Raton, FL: CRC Press.

Yule, G. (1903). Notes on the theory of association of attributes in statistics. *Biometrika 2*, 121–134.

Index